The Lung in
Its Environment

ETTORE MAJORANA INTERNATIONAL SCIENCE SERIES

Series Editor:
Antonino Zichichi
European Physical Society
Geneva, Switzerland

(LIFE SCIENCES)

The Lung in Its Environment

Edited by

G. Bonsignore

Institute of Respiratory Pathophysiology
University of Palermo
Palermo, Italy

and

G. Cumming

The Midhurst Medical Research Institute
Midhurst, West Sussex, United Kingdom

Plenum Press · New York and London

Library of Congress Cataloging in Publication Data

Main entry under title:

The Lung in its environment.

(Ettore Majorana international science series. Life sciences; v. 6)
"Proceedings of a symposium on the lung in its environment, held June 16–21, 1980,
at the Ettore Majorana Center for Scientific Culture, Erice, Sicily."
Includes bibliographies and index.
1. Lungs—Diseases—Congresses. 2. Lungs—Congresses. 3. Environmentally induced
diseases—Congresses. I. Bonsignore, G. II. Cumming, Gordon. III. Series.
RC756.L837   616.2′4   81-12004
                                                                     AACR2

ISBN 978-1-4684-3973-1   ISBN 978-1-4684-3971-7 (eBook)
DOI 10.1007/978-1-4684-3971-7

Proceedings of a symposium on the Lung in Its Environment,
held June 16-21, 1980, at the Ettore Majorana Center for
Scientific Culture, Erice, Sicily

© 1982 Plenum Press, New York
Softcover reprint of the hardcover 1st edition 1982

A Division of Plenum Publishing Corporation
233 Spring Street, New York, N.Y. 10013

FOREWORD

This volume documents the proceedings of a symposium on "Lung in its Environment" held at the Ettore Majorana Center for Scientific Culture, in Erice, Sicily, between 16th June and 21st June 1980. This was attended by about 200 participants drawn from Europe as a whole, but the majority were from Southern Europe.

The discussion was recorded either in English or Italian and the tapes were reduced to a verbatim typescript by the Ente Nazionale Interpreti Congresso. The verbatim typescript has been edited using a few guiding principles as follows:-

1.   Titles and honorifics have been eliminated unless the statement is addressed to a specific person.

2.   The style of the speakers in the discussion has been preserved as far as possible and not reduced to a strictly grammatical format.

3.   Where references to illustrations (e.g., on the blackboard) are made, the comments have been left unaltered and many are understandable. Removing them detracted from the sense.

4.   The air of informality in the proceedings has been preserved so far as possible.

5.   The responsibility for the discussion rests solely with the editors, and no contributor has had the opportunity of correcting what he said.

6.   No manuscript was received from two participants, but the discussion of their presentations has been included since it contains some points of substance.

7.   It is a pleasure to acknowledge the great assistance rendered by Guiliana de Furia in translation, by Corinne Wade in the typescript and by John Griffiths in the illustrations.

G. Cumming & G. Bonsignore

CONTENTS

PRE- AND POSTNATAL GROWTH OF THE HUMAN LUNG

D.G. Fagan

Histopathology Department
University Hospital
Queen's Medical Centre, Nottingham

Much more is known of the pre- and postatal development of the lung in experimental animals than in man. Although the pattern of development in man broadly parallels that of other mammals, there are important interspecies differences which render animal work of very limited use in Histopathology. To identify deviation from the norm, the normal - and range of normal - must be precisely delineated, not "paralleled in general".

Accordingly, this presentation will concentrate on data obtained from the study of human material rather than experimental animal studies, and will of necessity be dogmatic in parts as the appropriate studies have not been done on human material.

Normal Development

Intra-uterine development is classified[1] on simple subjective microscopic appearances into four phases:-

1. Embryonal: 0-5 weeks gestation
2. Pseudo-glandular: 6-16 weeks (approximately)
3. Canalicular: 17-24/27 weeks (approximately)
4. Terminal sac (alveolar)   27 weeks onwards

Each of these stages of lung growth has a set of typical histological appearances and there is little difficulty in classification. The details of these different stages have been exhaustively described in several publications and only a brief outline is given here.[2,3,4]

Embryonal Period In the first few weeks of development, a single

1

pouch develops in the laryngotracheal groove, lying along the ventral surface of the endodermal tube (gut). The lung bud divides into two as the endodermal epithelium pushes out into the mesenchyme investing the lung buds, forming a series of branching tubular structures. The ridges defining the laryngotracheal groove begin to hump up and fuse to separate the trachea from the oesophagus.

The endodermal branching continues rapidly, and by 5 weeks the trachea is completely separate from the oesophagus, and the lung bud mesenchyme has subdivided not only into left and right lungs, but also established the lobe pattern.

Pseudo-glandular Period The process of endodermal branching rapidly leads to the establishment of the basic segmental pattern within the lung lobes, and by about the twelfth week, the adult pattern of lobes and fissures has been established. Internally, the lung appears similar to an exocrine gland, with "ducts" lined by tall columnar endothelial cells invested in an undifferentiated mesenchyme. Cartilaginous differentiation begins in the trachea, and spreads centrifugally down the bronchial tree, particularly during the burst of bronchial proliferative activity at 10-16 weeks. Late in this period, aggregation of mesenchyme cells around the "ducts" begins to occur, and muscular differentiation commences. The bronchial tree appears to be defined at this time.

Canalicular Period The bronchi continue to divide, probably producing the generations of divisions destined to become the bronchioles, and later still alveolar ducts. These new generations of air passages are lined by a cuboidal epithelium. Centrifugal extension of cartilage, smooth muscle, and elastin differentiation continues during this phase of growth, and at the end of this period smooth muscle has reached the pre-terminal bronchioles. The epithelium in the bronchioles and acini remain cuboidal, but towards the end of the canalicular period the differentiation into terminal bronchioles and potential alveolar ducts occurs. Although the epithelium in both structures remains cuboidal, that lining the terminal acinus is becoming attenuated, and type I and II pneumocytes are becoming identifiable. Attenuation of the epithelium and differentiation between bronchiole and acinus is the mark of the onset of the next phase of development, the terminal sac phase.

Terminal Sac (Alveolar) Period It is of utmost importance that the reader is aware that many of the common experimental animals such as the rat, mouse or hampster are normally born with their lungs in a stage of development exactly equivalent to the end of the canalicular stage of development in man.[3] This means that the early POST natal development of the rat lung, so beautifully described by Burri,[5,6] is analogous to the events taking place in the human infant lung during the last trimester of pregnancy; that is, in man PRE natally.

It seems probable that this is the basis for the whole confusion over the question of the presence of alveoli in the human infant's lung. For example, in a recent, otherwise outstanding review, the author blurs the distinction between animals and man precisely on this point. Alveoli are clearly and unequivocally present in the newborn infant's lung.[7] Standard text books of Paediatric Pathology have illustrated them for many years.[8]

In the terminal sac period, the epithelium lining the terminal acinus rapidly attenuates, and capillaries develop along the sub-epithelial plane, protruding into the potential air space. A few areas of these 'air-blood barriers' develop during the canalicular period, but the main thrust of development is during the attenuation of the terminal sac epithelium.

At some point during the terminal sac phase, probably between 28 and 32 weeks in man, the "secondary crests" described by Burri in the rat lung begin to appear. They extend along the generations of alveolar ducts allowing definitive alveoli to be recognized. The terminal saccules commence alveolation[9] just before 40 weeks gestation in man, probably around 36-38 weeks. At full term, most terminal saccules are becoming alveolated, and the next phase of human lung development, the centripetal conversion of terminal bronchioles into respiratory bronchioles begins. This, and the next phase, the subdivision of existing alveoli, is well illustrated by Emery[9]. Figure 1.

It is emphasized that normal, full term delivery is not marked by any dramatic change in lung development. Normally, development continues uneventfully. This is far from true when the child is born prematurely, and this will be dealt with later.

Post Natal Development   The discerning reader will have detected the whiff of grape-shot in the account of the terminal sac phase of development, and in no field of study is there such confusion and uncertainty as in that of Post Natal lung growth in man.

The only generally accepted statements are these:-

1. That the newborn infant lung is not a miniature of
    the adult lung.

2. That growth in alveolar duct and alveolar numbers
    proceeds by two processes, the centripetal conversion
    of bronchioles, and the subdivision of large air
    spaces into smaller areas.

All other data is disputed. It is not established how many alveoli are present in the newborn infant's lung, neither how

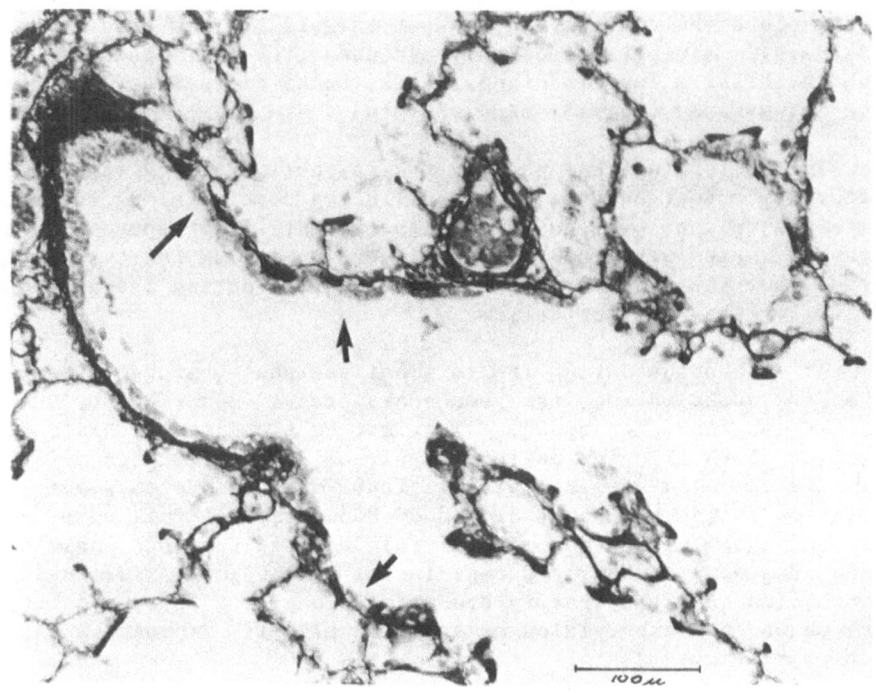

Figure 1   5/12 Infant. Necrotising Entero-colitis.

Shows a terminal bronchiole undergoing centripetal conversion into
an alveolated duct.   The fibro-elastin is demonstrated by a
Weigart-Picro sirius red stain.   The bronchiolar fibro-elastin
layer is seen breaking up into alveolar type condensations or
crests (arrow), while at this stage the surface epithelium is still
cuboidal.   Note that the gap between the newly formed crests is
slightly smaller than the gap between the established alveoli in
the adjacent ducts, and that the crests are only slightly thicker
than the other alveolar duct crests.   Contrast this with the
appearances in Figure 3.

rapidly they proliferate, nor when proliferation ceases, nor what the final adult population figure should be.

Before considering a lung model, the reasons for this confusion should be examined.

Firstly, the lung is an extremely variable organ. One investigator, having spent years exhaustively modelling several lung lobules, was driven to the conclusion that the branching pattern of each lung lobule was unique![10] It seems probable therefore, that modelling a few lung acini[11,12] and projecting that pattern of branching on to the rest of the lung, whether by sterology or computer modelling, is of limited value in Pathology.

In an attempt to overcome this problem, sterological methods have been devised in which random sampling methods are an essential.[13] Unfortunately this also requires assumptions as to the surface/cross sectional area ratio, the surface to volume ratio, and the ranges of size and shape of alveoli and alveolar ducts. These assumptions render use of these "shape constants" of limited value throughout the growth range, as the shape constants may alter with age. It has been shown in a young adult's lung that there is an underestimation of up to 85% in total populations of alveolar ducts between sterological estimates and reality as demonstrated by modelling.[14] This is far too great a gap to allow delineation of the limits of normal, let alone pathological deviation.

At the present time it seems more likely that alleged differences in lungs are due to different sterological assumptions and histological criteria than to reality.

However, we must have some skeleton on which to hang a model of lung growth.

Recent studies of alveolar population at birth have ranged from an extreme of none,[15,16] to 70×10^6 [17] with some authors finding low populations[18,19] of around 20×10^6. Comparison[20] of Dunnill's[18] figures for older children with the surprisingly reliable comparative data of Emery and Willcock[21] produce an answer for alveolar population around full term birth of almost exactly the number found by Thurlbeck and Angus[22] in a study which is methodologically the best so far undertaken. The following table is based largely on the work of Angus and Thurlbeck,[22] and Dunnill,[18] and seems to be the most likely pathway.

Full term birth	$60\text{-}70 \times 10^6$ Alveoli
Age 1 year	$130\text{-}200 \times 10^6$
Age 2 years	$250\text{-}350 \times 10^6$
Age 6 years	$350\text{-}400 \times 10^6$
Age 8 years	$375\text{-}450 \times 10^6$
Adult	$400\text{-}600 \times 10^6$

Mean air space diameter measurements suggest that size increases even faster than absolute numbers for the first 3-4 months post natal age,[19,20,21] then mean size reaches a plateau for several years, and from around the age of 6-8 years size begins to increase more rapidly than alveolar proliferation.

Authors holding the view that alveolar multiplication continues throughout childhood[21,22,23] at least have the merit of presenting direct evidence in favour of their view, while those authors opposing usually use an inferential argument.[18,19]

The proposition that alveolar replication would be expected to continue throughout the somatic growth period,[3] seems a priori sound, and the onus of refutation should be placed on the opposing authors to offer direct evidence, especially bearing in mind the possibility of inter-observer variation. It should also be remembered that not all children of the same post natal age are the same height, and alveolar surface areas and alveolar numbers are strongly related to physical stature.[24]

In view of the current interest in lung development in the first two years of life, I shall present another approach to the problem that is being evaluated now, that of the development of organisations rather than numbers and size alone. These studies are, amongst other objectives, designed to examine the extent to which environmental factors can modify alveolar development. Some of the data presented below is interim and as yet unpublished, but it illustrates an approach to the problem which seems hopeful.

The key to lung organisation appears to be the elastin network. That is not to say that it is the most important element from the lungs' point of view, simply that it is the best histological marker of lung development and organisation.

In the first study,[25] elastins were extracted from samples of peripheral lung tissue, free of pleura, large blood vessels and bronchi. The elastins were divided into alkali soluble and insoluble fractions, and the insoluble fraction was digested and the aminoacid composition determined. From this, the amount of "true" elastin (mature) was calculated, along with the amount of "polar contaminant" (foetal elastin). These values, expressed as a percentage of dry tissue weight, were plotted against post conceptional and post natal age.

The results showed a sharp increase in mature elastin in the last month of intra-uterine life, and the first four months of post natal life, to levels similar to older children and adults.

The increase in elastin clearly begins before birth, and continues across the birth period into the first four months of infancy. This suggested that normally the increase in elastin was predominantly genetically determined, as all the infants up to forty weeks post gestational age died within four days of birth. Some data which we did not publish due to lack of numbers, suggested that if the child were born prematurely, environmental factors such as air breathing stress could over-ride the genetic control, and induce the rise in mature elastin.

To investigate this point, lungs from 37 infants with physical growth parameters within the 50th percentile for gestational age were scored for the stage of development that the elastin had reached.[26]

The following series of steps were defined:-

1.   Elastin in bronchi.

2.   Elastin in bronchioles.

3.   Elastin in terminal bronchioles.

4.   Elastin at the bifurcation of alveolar ducts ("primary crests").

5.   Elastin along alveolar ducts ("secondary crests").

6.   Elastin in terminal saccules.

7.   Centripetal conversion of terminal bronchiole into respiratory bronchiole or alveolar duct.

8.   Sub-division of existing alveoli.

Preliminary investigation showed that the steps had been arranged in the correct order, correlating well with gestational age.

The normal 37 cases were then scored for the most advanced step of development reached, and a further 9 cases were then studied to test the hypothesis that premature birth with survival for two-to-six weeks produced accelerated elastin development. The results are shown in figure 2. It is re-emphasized that these are provisional, interim results. Fig 2 shows the elastin steps plotted against the post natal age in weeks.

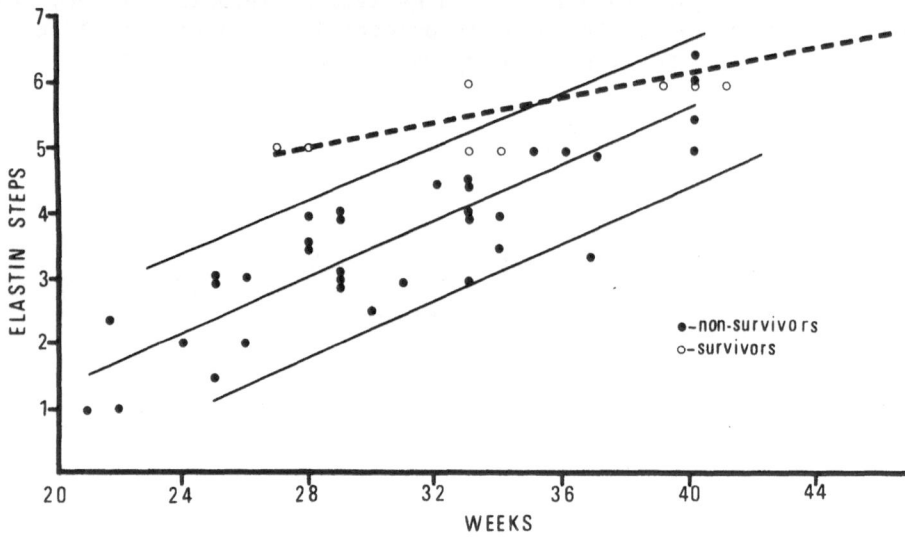

Figure 2

The results show that the main thrust of intra-uterine alveolar development lies between 32-40 weeks gestation, that centripetal conversion and alveolar sub-division are almost wholly full term and post natal patterns of development. The illustration of centripetal conversion at 30-32 weeks in the I.A.P. monograph "THE LUNG"[27] is a caption translocation. It should read ALVEOLATION. The production at full term gestation of an efficient, well alveolated air breathing lung is clearly completely genetically programmed for the infant which remains in utero until term.

Abnormal Development

    i) Post lung resection regeneration and other conditions.
The subject has been well reviewed by Thurlbeck[3] recently, and there is little more to say. The subject is one of considerable controversy, with little human data.

    ii) Pre-term delivery and intensive ventilatory support.
    As shown in figure   the premature infant exhibits markedly accelerated elastin development. The acceleration is the more dramatic the more immature the child. The change is most dramatic in the 28 weeks foetus, which has little, if any, elastin in the terminal acinus.

    An important point is that premature birth induces an acceleration of the normal steps of development, the lung proceeding at a faster rate along the normal sequence of developmental steps.

    Another important point is that the acceleration does not occur evenly throughout the lung. Acceleration of development only occurs in those areas subjected to the stress of air breathing. This suggests that the pattern of elastification, and extent, are pre-programmed, but the actual acini in which the acceleration occurs requires air breathing stress to trigger the induction.

    Alterations in the pattern of elastin deposition can occur, but these are at present only identified in association with the extreme distortion of lung geometry seen in post intensive ventilatory support infants with some form of hyperoxic or ventilator pressure damage, or perhaps asthma (figure 3).

    It may be considered of some interest that this shows that the pattern of alveolar growth can be modified by abnormal ventilation patterns and stresses. The extent of the effect with lesser degrees of abnormal ventilation is quite unknown.

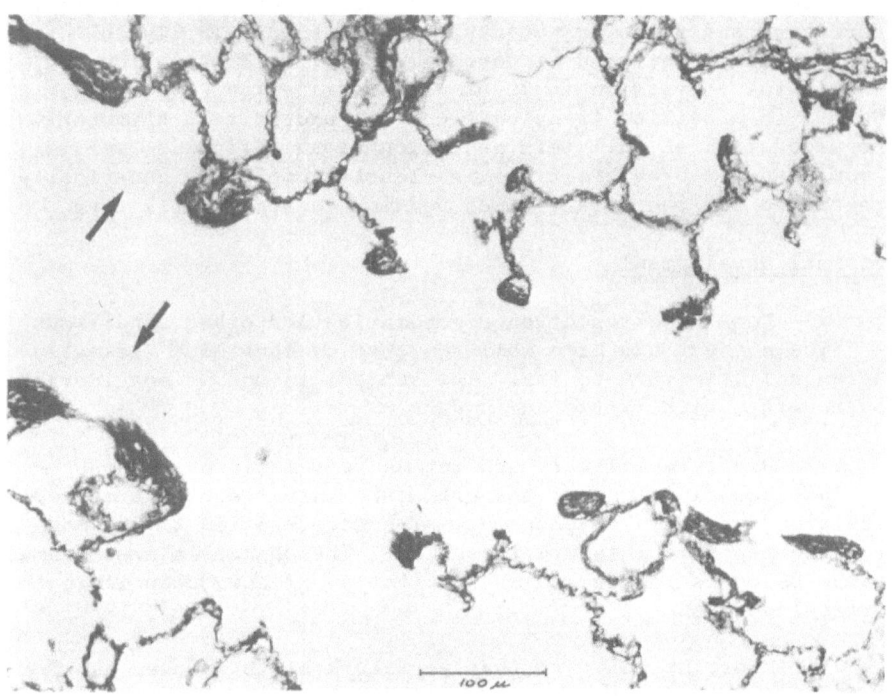

Figure 3A

Figure 3B - 11 years, Asthma. Shows centripetal conversion with the
new crests much further apart than the crests in other ducts, and
much thicker. (A) Central Lobular area.
              (B) Peripheral (sub-pleural) lobular area.

REFERENCES

1. Commission on Embryological Terminology, in "Nomina Embryologica"
    Leningrad 1970, B. Arey and H.W. Mossman, eds., Federation of
    American Societies for Experimental Biology, Bethesda, Md.

2. "The Anatomy of the Developing Lung", J.L. Emery, ed.,
    Heinemann/Spastics International Medical Publications,
    Lavenham (1969).

3. W.M. Thurlbeck, Postnatal growth and development of the lung,
    Amer. Rev. Respir. Dis., 111: 803 (1975).

4. Y. Kikkawa, Morphology and Morphologic development of the
    lung, in: "Pulmonary Physiology of the Fetus, Newborn and
    Child", E.M. Scarpelli, ed., Lea and Febiger, Philadelphia
    (1975).

5. P.H. Burri, J. Dbaly, and E.R. Weibel, The post natal growth
    of the rat lung. I. Morphometry, Anat. Rec. 178:711 (1974).

6. P.H. Burri, The postnataj growth of the rat lung. 111.
    Morphology, Anat. Rec. 180:77 (1974).

7. C.G. Loosli and E.L. Potter, Pre-and Post-Natal development of
    the respiratory portion of the human lung with special reference
    to the elastic fibres, Amer. Rev. Respir. Dis. 80:5 (1959).

8. E.L. Potter and I.M. Craig, Pathology of the fetus and infant,
    2nd and 3rd editions, Lloyd-Luke, London (1961 and 1975).

9. J.L. Emery, Connective tissue and lymphatics, in: "The Anatomy
    of the developing lung", J.L. Emery, ed., Heinemann, Lavenham
    (1969).

10. J.D. Davis, Anatomical Variations of the normal tracheobronchial
    tree, Proc. Staff Meet. Mayo Clin., 3:121 (1928).

11. J.E. Hansen, et al, Branching pattern of airways and air spaces
    of a single human terminal bronchiole, J.Appl.Physiol. 38:383
    (1975).

12. K. Horsfield and G.Cumming, Morphology of the bronchial tree
    in man, J.Appl.Physiol., 24:373 (1968)

13. E.R. Weibel, Morphometry of the human lung, Springer-Verlag,
    (1963).

14. J.E. Hansen and E.P. Ampaya, Lung Morphometry, a fa-lacy in
     the use of the counting principle, J.Appl.Physiol. 37:951
     (1974).

15. E.A. Boyden, The pattern of the terminal air spaces in a
     premature infant of 30-32 weeks that lived nineteen and a
     quarter hours- Am.J.Anat. 126:31 (1969).

16. L.Reid, The embryology of the lung in: "The Development of
     the Lung", A.V.S. de Reuck and R. Porter, eds., Churchill
     London (1967).

17. G. Hieronymi, Uber den durch das Alter bedingten Formwandel
     menschlichen Lungen, Engeb, Allg. Pathol.Anat., 41:1 (1961).

18. M.S.D. Dunnill, Postnatal growth of the lung. Thorax 17:329
     (1962).

19. G. Davies and L. Reid, Growth of the alveoli and pulmonary
     arteries in childhood, Thorax 25:669 (1970).

20. D.G. Fagan, "Structure and function of the human infant's
     lung" M.D. Thesis, Cambridge, 1974.

21. J.L. Emery and P.F. Willcock, The post natal development of
     the lung. Acta Anat. (Basel) 65:10 (1965).

22. W.M. Thurlbeck and G.E. Angus, Growth and ageing of the
     normal human lung, Chest, 67:35 (1975).

23. T. Nakamura, T. Takizawa and T. Morone, Anatomic changes in
     lung parenchyma due to ageing process, Dis.Chest, 52:518
     (1967)

24. G.E. Angus and W.M. Thurlbeck, Number of alveoli in the human
     lung, J.Appl.Physiol, 32:483 (1972).

25. F. Keeley, D.G. Fagan, and S. Webster, Quantity and character
     of elastin in developing human lung parenchymal tissues.
     J.lab.clin.med. 90:981 (1977).

26. P. Gray, Definition of the normal spatial and temporal growth
     pattern of elastin in infants and childrens lungs. B.Med.Sci.
     Thesis, Nottingham University, 1979.

27. C. Langston and D.G. Fagan, Recent advances in neonatal
     pulmonary disease, in: "The Lung" I.A.P. monograph 19,
     Williams and Wilkins, Baltimore (1978).

DISCUSSION

LECTURER: Fagan   CHAIRMAN: Cumming

CUMMING: Thank you. We will open this paper for discussion.
            Perhaps I may take the Chairman's perogative and
            begin by pointing out a logical difficulty in some of
            the slides. It is a very attractive hypothesis that
            structural changes are brought about by functional
            factors, and we have a slide in which there is a
            distribution of air space size and a distribution of
            elastin, the presumption being that the stress
            produced the increase in size and the elastin. The
            alternative hypothesis is that a biochemical change
            produced the elastin and that responded in its own
            way to the stress, and is in fact not the cause, but
            the result of the change. I would like a comment on
            that.

FAGAN: Yes!

CUMMING: You accept that as an alternative hypothesis.

FAGAN: Yes.   That is clearly a possibility.   The only
            problem is that I do not see how we can produce such
            a biochemical change by simple positive pressure
            ventilation.

CUMMING: I am sorry, I did not make myself clear. I am
            suggesting that the biochemical change has nothing to
            do with the pressure. It in fact occurs due to
            normal development, and as a result of this
            biochemical structural change it will respond to the
            pressure in the usual way. It is not the pressure
            that distends it, but the fact that it has more
            elastin makes it more susceptible to distention by
            normal pressures.

FAGAN: Well, I sort of regard this as putting the cart
            before the horse, since it is a very curious thing
            that you only get the mega alveoli appearing in
            children in whom one has applied a high positive
            pressure ventilation, so I find it difficult to
            understand how this child could have been going to
            get mega alveoli and that he of all people was
            selected by chance for positive pressure ventilation.

CUMMING: A fair point.

DENISON: There is a slide in which there were areas of lung
              that were distended and contained elastin, and other
              areas that were collapsed and did not. Are you
              argued that the elastin made the expanded areas more
              compliant. Very attractive. I am not sure how it is
              that if I take a tissue and then add elastic fibres
              to it and do nothing else to it, I make it more
              compliant.

FAGAN: Well, I would not say that you do nothing else to it.
              In order to become a large space, first you must pull
              the collagen apart and you must stretch it. So I
              suggest what you are looking at is a portion of lung
              which has the same sort of ventilatory properties as
              a paper bag. And that you take the paper bag and you
              put some rubber bands on the inside of it, not on the
              outside, and you will find that the rubber bands will
              hold the paper bag. And then, when you blow into it,
              it will quickly expand to its limit, checked by the
              rubber bands, and when you remove the inflation
              pressure it will very quickly collapse until it is
              held open by the elastin. So that is the paper bag
              model. The other part of the lung is more in keeping
              with a balloon. You have to blow hard in order to
              inflate that portion of the lung, and once you take
              the pressure off it just collapses down completely,
              because it has no elastin to hold it open. If you
              want to discuss what it is that chooses one lobule to
              develop into a paper bag and one lobule to develop
              into a balloon, it would take quite a long time but I
              am very happy to discuss that.

BIGNON: In your hypothesis you discussed the causes of the
              difference between your patient and others. She is a
              girl and we have studied such lungs in the new-born
              after long ventilation and we were surprised by the
              fact that there is a stenosis in the airways in the
              area where there is no ventilation, and it is not
              compression, but it is similar to an atelectasis. I
              think that the structure and the biochemistry of an
              unventilated area must be different from that of a
              ventilated area.

FAGAN: Are we talking about cause or effect? Because, what
              duration of ventilation are you talking about?

BIGNON: A long time; a few weeks to one month or more.

FAGAN: Yes, but I am talking about changes that you see in a
              few days and I would agree with you that once you go

secondary changes as well as teritiary changes, which is why I tried not to discuss too much the picture of bronchopulmonary dysplasia since the nature of the initial insult and the consequent reparative phenomena and compensatory growth followed by further damage is all too complicated. This is why I tried to deal with children that were born prematurely and perhaps showed the phenomenon described by Grunwald of some lobules showing low compliance and other lobules showing high compliance, and what happens if you ventilate a lung which shows grossly uneven ventilatory patterns between the lobules for only a few days. What I tried to show then was that if you do that for more than just a few days you run the risk of building this abnormal ventilatory pattern into the child's elastin framework, that the elastin development accelerates and may become abnormal depending on the pressures and the force with which you ventilate.

NESCI: I would like to know what other methods you used to determine which alveoli were functionally efficient and which were not. Did you use scanning electron microscopy or consider the ultrastructural aspects of the lining in both types of alveoli?

FAGAN: I hope I understood your question correctly. Basically we did not consider looking at these lungs ultrastructurally because they are not experimental animals, they are ordinary human autopsy cases. Therefore we were not able to obtain the tissues until ten or twelve hours post mortem. The ultrastructural examination is of limited value in this situation. What we did mainly was gross, overall assessment of the lung, in that we would inflate the lung to a pressure of 30 cm of air to obtain its maximum volume; then we would fix the lung either by instillation of formol saline or reluctantly by using formalin stream, and then carry out routine histological examination, sometimes using serial sections, thin sections, and so on. But no ultrastructure.

NESCI: Using a simple light microscope it is difficult to say what is a small alveolus and what is a large alveolus. I would think it is very difficult, if we only consider the elastin structures at the neck of the alveoli without any reference to the structural and ultrastructural aspects revealed by the scanning electron microscopy.

FAGAN: I am sorry, but I disagree. I am a classical
                pathologist and alveoli· were described by classical
                light microscopists. and that is the definition that
                I use. If you wish to develop an alternative
                definition of an alveolus, using the scanning
                electron microscope, then that is your perogative.
                But I am mostly concerned with ordinary light
                microscopy and ordinary pathology and I think that
                within those limitation you can define what is and
                what is not an alveolus.

CORRIN: It is always intriguing when a difference is found
                between the upper lobe and the lower lobe in human
                adults and these differences can often be related to
                postural differences. The foetus is twisting around
                until the last few months, so I wonder if you would
                be interested when you go home in attempting to
                correlate your observed differences with the
                obstetric history or whether there was a breach
                presentation, a vertex or a transfer lie.

FAGAN: I think that is a most interesting suggestion; I had
                not thought of that.

JEFFERY: I wonder whether you are perhaps being a little
                unkind to stereology and its principles - perhaps
                stereology is itself perfectly good but it is the
                criteria and that one puts on the observation and the
                structure that is under criticism - I see no
                discrepancy between defining clearly what you are
                counting and the stereological principles that you
                adopt and thereby adding more to the information
                which one gets out of the stereology.

FAGAN: I am sorry if you gain the impression that I am anti
                stereology, I am not. I think that stereology is a
                superb tool, and I agree with you entirely that what
                is happening is that we are asking it the wrong
                questions. This is why I wanted to try and refine
                what the question is that we are asking stereology to
                record or to measure for us. Unfortunately there are
                a number of assumptions that must be made in order to
                use the stereological method at the moment. The
                first assumption is that the lung grows equally in
                all three axes. I am not entirely satisfied that
                that is true. I think it is an assumption that we
                should examine, rather than just accept as being
                true. The second assumption that I am not happy
                with in stereology is the question of the limits in
                which one can apply the method. You see, in the new-

born lung that I showed you, I measured the first and third moments of distribution and I found that it came to 1.72. That is an incredible range of shape and size. The derivation of the formula does quite clearly state that you are dealing with structures of the same shape and of a similar size. I just wonder if the variation of the new-born lung is not too great to allow the same sort of stereological assumption to be made that you can quite safely make in the older child's lung. So, I do not challenge the application of stereology in an age range over about 6 or 7 years: the whole structure has settled down and it is much clearer what is an alveolus and what is a saccule and what is an alveolar duct. But I think there are terrible problems of interpretation in the new-born infant's lung.

JEFFERY: I would like to hear your comment in relation to my second point, and that is that perhaps we are labouring the point of alveolar number and we ought to be more concentrating indeed on the definition of an alveolus and more perhaps on surface area for which, again, there is a good stereological principle and the intercept method would tell us more of surface area, which after all is the important thing in terms of gaseous exchange.

FAGAN: I am delighted to hear you say that, because it is very much the conclusion that I have come to. I think that one can and does define an alveolus by one's own criteria and perhaps we should drop this search for an absolute number of alveoli as an answer to the number of alveoli in the lung at a certain age. Perhaps we should use the mean spherical analogue; perhaps we should concentrate on measuring the internal curvature of the lung and relating that to the internal surface area. Perhaps we should concentrate on demarcation into alveolated exchange surface and non-alveolated exchange surface. I am sure you have seen plenty of areas of gaseous exchange which are not obviously in alveoli. I think clearly we are on the same line.

CUMMING: I wonder if we might mention the dreadful word fractal in this connection implying that the surface area of any structure is determined solely by the magnifying power of the objective with which we examine it.

A VISITOR: Would there be any other relevant classical method or classical criterion for evaluating lung growth instead of using solely the staining method?

FAGAN: I am not quite sure whether this is a loaded question, because there are hundreds of other ways for assessing the growth of the lung: there are radiographical ways, there are pulmonary functional parameter methods, there are straightforward post-mortems measuring the height and the breadth of the lung. I think I am missing your point. If you mean histological methods, I cannot think of others off the top of my head.

CUMMING: Perhaps he is thinking of autoradiography and DNA conversion that one can look at in this way. Are you?

FAGAN: I am a simple histologist.

EMERSON: I was wondering if you have taken the opportunity to look at the collagen content of these lungs. You concentrated on the elastin, but did you look at, either histologically or biochemically, the amount of collagen?

FAGAN: Let me say we did not specifically look at the collagen content, except that the data might be retrievable in the biochemical context because of our method of purifying the elastin. We in fact threw the collagen away, but I think that we know the weights at each stage of the extraction process. So that in fact the fraction that was thrown away might be recoverable. The second thing is that the strain I was using quite clearly shows that while you get this exaggerated, heavy deposition of elastin you also get much exaggerated, heavy deposition of collagen. So in this particular situation the collagen and the elastin go hand in hand together.
            I strongly suspect also in the last case I showed you of bronchial asthma that those megadepositions also contain some smooth muscle, but there are still problems in demonstrating that.

EMERSON: Just another question: did you use any reticulin stain?

FAGAN: Not in this particular study. But yes, I have used reticulin stain.

EMERSON: I was just curious to see the amount of reticulin in those young lungs and possibly try and correlate that with a specific type of collagen.

FAGAN: I think that this is a most interesting field and I certainly have it at the back of my mind. Now that there are special staining methods available for fractionating the collagen components I certainly had it in mind as something to do.

HEATH: In considering the lung and its environment, there is no doubt that the size of the lung and the development of the lung are closely related to the partial pressure of oxygen in the alveolar space. I think that there is very convincing evidence, looking at populations of highlanders throughout the world, that if they develop in the hypoxic conditions of high altitude they develop larger lungs. Ten years ago it was shown that if you have three groups of growing rats, the hypoxic group develop lungs which are larger; the hyperoxic group develop lungs which are smaller than the normal group. Have you any idea how partial pressure of oxygen makes this important change to the structure of the lung and are there more alveoli there? If there are, do they come from centripetal conversion, or is this a peripheral thing with the subdivision of the existing alveoli?

FAGAN: This is absolutely true. All I can suggest is that the partial pressure of the oxygen does indeed alter the mechanics. If you have a hyperoxic environment then you have less nitrogen splinting, and if you have a low oxygen environment then equally the proportion of nitrogen will be higher, although perhaps if you are in a low partial pressure of atmosphere altogether then the mechanics may be different again. I would not like to suggest how this happens, but I would hope that it is explicable in terms of the normal growth pattern, since it is obviously an adaptation by man to a hostile environment. I could only say that it would be most interesting to look at some children's lungs from these areas, and then perhaps we could see what is happening.

NESCI: I would like to know whether you have any experience, from the pathologist's point of view, of the administration of cortisone to the mother to increase the maturity of the foetus, since from the 28th to the 32nd week we often see fetal immaturity and must make the mother deliver due to pathological reasons.

What is the action of cortisone on alveolar increase?

FAGAN: I do not think that there is any evidence that
                cortisone administration will increase the number of
                alveoli in the dose and the duration that it is given
                in order to induce the adult metabolic pathway of
                surfactant. I think, if there were any evidence, it
                would be very tenuous because of the susceptibility
                of the results to the assumptions as to what
                constitutes an alveolus. So, I do not think one need
                worry about the effects on the fetal lung, of a very
                short, high dose administration of steroids in order
                to induce the surfactant metabolism.

CUMMING: Thank you very much. I think we should call the
                first morning's session to a close. We now break for
                30 minutes and reassemble here at 11.30.

DISCUSSION

LECTURER: Cosmi   CHAIRMAN: Cumming

CUMMING: Thank you very much for your presentation. We now
                have a discussion period of about fifteen minutes.
                Would anyone like to ask the first question?

HEATH: One question I would like to ask is this: you did
                not appear to mention in your paper or show on your
                slides any reference to the argyrophilic cells, which
                we know occur in the airways. As pathologists we
                know that there exists in many species of animals
                such cells, - the APUD cells that are capable of
                putting out biogenic amines, which resemble small
                airway chemoreceptors or carotid bodies. You are
                talking about amines stimulating respiration and of
                the importance of chemoreceptors and then, strangely
                enough this group of cells which we know is in the
                bronchial airway and we know moreover that in many
                species these cells greatly diminish in number
                shortly after the perinatal period, which suggest
                that they must have a jolly important function around
                that time, and yet you did not mention them in the
                paper. Why was that?

COSMI: You are right. In a 30 minute talk it was not
                possible to go into the details of these cells ...
                the APUD system of the lung.

HEATH: Would you like in a few sentences now, just to tell
                us what your view is on the role of these cells?

COSMI: I think that they are very important, not only to
                explain respiratory changes in the fetus, but
                particularly the metabolic changes in the lung,
                referring both to the response to chemical stimuli
                and to the production of surfactant.

CUMMING: We have heard a particular view, so could I pursue
                this for a moment? Perhaps it will be unfair, (but I
                am noted for being unfair), to ask a pathologist to
                give a physiological view. So perhaps, Donald, you
                could tell us your view of the role of the argyrophil
                cells during the pre- and post-natal period?

HEATH: Well, of course there are various aspects; they are

called APUD cells, so we imagine that they must put
out biogenic amines. What sort of amines they could
put out I do not know. Histamine, of course, has
been very much in the news with regard to dilatation
and constriction of vessels. Histamine secreted by
mast cells is a vasoconstrictor. Now, I think in the
human lung, in contrast with this view, histamine has
a vasodilating effect on the pulmonary vasculature
and it may be that these conditions you described
with the opening of the pulmonary vasculature are
related to the secretion of some amine which has a
vasodilating effect. On the other hand, of course,
perhaps they are acting as small airway carotid
bodies. So there are two lines along which they
could act.

CUMMING: I should say first that APUD means Amine Precurser
              Uptake and Decarboxylation. Do you believe that the
              secretion of biogenic amines is notable for its
              effect on the pulmonary circulation or for its effect
              upon the smooth muscle of the conducting airways?

HEATH: I do not have any answer for that. That is why I
              have come all this way to Erice to find out.

CORRIN: We know that the lung contributes surfactant to the
              amniotic fluid and that at birth at least there is
              considerable absorption of fluid from the lungs. I
              wonder if Cosmi would like to tell us whether he
              considers the lung a net contributor to the amniotic
              fluid or whether they absorb more than they
              contribute. I wonder also whether you would like to
              consider the relationship of pathological conditions,
              such as hypoplasia of the lung to oligohydramnios -
              which is the cause and which is the effect?

COSMI: Although amniotic fluid phospholipids could be
              derived from other sources, like fetal urine and
              fetal skin, as is the case with other compounds like
              uric acid and certain hormones, there is no doubt
              that there is a direct transfer of phospholipids from
              fetal pulmonary fluid to the amniotic fluid. This
              evidence has been demonstrated in the past and in
              more recent studies.

CUMMING: There is liquid continuity between amniotic fluid and
              pulmonary water anyway, and diffusion and convective
              mixing could produce the results you described. Can
              you be more specific and say: is the flux of fetal
              fluid into the lungs or out of the lungs?

COSMI: In general, if you inject radio-opaque material as Potter did, into amniotic fluid you can recover it in the fetal pulmonary fluid. It seems that in general the flux is from the fetal lung to the amniotic fluid, and only during delivery there may be the reverse, i.e. amniotic fluid is aspirated into the fetal pulmonary fluid, and this of course explains certain conditions like meconium aspiration syndrome. in in fetal pulmonary fluid than in amniotic fluid. The viscosity also is different, that of the amniotic fluid being greater. So, under normal conditions the flux should be from the fetal pulmonary fluid to the amniotic fluid. But the movement of this inward and outward is minimal, about 1 ml under normal conditions. Then there are greater variations with grunting or gasping. So in general, in the healthy foetus, it should be from the fetal pulmonary fluid to the amniotic fluid.

CUMMING: Thank you Cosmi, that seems a fairly specific answer.

COSMI: I am sorry, you asked about pulmonary hypoplasia and oligohydramnios. This relationship certainly exists, and this is why fetal breathing movements may be important in the production of surfactant. In cases of diaphragmatic hernia or section of the phrenic nerve with no respiratory activity of the fetus then there is pulmonary hypoplasia, and there is a lot of evidence, as reported originally by Potter and Bolender, that cases of oligohydramnios and anuria, (and this is also the role of fetal urine), are associated with fetal pulmonary hypoplasia.

CORRIN: Would you like to say what it is at birth which causes the reversal of the direction of the fluid, from being a net secretor of fluid to having the lung as a net resorber of fluid. Can you explain this in terms of osmotic gradients or something like that?

COSMI: First of all there is an oncotic gradient, as you said, which explains part of resorption. Secondly, as certain alveoli are starting to be open, the angle increases and therefore the radius increases, and therefore there is less pressure required for this movement of which will favour the outward flow of the fetal pulmonary fluid. And third there is an opening of interstices in the capillaries, in the pores, in the alveoli which may favour this transfer of fetal pulmonary fluid into the lymphatic circulation. It could be the high PO_2 in these bubbles which decreases the pulmonary vascular resistance and therefore increase pulmonary flow.

CORRIN: In terms of the Starling hypothesis, I think you would expect the forces for transudation from the pulmonary vessels to the lung fluid to be greater after birth because the pulmonary circulation opens up. I do not know what happens to the capillary pressures, but I imagine that they are greater after birth.

COSMI: It will depend, I think, a lot on the initial oncotic pressure.

CORRIN: Why should that change?

COSMI: You mean in the blood?

CORRIN: Yes.

COSMI: Because there is a change in the circulation and maybe even a shift of fluid, at birth, into other compartments. I think we have to consider the change in outputs which take place at birth in the pulmonary circulation. Once the pressure in the fetal pulmonary artery drops from about 75 mm to around 55 mm, there is a period in which the flow through the foramen ovale or ductus arteriosus may be bidirectional or maybe preferentially from left to right, and therefore there is this dynamic change which maybe also influences fluid movement through the compartments.

CORRIN: Do we know what the capillary pressure is in the lungs before and after birth? If it actually drops with birth then you can explain it all very simply.

COSMI: Yes, it does.

CORRIN: Thank you.

NESCI: I would like to know what is the action of bradykinins. We know that when the umbilical cord is clamped bradykinins are released. They would seem to be the "primum movens" for the opening of pulmonary vessels. Secondly you spoke of neonatal temperature, that is the passage from $37^{\circ}C$ to ambient temperature, that if a new-born is plunged into water at $40^{\circ}C$ respiration is arrested. This is true, but only to a certain extent, because we know that if a premature baby or a neonate with respiratory distress is left at room temperature then thermogenetic, glycogenetic and glycolytic phenomena occur which cause the pH to diminish considerably and the initial respiratory act also decreases. Thank you.

COSMI: As far as bradykinins are concerned, these have always been considered as mediators. For example, the effect of hyperoxia or hypoxia has always been ascribed to such mediation especially in the case of high PO_2. The reduction in vascular resistance has been attributed to bradykinin and more recently to prostaglandin I, G2 etc. Undoubtedly they play an important role as mediators in the reduction of vascular resistance and therefore in the increase in flow at the pulmonary level but also at the level of the ductus arteriosus. It is known that this prostaglandin is produced together with bradykinin, histamine and so on. Some of the drugs are used to treat neonatal conditions such as the persistence of fetal circulation, pulmonary hypertension and so on. Therefore their importance is beyond any doubt. As far as temperature is concerned, first of all it is difficult to discriminate between thermal receptors, proprioceptors, pain receptors and so on, because, among other things, the same fibres are involved in transmission. So it is difficult to differentiate the various impulses. Certainly thermal stimulation is important, though it is not essential, because there are situations where, for example in the fetus of sheep, if the circulation is not impaired, respiration is stimulated by low temperature. Then, without any reactions involving PO_2 or PCO_2, if it is heated respiration is arrested.

WIDDICOMBE: May I raise a semantic point? Early in your talk you described the grunting of the fetus and yet as far as I know if you have your airways full of liquid, of amniotic fluid, you cannot grunt, because grunting is a sound which you only have with gas in your airways.

COSMI: The phenomenon is similar to grunting.

WIDDICOMBE: Your slide was also very similar to a cough or an expiratory muscle contraction, and I think this is an interesting point because we are told that the new-born baby, if it has respiratory distress syndrome, can grunt. This is thought to be due to laryngeal constriction. I wonder if you would try to explain the mechanism in the records which you called grunting. Are they, for example, expiratory muscle contractions or are they expiratory muscle contractions plus laryngeal constrictions? And if they are, is that something like a cough, which we are told that the new-born baby has not got? And the last point that goes along with that: do you think the presence of a catheter in the trachea of the fetus might be stimulating something like cough receptors?

COSMI: That point is correct; as for grunting, of course it is similar to grunting. If we believe in the pharyngo-laryngeal outlet, the functional sphincter that prevents the fetal pulmonary fluid from entering the amniotic fluid except during swallowing or communication, then you could explain part of this. Another point could be the intensive stimulus we applied, in this case it was electrical and there was muscle contraction which could explain in part the response due to the role played by the superior laryngeal nerve. When you introduce normal saline there is no response, whereas if you introduce distilled water or heterologous milk, there is a feed-back response. That could in part explain the reflex during deglutition of the new-born, certainly in case of apnea of prematurity or sudden infant death.

CUMMING: Thank you very much Cosmi. There are many people who want to ask questions, but since the time has now arrived for the next speaker we must postpone it. I suspect very much that the questions you wish to ask now will be equally relevant. Thank you. Can I call on Dr. Fagan to tell us something about the growth of the human lung.

PROTECTIVE MECHANISMS OF THE RESPIRATORY TRACT

P.S. Richardson

Department of Physiology,
St. George's Hospital Medical School
Cranmer Terrace, London, SW17 0RE, U.K.

INTRODUCTION

To say that the respiratory tract has protective mechanisms
implies that there are hazards against which it needs protection.
This paper will start by outlining the principal threats to the lung
from the environment and go on to describe the mechanisms for the
defence of the airways.

THREATS TO THE RESPIRATORY TRACT

Of all the organs in the body the lungs present the largest
surface, some 70 m^2, in contact with the environment. The air in the
lung is constantly changing as we breathe: the respiratory tract is
exposed to about 10^4 litres of air each day, about two thirds of which
penetrates to the respiratory exchange regions of the lung and one
third only to the conducting airways. The principal threats are:

(1) The air inhaled is almost invariably cooler than body core
temperature and is not saturated with water vapour. This gives
inspired air the potential to cool and dry both the delicate
respiratory surface of the alveoli where gas exchange occurs and the
ciliated surface of the conducting airways which play an important
role in lung cleaning.

(2) The dust which inspired air contains poses the next threat to
the well-being of the lungs. Even people who inhabit clean environments inhale about 10^{10} dust particles every day and workers in dirty
atmospheres (e.g. miners, cement and grain workers) may breathe in
a hundred times that number (Brain and Valberg, 1979). The accumulation of sufficient dust would obviously clog the airways and

prevent gas exchange. A person living in a clean environment may
inhale 100 g of dust in a lifetime. It has been estimated that in
the course of a working lifetime a miner from a dirty mine might
inhale 4 kg of particulate matter.

(3) Many of the dust particles contain living organisms, viruses,
fungi, bacteria, mycoplasma, which could colonize the lungs. One
important function of the lung defences is to neutralize and dispose
of potential pathogens.

(4) The inhalation of noxious gas and smoke is another threat to
the airways. A wide variety of chemicals used or produced by
industry, as well as tobacco smoke, fumes from fires and volcanoes
and riot control agents, can potentially harm the lungs. The
respiratory tract has mechanisms for removing many of the irritants
from inspired air before they reach the lungs and, when they do
penetrate the lower respiratory tract, of diluting and disposing of
them.

(5) The final hazard discussed here comes from the inhalation of
matter from the alimentary tract; saliva, food, drink and vomit.
The reflexes involved in swallowing and vomiting normally stop
breathing and close the larynx to prevent entry of foreign material
into the airway, but sometimes these protective reflexes work poorly
and food, etc. spill into the lower airway where, if they are not
cleared effectively, they can cause pneumonia.

THE DEFENCES OF THE AIRWAYS

There are many defence systems in the respiratory tract and
some are considered in detail in accompanying papers in this volume
(viz. the defensive reflexes, Sant'Ambrogio; nasal mechanisms,
Widdicombe; humidifying and warming of the inspired air, Cumming;
immune mechanisms in the lung, Molina, Turner-Warwick; ciliary action,
Sanderson; alveolar defence mechanisms, Corrin). This paper will
briefly outline the principal protective mechanisms at each level of
the respiratory tract and discuss how these defend the lungs against
the threats described above. The role that mucus plays in lung
protection will be emphasised because none of the other papers deals
with this aspect of the subject.

The Nose

Inspired air leaves the nasal cavity cleaner, warmer and more
humid than it entered. The anatomy of the nose is adapted to allow
it to fulfil the role of air conditioner for the lungs in several
ways (Proctor, 1977). The nostrils contain a mesh of hairs which
filter the very large particles of dust and fluff from the airstream.
As air passes from the nostril to the vestibule of the nose it is
forced through the narrow ostium internum and simultaneously

deflected so that the direction of flow changes from upwards to backwards. The velocity of the airstream increases in the same way that a river runs faster as it enters a narrow gorge. The larger dust particles (virtually all of those greater than 10μ diameter and many between 2 and 10μ) have sufficient momentum to prevent them from following the twist of the nose at this speed and are thrown against the wall of the nasal cavity where they are caught by the layer of mucus. The turbinate bones project into the nasal cavity and disturb the inspiratory airflow further, hurling more dust particles against the airway wall and ensuring intimate contact between all parts of the airstream and the walls of the nose. The mucosal tissue overlying the turbinates has a particularly rich blood supply and contains numerous mucous glands. The warm, moist wall of the nasal cavity, coupled with the turbulent airflow, ensures that there is efficient transfer of heat and water vapour to the inspired air. Exactly the same features allow the nose to remove the soluble gases from inspired air (Anderson et al., 1974). Rather surprisingly the nose becomes more efficient at removing soluble gases when airflows are faster, perhaps because the greater turbulence increases contact between gas and the nasal lining (Aharonson et al., 1974). The efficiency of the nose at cleaning inspired air is beyond question (Proctor, 1977) but the degree to which it completes the warming and humidifying process is still a matter for dispute (Proctor and Swift, 1977; Cumming, 1980).

The Larynx

The oropharynx acts as a conduit for food, drink and air but the larynx is the first part of the lower airway restricted to air. The larynx evolved as a valve to exclude material in the alimentary tract from the lower airway (Negus, 1949). The swallowing and vomiting reflexes involve closure of the glottis so that food etc. cannot penetrate it. There is simultaneous apnoea. Small quantities of water which flow into the larynx during normal breathing will elicit the swallowing reflex (Storey, 1968). Any more powerful stimulus, either mechanical or chemical, causes coughing (Boushey, Richardson and Widdicombe, 1972). It is easy to see how a cough would dislodge foreign matter from the larynx but it is less obvious how it would expel gaseous irritants. One facet of the laryngeal cough reflex which might help the respiratory tract rid itself of soluble gases is the simultaneous increase in mucus secretion from the trachea (Phipps and Richardson, 1976).

Defence of the Tracheobronchial Tree

Much of the dust with a diameter less than 3μ, and some larger particles, evade the dust-trapping mechanisms of the nose and penetrate to the tracheobronchial tree. What happens to dust particles here depends partly on their size, partly on chance

(Hounam and Morgan, 1977; Brain and Valberg, 1979). The larger
particles are likely to impact upon the airway walls, particularly
where the airstream changes direction at the bifurcations of the
larger airways. In the small airways, however, air movement
becomes sluggish and the chance of particles colliding with the
walls lessens. Many particles here, however, settle under the
influence of gravity (sedimentation). With the very small diameter
dusts (< 0.1μ) another process becomes more important : the random
jostling of air molecules imparts sufficient movement to these that
they are likely to hit the bronchiolar walls (Brownian movement).
Many of the intermediate sized particles (0.1 - 1.0μ) avoid being
caught by any of these mechanisms and are breathed out without
harming the lungs in any way.

There are two principal mechanisms for removing dust which
settles in the conducting airways : mucociliary clearance and cough.
Ciliated epithelium lines the conducting airway from the trachea to
the terminal bronchioles. For the cilia to move particles it is
essential that the dust is embedded in mucus. Naked cilia, with no
mucus covering, beat ineffectively and fail to move dust particles
(King et al., 1974). The large airways (bronchi and trachea) have
numerous secretory cells in their walls which provide sufficient
mucus to blanket the epithelium (Yoneda, 1976). In the bronchioles,
however, there are no submucosal mucus glands and the epithelial
goblet cells are sparse (Reid, 1954). How do the ciliated cells
function with so few mucous cells? It is possible that two other
cell types furnish the secretions necessary for efficient ciliary
action : the Clara cells, abundant in the bronchioles, contain
secretory granules though they do not stain like those in the
classical mucous cells (Thurlbeck, 1977), and the ciliated cells
themselves have a layer of glycoprotein on their luminal border
(Spicer et al., 1971) which they may liberate into the airway lumen
in the form of mucus (Gallagher et al., 1978).

There is growing evidence that acute airway irritation increases
the production of airway mucus but slows mucociliary transport
somewhat (Carson, Goldhamer and Carpenter, 1965; Pavia, Thomson
and Pocock, 1971). Increased mucus production dilutes the irritant,
and the augmented secretion together with the slowed transport would
increase the thickness of the mucus lining to the airway. It is
debatable whether the slowing of mucociliary transport is a protective mechanism despite the thickening of mucus coating it produces.
Airway reflexes acting through cough receptors and parasympathetic
nerves can explain part of the increase in secretion (Phipps and
Richardson, 1976) but irritants also increase mucus output by a
direct action even in the denervated airway. The direct action may
involve the liberation of chemical mediators such as prostaglandins
(Richardson et al., 1978) and the leakage through the damaged
capillaries and epithelium of serum proteins, the presence of which

in the airway lumen stimulates secretion (Hall, Peatfield and Richardson, 1978).

Camner, Helström and Philipson (1973) showed that the inhalation of inert dust hastened the rate at which cilia removed radioactive markers from the lung. We have recently found that similar dust enhances mucus secretion from the airway (Peatfield and Richardson, unpublished). Reflexes account for part of this effect. There is evidence then that both inert dusts and irritant chemicals can summon the secretions required for their removal by cilia.

Cough is the other mechanism by which the airways rid themselves of dust, irritants and pathogens. In a cough the patient will inspire, close his larynx and build up a pressure against this before he suddenly opens it to emit a blast of air. The air velocity needed to shear mucus from the bronchial wall can only be attained in the larger airways, perhaps the first six or eight generations of bronchi (Leith, 1968) so the small bronchi and bronchioles probably depend solely on the cilia to clear them. The non-newtonian physical properties of mucus are appropriate to allowing a blast of air to shear it from the airway wall. A blob of mucus will bend rather than flow in response to a gentle shearing force, because stronger forces are required to break the chemical bonds which link adjacent glycoprotein molecules in the mucus scaffolding. A greater force will, however, cleave the bonds and induce flow. Once the blast of air from a cough has broken the mucin scaffolding in one plane it will shear easily along that plane and join the expired airstream (Clarke, 1973).

There are important changes in the airway during coughing. Bronchial smooth muscle contracts as part of the cough reflex, narrows the airway and increases air velocity. There is dynamic compression of part of the airway as intrathoracic pressure rises and again this will ensure a rapid airflow in this segment (Ross Gramiak and Rahn, 1955). Coughing is only effective providing that there is a sufficiently thick layer of mucus on the airway wall, as the normal thin layer is almost impossible to shear (Yeates et al., 1975). It is interesting to find that the sort of airway irritation which provokes coughing will also, through a reflex which involves the same afferent nerves, elicit mucus secretion (Phipps and Richardson, 1976).

This paper has concentrated on the role of mucus in removing dirt and irritants from the lung but lung secretions have direct actions against pathogens as well. They contain immunoglobulins, principally lgA (Kaltreider, 1976), which is mainly synthesised locally by plasma cells in the submucosal tissue, joined in a dimeric form and secreted across the epithelium.

Defences of the Lung Periphery

Coughing is ineffective in small airways and mucociliary transport is absent beyond the terminal bronchioles, so other cleaning mechanisms are needed for the respiratory regions of the lung. The alveolar macrophages, phagocytic cells of the reticulo-endothelial system, scavenge dust and bacteria which settle on the alveolar surface or respiratory bronchioles (see recent review by Green et al., 1977). The phagocytosis and killing of bacteria are important functions of the healthy macrophage, but these can fail in disease (Green et al., 1977), and it is likely that this failure is one of the causes of illnesses such as pneumonia, tuberculosis and cystic fibrosis. Once macrophages are full of ingested particles they leave the alveolar surface, and join the muco-ciliary escalator or the lymphatics of the lung. Many macrophages cross the respiratory bronchiole surface to reach the terminal bronchioles where the mucociliary escalator begins, but others probably penetrate the alveolar lining to travel via the interstial space of the alveolar membrane and the lymphatic channels of the respiratory bronchioles before re-crossing the epithelium into the lumen of the terminal bronchioles. Other macrophages remain within the lymphatics and either stick in the lymph nodes or enter the circulation via the thoracic duct.

When macrophages ingest certain particles such as asbestos and silica they release some factor which causes fibroblasts to lay down collagen. This is almost certainly the mechanism of pulmonary fibrosis in the dust diseases (Allison, 1977).

CONCLUSIONS

This paper had described the major threats from the environment to the lungs and respiratory passages and outlined the first line non-specific defence mechanisms which generally overcome these threats by removing them. Once pathogens have started to multiply in the lungs there are pathogen-specific immunological defence mechanisms which re-inforce the processes described here. If these fail the lungs enlist the general immune defences of the body.

REFERENCES

Aharonson, E.F., Menkes, H., Gurtner, G., Swift, D.L. and
    Proctor, D.F., 1974, Effects of respiratory airflow rate on
    removal of soluble vapors by the nose, J.appl.Physiol.
    37:654-657.
Allison, A.C., 1977, Mechanisms of macrophage damage in relation to
    pathogenesis of some lung diseases, in: "Respiratory Defense
    Mechanisms, Part II", ed. Brain, Proctor, Reid.
    Marcel Dekker, New York.

Andersen, I., Lundqvist, G., Jensen, P.C. and Proctor, D.F., 1974,
    Human response to controlled levels of sulfur dioxide,
    Arch Environ Health., 28:31-39.
Boushey, H.A., Richardson, P.S. and Widdicombe, J.G., 1972, Reflex
    effects of laryngeal irritation on the pattern of breathing
    and total lung resistance, J. Physiol., 224:501-513.
Brain, J.D. and Valberg, P.A., 1979, Deposition of aerosols in the
    respiratory tract, Am.Rev.resp.Dis., 120:1325-1373.
Camner, P., Helström, P.A. and Philipson, K., 1973, Carbon dust
    and mucociliary transport, Arch.Environ Health., 26:294-296.
Carson, S., Goldhamer, R. and Carpenter, R., 1966, Responses of
    ciliated epithelium to irritants. Mucus transport in the
    respiratory tract, Am.Rev.resp.Dis., 93:86-92.
Clarke, S.W., 1973, The role of two-phase flow in bronchial
    clearance, Bull. Physio-pathol. Resp. 9:359-376.
Cumming, G., 1980,

Gallagher, J.T., Hall, R.L., Jeffery, P.K., Phipps, R.J. and
    Richardson, P.S., 1978, The nature and origin of tracheal
    secretions released in response to pilocarpine and ammonia,
    J. Physiol, 275:36-37P.
Green, G.M., Jakab, G.J., Low, R.B. and Davis, G.S., 1977,
    Defense mechanisms of the respiratory membrane, Am.Rev.resp.Dis.
    115:479-514.
Hall, R.L., Peatfield, A.C. and Richardson, P.S., 1978, The effect
    of serum on mucus secretion in the trachea of the cat.
    J. Physiol, 282:47-48P.
Hounam, R.F. and Morgan, A., 1977, Particle deposition in: Respiratory
    Defense Mechanisms, Part I, pp125-156, ed., Brain, Proctor,
    Reid. Marcel Dekker, New York.
Kaltreider, H.B., 1976, Expression of immune mechanisms in the
    lung, Am.Rev.resp.Dis., 113:347-379.
King, M., Gilboa, A., Meyer, F.A. and Silberberg, A., 1974,
    On the transport of mucus and its rheologic simulants in
    ciliated systems. Am.Rev.resp.Dis., 110:740-745.
Leith, D.E., 1968, Cough, J.Am.Phys.Ther.Assoc., 48: 439-447.
Negus, V.E., 1949,"The comparative anatomy and physiology of the
    larynx," Hafner, New York.
Pavia, D., Thomson, M.L. and Pocock, S.J., 1971, Evidence for
    temporary slowing of mucociliary clearance in the lung
    caused by tobacco smoking. Nature, 231:325-326.
Phipps, R.J. and Richardson, P.S., 1976, The effects of irritation
    at various levels of the airway upon tracheal mucus
    secretion in the cat. J. Physiol., 261:563-581.
Proctor, D.F., 1977, The upper airways, 1. Nasal physiology and
    defense of the lungs. Am.Rev.resp.Dis., 115:97-129.
Proctor, D.F. and Swift, D.L., 1977, Temperature and water vapour
    adjustment, in: "Respiratory Defense Mechanisms, Part I"
    pp95-124. ed., Brain, Proctor, Reid. Marcel Dekker, New York.

Reid, L., 1954, Pathology of chronic bronchitis, Lancet, (i).
    275-278.
Richardson, P.S., Phipps, R.J., Balfre, K. and Hall, R.L., 1978,
    The roles of mediators, irritants and allergens in causing
    mucin secretion from the trachea. Ciba Symposium 54 (new
    series), Respiratory Tract Mucus, 111-126, Elsevier,
    Amsterdam.
Ross, B.B., Gramiak, R. and Rahn, H., 1955, Physical dynamics of
    the cough mechanism, J.appl.Physiol., 8:264-268.
Spicer, S.S., Chakrin, L.W., Wardell, J.R. and Kendrick, W., 1971,
    Histochemistry of mucosubstances in the canine and human
    respiratory tract, Lab.Invest, 25:483-490.
Storey, A.T., 1968, Laryngeal initiation of swallowing,
    Exp.Neurol., 20:359-365.
Thurlbeck, W.M., 1977, Structure of the lungs, in: "International
    Review of Physiology, Vol.14, pp1-36, ed., Widdicombe, J.G.,
    University Park Press, Baltimore.
Yeates, D.B., Aspin, N., Levison, H., Jones, M.T. and Bryan, A.C.,
    1975, Mucociliary transport rates in man. J.appl.Physiol.,
    39:487-495.
Yoneda, K., 1976, Mucous blanket of rat bronchus, an ultra-
    structural study. Am.Rev.resp.Dis., 114:837-842.

D I S C U S S I O N

LECTURER: Richardson   CHAIRMAN: Cumming

DENISON: How does an alveolar macrophage know when it is full?
                And how does it know where to go once it is full?

RICHARDSON: A challenging question sure enough. They probably do
                not know when they are full, in fact there is quite a
                lot of evidence that an alveolar macrophage can
                become too full and then die. If you give coloured
                dusts into the lung you find that they are taken up
                in macrophages when you section the lungs, and that
                some collections of dust are not in any obvious
                macrophage. The explanation for this that has been
                put forward is that the dust has been taken up in
                macrophages which have become too full and died,
                depositing all the dust that they contain on the
                alveolar surface. How then do they know how to get
                into the ciliated airway? That is very difficult to
                answer. One explanation is that there is a screen of
                alveolar surfactant which makes its way from the type
                II cells over the surface of the alveoli and finally
                into the conducting airways. I think that there is
                good evidence that there is surfactant, at least in
                the smaller conducting airways, because it can be
                seen there under the electron microscope, so that is
                a possible explanation. They may be carried there on
                an extension of the mucociliary escalator which works
                on the principle of surfactant flow.

LAVAL: I would like to know whether the paranasal sinuses
                have not been mentioned because their activity is
                neglectable or because they are included as a part of
                the nasal mucosa. Secondly, I would like to know
                something more about what you think is the function
                of Clara cells. The third question is about the
                relationship between the number of cilia and the
                amount of mucus produced. Dr. Jeffrey and myself, in
                1974 in Marseille, observed - Jeffery in the rat and
                myself in man - that ciliated cells have a remarkable
                number of microvilli, so ciliated cells would seem to
                be involved in the process of resorption, as it seems
                that the amount of mucus produced exceeds the
                requirements, and this turnover -
                production/resorption - is necessary to guarantee the
                optimum quantity of mucus. The last comment is about

macrophages. Most macrophages are not eliminated by
the bronchi but cross the alveolar septum again where
there is some mechanism of lysosomal activity, all
the material collected at the alveolar level being
found at the level of lymphnodes.

RICHARDSON: Thank you for those four questions, I shall try to
answer them. About the paranasal sinuses, I think
that they are chiefly important in modifying the
sound of speech. They also act as a source of mucus
which will flow over the surface of the nose, but the
inspired air stream does not enter the paranasal
sinuses to any great extent, so they cannot modify
this. About the Clara cells, it is still a mystery
what they produce. Widdicombe is really the person
to ask about this, because he is collecting Clara
cells secretions at the moment and he has discovered
that in the mouse airway there are many Clara cells,
even in the largest of the mouse airways. He is able
to cannulate a mouse trachea and collect the
secretions and he hopes to be able to analyse them.
My best guess at the moment as to what the Clara
cells produce is that they give out something like
mucus which enables the cilia to beat, which can form
a roof over the cilia and which can take out dust.
The third point there certainly are macrophages which
enter the lymphatics. The best evidence we have from
Green and his co-workers, he thinks that the majority
of the macrophages enter the ciliated airway and are
removed from other ciliated airways, but some do
enter the lymphatics. The numbers which leave the
lung by each route, I do not know. But the evidence
that the majority do actually exit from the lung
comes from that sort of study. Most of the
radioactivity does eventually leave lungs rather than
remaining in the lymphnodes. I think that the
majority leave the lung by the ciliated airway; the
minority, which is taken into the lymphnodes, is very
important for the immunological recognition of the
antigens which have penetrated the lungs. The fourth
point: do the ciliated cells absorb fluid from the
lungs? I think it is very likely that they do.
Almost certain, in fact. The problem with the lungs
is that there is a tremendous surface area of
peripheral airways, whose secretions converge upon
the central airways. If you imagine anything even
approaching a mucous blanket in the peripheral
airways, then all those secretions have to converge
upon the central airways and they would completely
block the trachea. So, either there must be very few

secretions from the peripheral airways or else they have got to be reabsorbed on the way out. The problem is that no one has yet proved that anything apart from water can be absorbed from the airways, but I think that it is almost certain that it does happen and that the microvillous border of the ciliated cells is a very convenient place where it could happen.

PRODI: I wanted to ask you how do you rate in vivo solubility as a clearance mechanism. In very long particles a correlation has been shown between the solubility in body fluids and the clearance half times, but you did not talk about that. Can you comment?

RICHARDSON: Solubility is certainly very important in clearance. In the nose it is known that soluble gases are removed from the inspired airstream before they reach the lower airways. For instance, when sulphur dioxide is breathed through the nose at quite a high concentration, as high as 25 ppm, it is virtually completely removed from the inspired airstream by being dissolved in the surface mucous before it reaches the pharynx. So, solubility is very important to allow the nose to clean the gas. You were talking of particles which actually deposit in the lung. I am afraid I do not know too much about that. Presumably, if the particle is soluble and it is depositing on a fluid surface, which is air, it will dissolve and can be removed through the blood stream.

CUMMING: Do you make any distinction between soluble organic and soluble ionic compounds in your question?

PRODI: I am talking about chemically quoted insoluble particles, for instance plutonium oxide, which is of interest in the nuclear industry and particles that behave like these with clearance half times of the order of years, the so-called Y group in the international classification. So, can you comment on this correlation that has been put forward between some solubility in body fluids, which is now called transferability, and clearance half times?

RICHARDSON: What is certain is that the sort of picture you see with a rapid removal of particles from the alveoli, does not apply to all substances. It does not apply for instance to plutonium or tantalum, where the half lives for clearance of the alveolar fraction, are not measured in days or months, but as you say in years. Presumably there might be some sort of recognition mechanism implied. If a particle is completely insoluble it might be very difficult for the macrophages to recognize that it is anything different from normal lung and so make it very difficult for ingestion because it does not know that it is there. It has only chemo-sensors and so it has to have something to sense and dissolved substances, I suppose, are most likely to be sensed.

CANDURA: I would like to know your opinion on the mechanism through which the lysis of macrophages that have phagocytized silica particles occurs. I am asking this because if the macrophage were actually able to identify what is going to phagocytoze, perhaps it would not phagocytoze crystalline silica knowing that this is going to bring about its lysis.

RICHARDSON: This is a very important aspect which you have raised. When people inhale silica dust, there is this terrible reaction to it, with the laying down of fibrous tissue. You are referring I think to Allison's story. He finds that fibroblasts, when exposed to silica, do not lay down fibrous tissue. However, when there are also macrophages in the system, the macrophages take up the silica particles and release something and what the macrophages release is what causes the fibroblasts to lay down fibrous tissue and cause the sort of reaction in the lungs that you saw in the final slide that I showed. I do not know quite what it is about silica that is so damaging to macrophages, I am sorry, I cannot be very helpful in answering your specific question.

PAOLETTI: You say that in bronchitis, after cough, there is an increase in clearance, when evaluated by radio aerosol technique. I do not think so, because in bronchitis there is a decrease in clearance. I think there is a technical reason, because there is a rapid change of activity in the lung when you study it in this way. So, you cannot speak of cough and clearance because now a study of clearance with this technique takes into account the changes in this activity and so there is a correction for cough. In fact, if you study clearance in normal subjects after having cough, there is a rapid increase in clearance, but this is not clearance, it is a change of radioactivity in the lung.

CUMMING: I think to put that question with your comment you
              have to be much more specific about what you mean by
              a change in radioactivity. You are not suggesting
              that there is a change in the molecule. Can you be
              more specific?

PAOLETTI: If you cough and follow the radioactivity for a long
              time with a gamma camera, when there is cough you see
              the change of position of the bolus of mucus, not an
              increase of clearance, because in bronchitis there is
              a long period of clearance, longer than in normal
              subjects. The change of radioactivity is due only to
              the technique.

RICHARDSON: I do not think you are disputing what I was trying to
              say, though you have misunderstood. I would not
              claim that bronchitics have a more rapid lung
              clearance than normal people because of coughing.
              They have a slower ciliary clearance, because they
              have fewer ciliated cells which are attempting to
              move a greater burden of mucus but it is difficult to
              portray this on a slide. This is the ciliary
              clearance for a bronchitic, much slower than for a
              normal person. However, when they cough, there is a
              very real increase in the rate at which they clear
              radioactivity from their lungs as you can see and
              there is no possible artefact that could explain a
              step like that, which occurs immediately on coughing.
              Either the radioactivity is there in the chest or it
              is being removed from the chest and spat out or
              swallowed. In fact one can measure the radioactivity
              of the mucus which has been brought up into the mouth
              and expectorated, so it is a real clearance
              mechanism. In normal subjects coughing produces very
              little sputum. I challenge you to cough now and spit
              out something.

CUMMING: If I may comment, ladies and gentlemen, I think there
              is no issue with regard to the data, there is an
              issue with regard to the meaning of the words.
              People are using clearance in two different senses.
              If we think in terms of clearance in relation to what
              the body does, Paoletti is correct. If we are
              thinking of clearance and whether radioactivity
              leaves or changes its position in the gamma camera,
              then Paul is correct.

LENZINI: I think it is important at this point to throw some
              light on a particular point with reference to

pulmonary macrophages. It is important here to
consider pulmonary macrophages as belonging to the
system of mononuclear phagocytes which, as in all
systems, has its own kinetics. The kinetics of the
pulmonary macrophages, are linked to the bone marrow
in its three compartments: bone marrow, blood and
pulmonary tissue. Among the pool of the pulmonary
macrophages there are conceivably subpopulations of
pulmonary macrophages with their own biological
properties. There are pulmonary macrophages that are
equipped only for non-immune pinocytosis or
phagocytosis, there are pulmonary macrophages that
are equipped for immune phagocytosis, i.e. they are
"armed" with receptors for complement and
immunoglobulins and act against some bacteria, and
there are pulmonary macrophages that are especially
equipped for biosynthesis. This is important because
the death of each type of macrophage and its
elimination are effected through different routes,
which are presumably linked to the individual
properties. The scavenger macrophages, when they
die, probably fall within the airways, but all the
macrophages engaged in immune phagocytosis are
assisted by the T or B subpopulations of lymphocytes,
and their turnover is different as well as their
elimination. Another important point - I was
stimulated by my colleague wondering why macrophages
do not recognise silica - the scavenger macrophages
recognise its antigen with the collaboration of
lymphocytic populations. Finally, some macrophages
are involved in the production of substances such as
interferon and chemotactic factors - and probably
they have a certain tropism for the interstitium and
are linked to the lymphatics. I think this is
important, because among the defence mechanisms, as
Richardson said, there are pneumocytes I, pneumocytes
II and Clara cells, whereas that of macrophages is
totally different belonging to the system of
mononuclear phagocytes and therefore the pulmonary
macrophages should not be considered in isolation and
we must stress that these cells have their origin in
the bone marrow.

TURNER-
WARWICK: The last contribution is a very important one, and I
                would like to ask the question: what evidence is
                there that the pulmonary macrophages, as described
                just now, have any other origin that from the bone
                marrow? Many people accept that mononuclear cells
                have a very adaptable function depending on the

environment that they face. But that does not mean necessarily that they have a different origin.

EMERSON: I think one can get into great difficulty when trying to look into the origin of the alveolar macrophage. I believe that the bone marrow, at the promonocyte level, is the origin of the alveolar macrophage. The work of Altman has shown that there may not be a pool of monocytes in the lung. However, when the lung is insulated there is a massive influx of monocytes from the circulation into the lung and these eventually mature into the alveolar macrophage. Granted there may be subpopulations of the pulmonary macrophage: one could be the alveolar macrophage itself in the air space, a second is the interstitial macrophage, and a third would be the macrophage seen in the conducting airways. I would not want to make subcategories beyond that. Whether the macrophages within each of those areas are the same or not is difficult to say. There are as yet no good markers to prove this. I think some people are working on monoclonal antibodies to look specifically at the different types of macrophages in the lung. That would be the ideal situation. I think one can get into difficulties in looking for subpopulations of pulmonary macrophages without specific enough techniques to identify these cells. But I hope that no-one would leave here thinking that the alveolar macrophage is not derived from the bone marrow.

CUMMING: Thank you very much. The function of the macrophage of course will form much of the discussion this week. One of the principle functions of course is ingestion, one of the functions we should have in mind for the moment is the ingestion of food by the members of the course so we will adjourn for lunch.

ALVEOLAR CLEARANCE AND LUNG LYMPHATICS

Bryan Corrin

Cardiothoracic Institute
Brompton Hospital
London, SW3 6HP

Alveolar clearance concerns the effective removal of both inhaled extraneous matter and fluid transudates from the lung, and these will be dealt with in turn.

Consideration of the clearance of inhaled particulates first requires an understanding of those factors which govern deposition in the lung. Deposition may be by inertial impaction, sedimentation or diffusion and is influenced particularly by the size, shape and density of the particles (Stuart, 1973), and by the pattern of breathing (Stahlhofen, 1980). Solubility is also important in relation to the ultimate fate of inhaled particulates. All the heavier particles are deposited in the conductive air passages by impaction or sedimentation and are removed by muco-ciliary activity. Alveolar deposition is by sedimentation of particles of 1 to 5 μ Stokes diameter and by diffusion of very small particles (Fig.1.). Stokes diameter is a theoretical concept applied to irregular particles to indicate the equivalent size, in terms of sedimentation, of spheres of unit density. Marked irregularities in particle shape are encountered with asbestos, and in density with heavy radioisotopes. Asbestos fibres of up to 100 μ in length have a Stokes diameter in the respirable range and therefore reach the alveoli. Straight amphibole fibres such as crocidolite (blue asbestos) penetrate the air passages more easily than curly serpentine fibres such as crysotile (white asbestos).

Alveolar clearance of inhaled deposits is effected very largely by the alveolar macrophages. These cells are avidly phagocytic (Fig.2)(Corrin, 1969) and this property is facilitated by surfactant (La Force et al., 1973) and by immune mechanisms, the macrophages

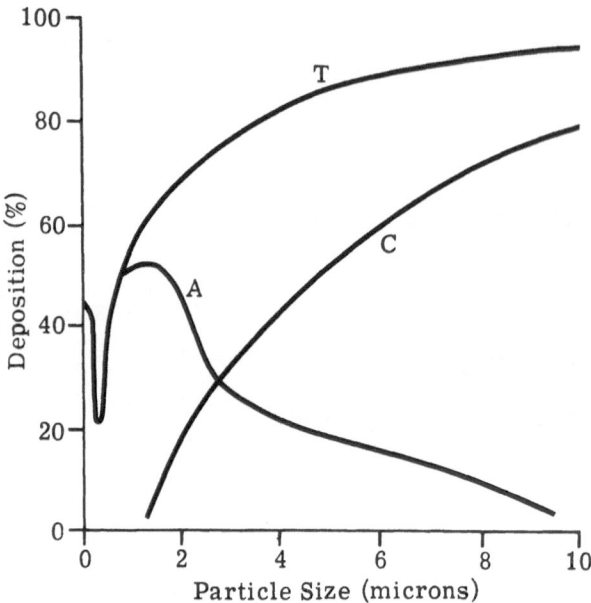

Fig. 1. Total (T) and regional dust depositions related to
particle size. A, alveolar; C, conductive air
passages.
Redrawn from Hatch and Hemeon, 1948 and Nagelschmidt,
1963.

having surface receptors for immunoglobulin and complement
(Reynolds et al., 1975). Microbial invaders are killed by
macrophages by a variety of mechanisms, including lysosomal acid
hydrolases, peroxidation and lysozyme (Corrin et al., 1969; Gee et
al., 1971; Klockars and Reitamo, 1975). These enzymes not only
attack phagocytosed microbes but are secreted from the cell (Unanue,
1976). Green and colleagues have shown that several factors impair
the bactericidal properties of the macrophage, notably acidosis,
hypoxia, cold, ethanol, tobacco smoke and virus infection (Green and
Kass, 1965; Green and Carolin, 1967; Goldstein et al., 1970; Jakab
and Green, 1976) whilst Allison et al., (1966) have shown that silica
particles damage phagocytic cells by attacking the phagolysosomal
membrane.

     Alveolar macrophages are derived from the bone marrow via the
blood and alveolar interstitium (Brunstetter et al., 1971; Godleski

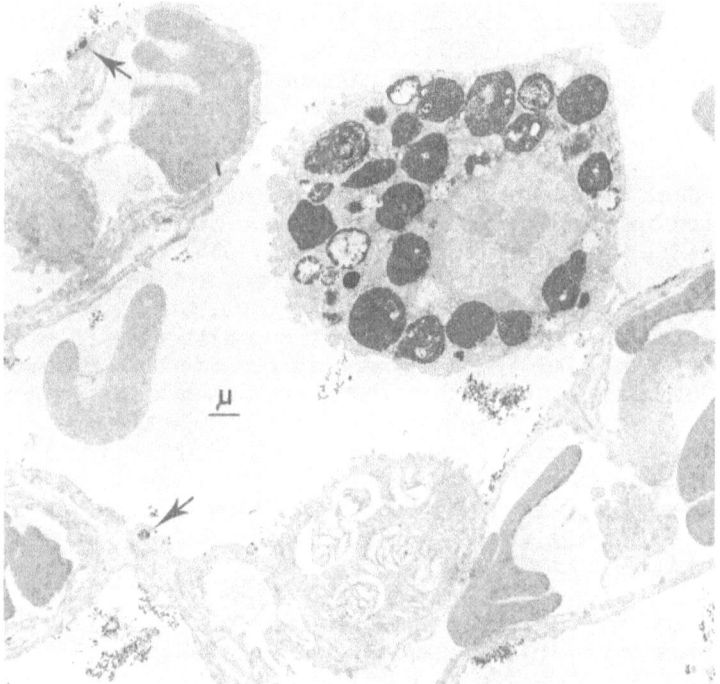

Fig. 2. Unstained electron micrograph of rat lung into which
finely divided thorium has been introduced via trachea.
A little free thorium is seen in the alveolar lumen but
most has been taken up by an alveolar macrophage where it
is seen in large phagosomal vacuoles. Arrows indicate
small phagosomes in the type I alveolar epithelium. A
type II cell (centre bottom) is virtually devoid of
thorium. Reproduced from Corrin (1969) by permission
of the Editor of Thorax.

and Brain, 1972; Bowden and Adamson, 1972) and once in the alveolus
the great majority never return to the tissues (Gross and Westrick,
1954). In the alveolus they are present in the hypophase of the
alveolar lining film and gradually converge on the centre of the
acinus where they reach the ciliated bronchiolar cells and are
removed to the pharynx. It is uncertain whether they move actively
in the alveolar lining film or travel passively as this fluid
replaces that cleared from the bronchioles. Their mobility does
however appear to be influenced by movement of the lung tissue for
they accumulate in those alveoli which abut onto the more solid
elements of the lung, in particular the connective tissue of the
bronchiolo-arterial sheaths, the interlobular septa and the pleura.
In such alveoli the underlying epithelium may be eroded and dust-

laden macrophages then regain the interstitium (Policard et al., 1957; Gross and Hatch, 1962). The vast majority of dust cells however are removed via the air passages. It is estimated that up to five million macrophages leave the lungs in this way every hour (Brain, 1970).

The dust cells which erode the alveolar epithelium have already accumulated in focal groups at the centres of the acini (Gross and Hatch, 1962), where they form Macklin's (1955) dust sumps. In the interstitium they are bound together by connective tissue, the amount of which depends upon the quantity and particularly, the fibrogenicity of the dust. Such dust deposits appear to be relatively stable but in fact there is considerable movement and turnover of macrophages within them, even when there is considerable fibrosis, as shown by Heppleston's (1958) double dusting experiments with different coloured dusts. Dust cells are continually dying and their dust load is taken up repeatedly by new cells. Removal of dust from the dust sumps is a continuous but slow process, as shown by the persistance of these focal dust aggregates long after employment in a dusty environment has ceased.

Compared to the macrophages, the alveolar epithelium phagocytoses only miniscule amounts of inhaled particles (Fig.2) (Corrin, 1969). This may be sufficient to elicit an immune response but is of relatively minor importance in regard to the pneumoconioses. It must be recognised however that the epithelium has some, if limited, phagocytic capacity and that some dust particles may pierce the alveolar epithelium without being phagocytosed, notably very small needle-like asbestos fibres. Some of the dust particles entering the epithelium are passed on to the interstitium where they may be taken up by interstitial mononuclear phagocytes. Ingested or free dust particles move in the interstitial fluid to the lymphatics, which begin near the dust sumps, and are removed to the lymph nodes. Brundelet (1965) and Tucker et al. (1973) both reported excretion of some dust cells from the interstitium into the bronchiolar lumen.

Lymphatic clearance is concerned not only with dust transport but is particularly important in preventing pulmonary oedema. Compared to the alveolar epithelium the capillary endothelium is relatively permeable. Whereas the epithelial cells are joined to each other by tight junctions, the endothelial cells adhere by ordinary desmosomal connections (zonulae adherens rather than occludentes). Considerable volumes of fluid leave the capillaries but the alveolar space is prevented from flooding by its epithelial lining, thin though this barrier is. Any fluid not reabsorbed by the blood vessels traverses the interstitium to reach the lymphatics. These are not found at the alveolar level but begin near the terminal bronchioles at the centres of the acini. Fluid loss from capillaries occurs mainly between the endothelial cells, but

pinocytosis permits the transport of larger molecules (Feldmann et
al., 1973). Increased permeability may be brought about by
hydrostatic or cytotoxic mechanisms but there is no distinction in
regard to the route or molecular size of the extravasated solutes
(Schneeberger and Karnovsky, 1971; Pietra, 1978). Capillary
endothelial cell junctions are mainly found on the thick side of the
air-blood barrier and fluid leaving the blood stream gains access
directly to the interstitium of the interalveolar septum. Basement
membranes appear to offer no barrier to fluid movement. Although
fluid leaving the capillaries first enters the connective tissue of
the interalveolar septum, this is relatively limited in amount and at
the light microscopic level interstitial oedema is first evident in
the thicker sheaths of connective tissue about the airways and blood
vessels. Interstitial oedema is always present before alveolar
oedema develops. However interstitial oedema only develops when
lymphatic clearance is overloaded and the lymphatics have
considerable reserve, being able to increase their capacity tenfold

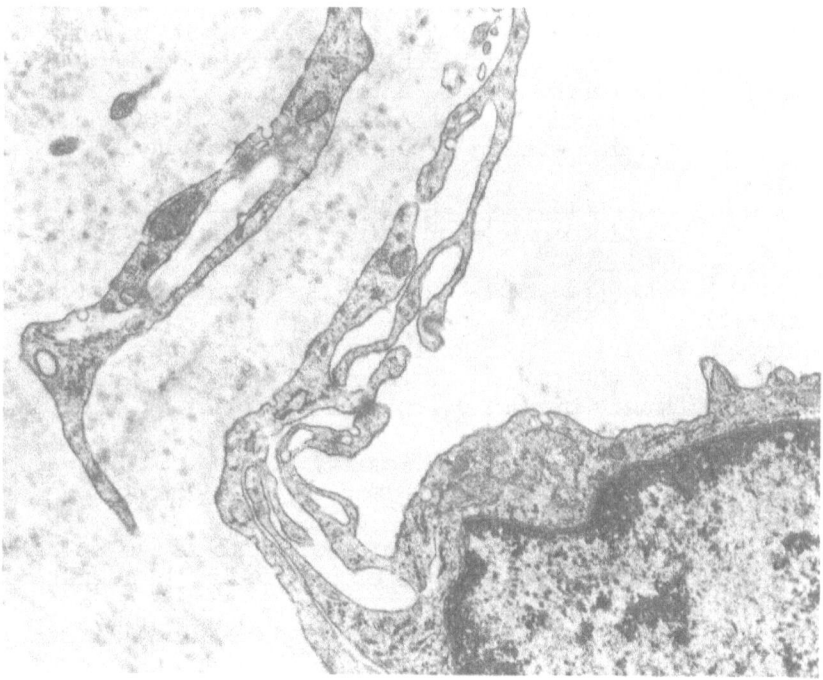

Fig. 3. A lung lymphatic showing overlapping endothelial cells
        with poorly developed junctions and basement membrane.
        Photograph by courtesy of Mrs. D. Bowes.

(Staub, 1970). At birth considerable volumes of fluid are absorbed from the alveoli and removed by the lymphatics (Ahearne and Dawkins, 1964).

Lymphatics differ from blood capillaries in structure as well as position (Fig. 3) (Lauweryns, 1971). Endothelial cell junctions and basement membranes are poorly developed in lymphatics and anchoring fibrils serve to maintain patency of the lumen when oncotic pressure in the surrounding interstitium increases as fluid accumulates. Some endothelial cells merely overlap so there is no true junction but a small intercellular gap through which interstitial fluid may easily enter the vessel. Larger lymphatics are valved to ensure that flow is unidirectional.

In conclusion, alveolar macrophages are of prime importance in clearing the alveoli of inhaled foreign material, whilst the pulmonary lymphatics are important in maintaining fluid balance in the lung.

REFERENCES

Aherne, W., and Dawkins, M.J.R., 1964, The removal of fluid from the pulmonary airways after birth in the rabbit, and the effect on this of prematurity and pre-natal hypoxia, Biol. Neonat.,7:214.

Allison, A.C., Harington, J.S., and Birbeck, M., 1966, An examination of the cytotoxic effects of silica on macrophages, J.Exp.Med. 124 (2):141.

Bowden, D.H., and Adamson, I.Y.R., 1972, The pulmonary interstitial cell as immediate precursor of the alveolar macrophage, Amer. J.Path., 68:521.

Brain, J.D., 1970, Free cells in the lungs, Arch. Intern. Med., 126:477.

Brundelet, P.J., 1965, Experimental study of the dust-clearance mechanism of the lung 1. Histological study in rats of the intra-pulmonary bronchial route of elimination, Acta Path. Microbiol. Scand. Supp.175.

Brunstetter, M-A., Hardie, J.A., Schiff, R., Lewis, J.P., and Cross, C.E., 1971, The origin of pulmonary alveolar macrophages, Arch. Int. Med., 127: 1064.

Corrin, B., 1969, Phagocytic potential of pulmonary alveolar epithelium with particular reference to surfactant metabolism, Thorax 24:110.

Corrin, B., Clark, A.E., and Spencer, H., 1969, Ultrastructural localisation of acid phosphatase in the rat lung, J. Anat., 104:65.

Feldmann, G., Chahinian, P., Leturcq, E., and Bignon J., 1973, Localisation ultrastructurale par anticorps couplés à la peroxydase de l'albumine extra-vasculaire dans le poumon de rat, Compt.Rend. Acad. Sci., 277:251.

Gee, J.B.L., Vassallo, C.L., Vogt, M.T., Thomas, C., and Basford, R.E., 1971, Peroxidative metabolism in alveolar macrophages, Arch. Intern. Med, 127:1046.

Godleski, J.J., and Brain, J.D., 1972, The origin of alveolar macrophages in mouse radiation chimeras, J.Exp.Med., 136:630.

Goldstein, E., Green, G.M., and Seamans, C., 1970, The effect of acidosis on pulmonary antibacterial function, Lab. Clin. Med., 75:912.

Green, G.M., and Carolin, D., 1967, The depressant effect of cigarette smoke on the in vitro antibacterial activity of alveolar macrophages, New Eng. J. Med., 275:421.

Green, G.M., and Kass, E.H., 1965, The influence of bacterial species on pulmonary resistance to infection in mice subjected to hypoxia, cold stress, and ethanolic intoxication, Brit. J. Exp. Path., 46:360.

Gross, P., and Hatch, T., 1962, Pneumoconiosis: the requirements for lymphatic dust transport, Int. Archiv Gewerbepath. Gewerbehyg., 19:660.

Gross, P., and Westrick, M., 1954, The permeability of lung parenchyma to particulate matter, Amer. J. Path., 30:195.

Hatch, T., and Hemeon, W.C.L., 1948, J. Ind. Hyg. Toxicol., 30:175.

Heppleston, A.G., 1958, The disposal of coal and haematite dusts inhaled successively, J.Path Bact., 75:113.

Jakab, G.J., and Green, G.M., 1976, Defect in intracellular killing of staphylococcus aureus within alveolar macrophages in Sendai virus-infected murine lungs, J.Clin.Invest., 57:1533.

Klockars, M., and Reitamo, S., 1975, Tissue distribution of lysozyme in man, J. Histochem. Cytochem., 23:932.

LaForce, F.M., Kelly, W.J., and Huber, G.L., 1973, Inactivation of staphylococci by alveolar macrophages with preliminary observations on the importance of alveolar lining material, Amer. Rev. Resp. Dis., 108:784.

Lauweryns, J.M., 1971, The blood and lymphatic microcirculation of the lung, Pathology Annual, 5:365 Sommers, S.C., ed., Appleton-Century-Crofts, New York.

Macklin, C.S., 1955, Pulmonary sumps, dust accumulations, alveolar fluid and lymph vessels, Acta Anat., 23:1.

Nagelschmidt, J.G., 1963, Dust parameters and their biological significance. In: "Mineral Dust in Industry", Dept. of Scientific & Industrial Research, Her Majesty's Stationery Office, London.

Pietra, G.G., 1978, The basis of pulmonary edema, with emphasis on ultrastructure. In: "The Lung", W.M. Thurlbeck and M.R. Abell, eds., Williams and Wilkins, Baltimore.

Policard, A., Collet, A., and Pergermain, S., 1957, Étude au microscope électronique due granulome pulmonaire silicotique experimental, Presse Medicale, 65: 121.

Reynolds, H.Y., Atkinson, J.P., Newball, H.H., and Frank, M.M.,
    1975, Receptors for immunoglobulin and complement on human
    alveolar macrophages, J. Immunol. 114:1813.
Schneeberger-Keeley, E.E., and Karnovsky, M.J., 1968, The
    ultrastructural basis of alveolar-capillary membrane
    permeability to peroxidase used as a tracer, J.Cell. Biol.
    37:781.
Stahlhofen, W., 1980, Experimentally determined regional deposition
    of aerosol particles in the human respiratory tract, Clin.
    Resp. Physiol, 16:145.
Staub, N.C., 1970, The pathophysiology of pulmonary edema, Human
    Path., 1:419.
Stuart, B.O., 1973, Deposition of inhaled aerosols, Arch. Intern.
    Med., 131:60.
Tucker, A.D., Wyatt, J.H., and Undery D., 1973, Clearance of inhaled
    particles from alveoli by normal interstitial drainage pathways,
    J.Appl.Physiol. 35:719.
Unanue, E.R., 1976, Secretory function of mononuclear phagocytes -
    a review. Amer. J. Path., 83:396.

D I S C U S S I O N

LECTURER: Corrin   CHAIRMAN: Cumming

SMITH: Can I ask you to what extent you think that the
                  uptake of particulate material by the alveolar wall
                  is an active phagocytic process and to what extent it
                  may be a passive or even accidental uptake of
                  material, perhaps by the micropinocytotic vesicles
                  which normally are very numerous in the alveolar
                  epithelium?

CORRIN: I cannot understand the distinction between
                  accidental and passive uptake.

SMITH: As I understand it, there is supposed to be an uptake
                  of fluid material, draining fluid from the alveolar
                  space back into the interstitium. I was just
                  wondering whether the particulate material is sort of
                  taken up purely accidentally as a result of this
                  mechanism.

CORRIN: As you said, pinocytotic vesicles are extremely
                  numerous throughout the epithelium and the
                  endothelium, and if you can cast your mind back to
                  that electron micrograph where I showed two widely
                  separated - what I term phagosomal vacuoles, type I
                  cells - I think that suggests they are not just
                  entering pinocytotic vesicles. I envisage that as an
                  active phagocytosis, as I emphasised, in comparison
                  with the activity of the free alveolar macrophages
                  that it is relatively insignificant. Nevertheless I
                  think it is probably important in the carrying of
                  antigenic material into the interstitium.

JEFFERY: I was interested when you made the distinction
                  between the release of lysosomal enzymes during
                  phagocytosis and the release of elastase following
                  phagocytosis. Is anything known of the mechanism of
                  release of elastase? Obviously this is of great
                  importance in the generation of emphysema.

CORRIN: I think that the mechanism would merely be a fusion
                  of the appropriate vacuoles with the cell membrane
                  rather than with the phagosomal membrane. That would
                  be the mechanism.

JEFFERY: But is this in response to phagocytosis or is it
                  going on all the time?

CORRIN: If you remember the second list of the acid
               hydrolases, that really escape during phagocytosis,
               lysozyme is secreted continually and the neutral
               collagenases to which you refer have an intermediate
               position. They are being secreted in small amounts
               all the time, but the secretion is augmented,
               promoted by phagocytosis. To develop your own ideas,
               perhaps, the more dust there is for the alveolar
               macrophages to grapple with the greater the release
               of these neutral proteases that one would expect.

JEFFREY: So one can presume that the release of elastase is
               not due to the messy eating.

CORRIN: I think we have to go beyond this idea of messy
               eating particularly in the case of lysozyme, they are
               secreting this material all the time and, I did not
               illustrate any macrophages which have morphological
               features other than phagocytic, but in certain
               situations we know these cells transform and the
               dense bodies are scanty, they do not seem to be so
               phagocytic and they contain many clear vesicles and
               abundant endoplasmic reticulum and possibly those are
               the cells which are switched on to secrete rather
               than phagocytose.

HEATH: I wonder whether you think that the lung has any
               difficulty in getting rid of fibres of the size of
               asbestos fibres. People exposed to asbestos have
               retained these fibres in the lung for many years,
               perhaps 40, and have then developed malignant
               mesothelioma. It was believed that this was related
               to the chemical nature of asbestos fibres. But now
               it has shown that in Turkey, where there is no
               asbestos that zeolite fibres, which have exactly the
               same shape and size as asbestos fibres, have also
               been retained in the lungs for many years and have
               given rise to pleural mesothelioma. I ask you this
               because the British Government has just commissioned
               the new telescope on the top of a volcano in Hawaii
               and Professor Elmes has just looked at the dust on
               this volcano and has found, horror of horrors, that
               the dust contains zeolite fibres. If these
               astronomers are going to breathe in fibres of this
               particular shape and size, do you think they are at
               any risk in developing pleural mesothelioma?

CORRIN: There are many questions there. The first one was:
               do I think the lung has any difficulty in removing

fibrous particulates and I would say: yes, because the mechanism I described for concentrating the dust which has been phagocytosed on the dust sumps leads to the focal process which we recognise in coal workers pneumoconiosis and silicosis. But, as you know, asbestosis is a diffuse fibrosis and the asbestos bodies and fibres are not congregated in this focal manner, as are more spherical particulates. I think the reason why the lung finds it difficult to handle these fibres is not their chemical nature, but their shape. It is of course interesting that they enter the lung at all. These are fibres that are often in excess of a hundred mu in length. When Paul Richardson was talking of particulate size he was probably talking of spheres of unit density and we can relate particles which depart from this in both shape and density to spheres of unit density by referring to their standard diameter or aerodynamic equivalence, and these fibres probably enter on the airstream like zeppelins borne along by the wind. Once they have settled, it is up to macrophages to remove them, and a macrophage is 10, 15 or 20 mu and the fibre may be 100, and we see two or three macrophages trying to grapple with these fibres and they are unable to.

HEATH: There is of course an important corollary and this is why I am taking up time for a second go; to avoid the development of mesothelioma, people have used substitutes which are not chemically like asbestos but have the same physical characteristics. So, is the message that when you substitute a synthetic to replace asbestos it must have the same shape and size, otherwise you would be in just the same predicament?

CORRIN: I think there is a series of explanations. It does not matter so much what is the chemical nature of the fibre, but it is its physical dimensions. It is claimed that anything over 5 mu in length and under 0.3 mu in diameter is carcinogenic.

BIGNON: First a comment. I think I will bring some data in the next few days about fibres which may shed some light. I want, however, to come back to your distinction between the release of enzyme during phagocytosis and after. I am surprised that you said that elastase was released after phagocytosis. It is a subtle distinction and in the paper you published with Robert in the Journal of Pathology describing the elastase from alveolar macrophages you could obtain a release in culture during phagocytosis of latex. What is the distinction between, during and after phagocytosis?

CORRIN: The distinction was that the secretion of the neutral proteases is increased if the cell is given something to phagocytose, so there is always a low grade secretion of them, but secretion is augmented if the cell is stimulated.

BIGNON: You mean that the stimulation by a particle increases first the synthesis and then there is a release of elastase? Yes? O.K.?

CUMMING: A point that occurred to me with regard to the difference in chemical nature between zeolite on the one hand and asbestos on the other and the prevalence of mesothelioma, is firstly that the chemical structure of these minerals is based on the amphibole chain in which there is a molecule of silica linked to four atoms of oxygen at the six corners of a hexagon, and hexagons fuse together to form a long chain, and that is common to all. What is not common to all are the lateral side linkages and the atoms which are used, magnesium, iron and so on. It may be the size or it may be something else, and I am not at all clear which of these two things is correct. About mesothelioma, although the amphibole chain is characteristic of all the asbestos minerals, the prevalence of mesothelioma varies enormously with the particular type of mineral. Margaret disagrees with me, excellent.

TURNER-
WARWICK: I think, correct me if I am wrong, Chris Wagner also performed his experiments demonstrating the development of mesothelioma using fibre glass of completely different chemical structure but of the critical dimensions that Corrin mentioned. Diameter of about 0.5 mu and a length of about 10 mu. So, I think it is dimension that seems to be the important factor, not the chemistry.

CUMMING: I think there are two conflicting observations. It is perfectly true that particles of a certain dimension can produce it, but then when you have particles of the same dimension and different chemistry their ability to produce it differs markedly. Chrysotile and chrysodolite, I cannot understand how these two things differ.

CORRIN: In Wagner's dusting experiments, when he uses an
             amphibole, the dust load increases as the dusting
             experiment goes on. But with the curly, serpentine
             chrysotile fibre a certain level is reached and they
             plateau out, which suggests that the chrysotile is
             landing on a muco-ciliary mechanism and is not
             reaching the alveoli. And in agreement with this is
             the electron analysis of asbestos bodies. He very
             rarely finds chrysotile inside an asbestos body, it
             is the amphiboles. I think the reason for this is in
             the aerodynamic characteristics of asbestos fibres.
             Let us talk about the length and thickness of an
             asbestos fibre, and that is appropriate to an
             amphibole, which is a straight one. But when we come
             to a serpentine or curly fibre of chrysotile the
             length may be the same, the thickness may be the
             same, but the aerodynamic functions are given by this
             distance here and so its comparative thickness would
             be enormous and it is quite likely, I think, that the
             majority of chrysotile fibres do not reach the
             alveolar level, that brings us to the disease in a
             chrysotile miner, perhaps.

CUMMING: That seems to be illuminating that particular
             question.

PRODI: If I may comment about the different fibre shape.
             The different behaviour is in terms of tumbling of
             fibres. If you have a straight fibre it tends
             generally to align in a shear flow and it tends to
             travel all the way down. On the other hand a curly
             fibre does not show that behaviour and it tends to be
             captured in the upper airways. So this is
             essentially one difference in the behaviour, why the
             fibres are airborne and why they deposit in the
             airways. On the other hand it can be a difference in
             the behaviour during clearance, in the sense that
             straight fibres may travel through the tissues and
             can be found even in the periphery, in pleural areas,
             while curly fibres cannot. So, this is just a
             difference in the physical behaviour, both in
             deposition and clearance.

CORRIN: You expressed my own views better than I could.

CUMMING: We seem to have answered Heath's point that merely a
             change in the chemical constitution will not have the
             effect that is looked for, but that the physical

dimension, the spear, is what we must avoid. Thank
you, Brian, we will finish the session at that point
and ask for the next presentation by Peter Jeffery.

THE NORMAL STRUCTURE OF BRONCHIAL EPITHELIUM

Peter K. Jeffery

Cardiothoracic Institute
Brompton Hospital
London, SW3 6HP

The airways which supply the respiratory portion of the lung
are responsible for the clearance of inhaled particulate and gaseous
pollutants deposited on their mucosal lining. To this end, the
lining mucosa produces a mucous secretion in which pollutants become
dissolved or entangled, and which is then moved by cilia to the
pharynx, where it is normally swallowed. The mucous secretion so-
produced (normally between 10 and 100 ml/day) is a complex mixture
of about 95% water and 5% carbohydrate, protein, lipid and inorganic
material – usually in the form of a number of different glycoproteins,
each differing in the ratio of its protein to carbohydrate component
and its degree of acidification. The physical (i.e. viscoelastic)
properties of the secretion are largely due to these glycoproteins
which are produced by a variety of cell types comprising the airway
wall.

In man, bronchial mucus may come from two anatomical sources –
the secretory cells of the surface epithelium and those of the
submucosal glands (Reid 1954; McCarthy & Reid 1964). Epithelial
secretory cells are abundant in the trachea and bronchi but are
rarely found in bronchioli less than 1 mm in diameter. Submucosal
glands are present where there is cartilage in the bronchial wall –
the mean gland to wall ratio being normally 1:3(i.e. a Reid index
of 0.33). In regard to amount and distribution of mucus-secreting
tissue, there are species differences. In the specific pathogen-
free (SPF) rat, used for studies of experimental bronchitis, the
epithelial secretory cells are sparse at all levels of the
respiratory tract and submucosal glands are restricted to the
cranial portion of the trachea (Jeffery & Reid 1977). In the cat, a
much favoured animal for physiological studies, epithelial secretory
cells are numerous and are still found in intrapulmonary bronchi of

0.5 mm diameter: submucosal glands occupy a significant proportion
of the tracheal and hilar airway walls and are still found in the
most distal bronchi albeit to a lesser extent. The mean gland to
wall ratio in the trachea is 0.70, over twice that for normal man
(Jeffery 1978).

Ciliated cells are present throughout the whole of the
tracheobronchial tree and have been shown to be present in the most
distal respiratory bronchioli of the rat. Each beats at about 1000
times per minute in a fluid layer of low viscosity. The presence of
such a vast array of beating cilia has given rise to the term
"ciliary escalator" although it is now recognised that the escalator
may not be continuous but rather organized as individual fields of
cilia, each field separated by non-ciliated areas the extent of which
varies with species and with disease. (fig. 1). The consequences of
such an organization is that while the cilia of each cell in a given
field beat (i.e. effective stroke) in sequence and in a similar
direction to those of its immediate neighbour, distinct fields beat
asynchronously and in slightly differing directions. Normally the
resultant force moves mucus and pollutants to the throat where they
are normally swallowed.

Fig. 1.   A scanning electron micrograph of rat trachea to show
          surface features of the lining epithelium. Fields of cilia
          (arrows) are separated by nonciliated areas. The longer
          extensions are cilia the shorter microvilli.
          Glutaraldehyde fixed: critical point dried and gold coated

Thus mucus flowing from gland ducts or released by discharge of epithelial secretory cells comes to lie above the epithelium and is then moved cranially by the tips of the cilia. Whether the mucus so-produced normally lies as a continuous blanket (Sturgess 1977) or as discrete droplets or flakes (van As & Webster 1974) is still unresolved.   If it is a blanket throughout the tract then much of it must be reabsorbed en route from bronchi to trachea otherwise, with a progressive reduction in airways' surface area, the thickness of the blanket would increase to unmanageable proportions.

Cell types

The epithelial lining is complex with up to 11 cell types now distinguished (fig.2). Species differences in anatomical distribution exist with all 11 cell types found in the cat, 10 in the rat and 9 in the human (Table I). The ultrastructural detail of each has been recently reviewed (Jeffery & Reid 1975, Breeze et al 1976); and the following special points are noteworthy:

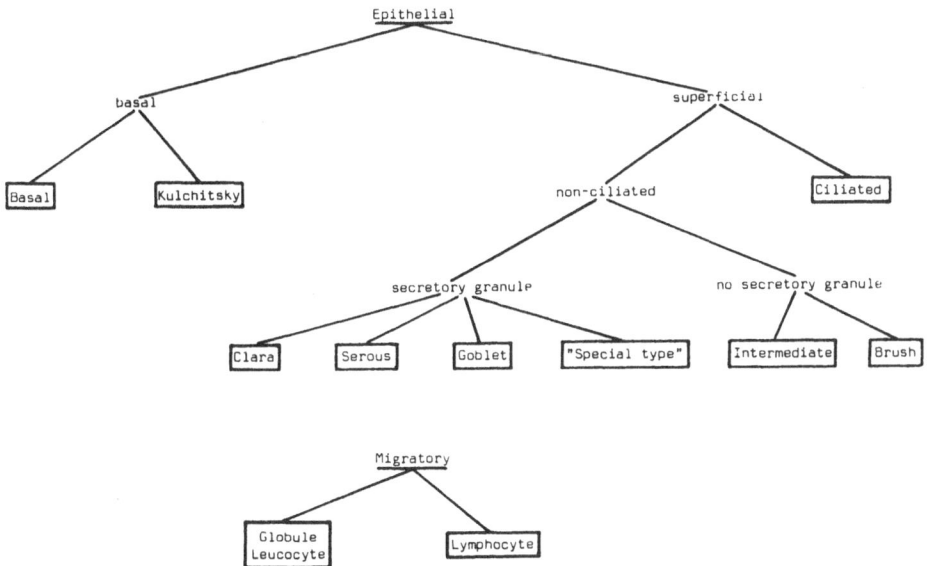

Fig. 2. Cell types found in tracheo-bronchial epithelium.

Table I

Epithelial cell types identified by electron microscopy in various species

	Man	Monkey	Dog	Cat	Pig	Cow	Rat	Mouse	Hamster	Rabbit	Certain birds
Ciliated	+	+	+	+	+	+	+	+	+	+	+
Goblet (lucent granule)	+	+	+	+	+	−	+	+	+	+	+
Serous (dense granule)	+	−	−	+	−	−	+	−	−	−	+
'Special type'	+	−	+	+	−	−	−	−	−	−	−
Clara	+	+	−	+	+	−	+	+	?	+	−
Intermediate	+	−	−	+	+	−	+	−	−	−	+
Brush	?	−	−	+	+	−	+	−	−	?	−
Basal	+	+	+	+	+	+	+	+	+	+	+
Kulchitsky	+	−	−	+	−	−	+	+	−	+	+
Globule leucocyte	−	'migratory'	+	+	−	−	+	−	−	−	−
Lymphocyte	+	'migratory'	−	+	−	−	+	−	−	−	−
Intraepithelial nerve	+	−	−	+	−	−	+	+	−	+	+

+ identified; − not yet identified; ? unconfirmed.

1)   Ciliated cells. Remarkably the organisation of the internal
microtubules (i.e. axoneme) of each cilium, which are responsible
for its movement, has remained constant throughout evolution
(fig.3a & b). The central pair of tubules imparts bilateral
symmetry to the cilium, the plane of which coincides with the plane
of effective ciliary motion. The 9 peripheral doublets have
attached 'arms' composed of dynein with ATPase activity and the
bending motion is thought to be associated with mechano-chemical
interaction of these arms with the other tubular components of the
axoneme, the details of which are well reviewed by Sleigh (1977).
The tip of the cilium is characterized by a reduction in diameter
and by the presence of claw-like projections which probably couple
cilium and overlying mucus: these have now been observed in a
variety of species including man (fig.4) (Jeffery - in preparation).

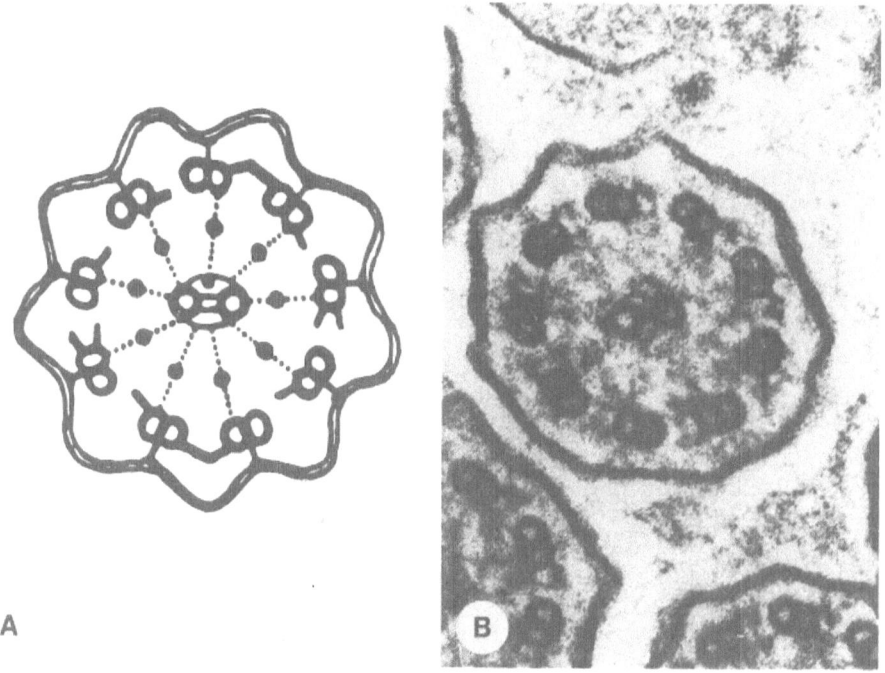

A   B

Fig.3. Transverse sections through the mid region of a single cilium
       a) a diagrammatic representation of the organization of
       internal microtubules showing 9 peripheral doublets
       surrounding 2 single central and b) a typical electron micro-
       graph of a similar area. Glutaraldehyde osmium tetroxide:
       uranyl acetate + lead citrate X.

At the base, each cilium is anchored within the cell by a 'rootlet'
the structure of which shows species variation: in man and cat it
is striated while in the rat it is not.   The reasons for this
difference is not understood.

2)   Secretory cells.   Four types of secretory cell have been
described - the mucous, serous, and "special type", these three
restricted to the most proximal airways, and the Clara cell
typically found in the distal.   The function and nature of secretion
of the "special type" is unknown.

    In proximal airways of the normal SPF rat most of the secretory
cell population are serous, each cell with discrete homogeneously
electron-dense granules and rough endoplasmic reticulum (fig. 5).

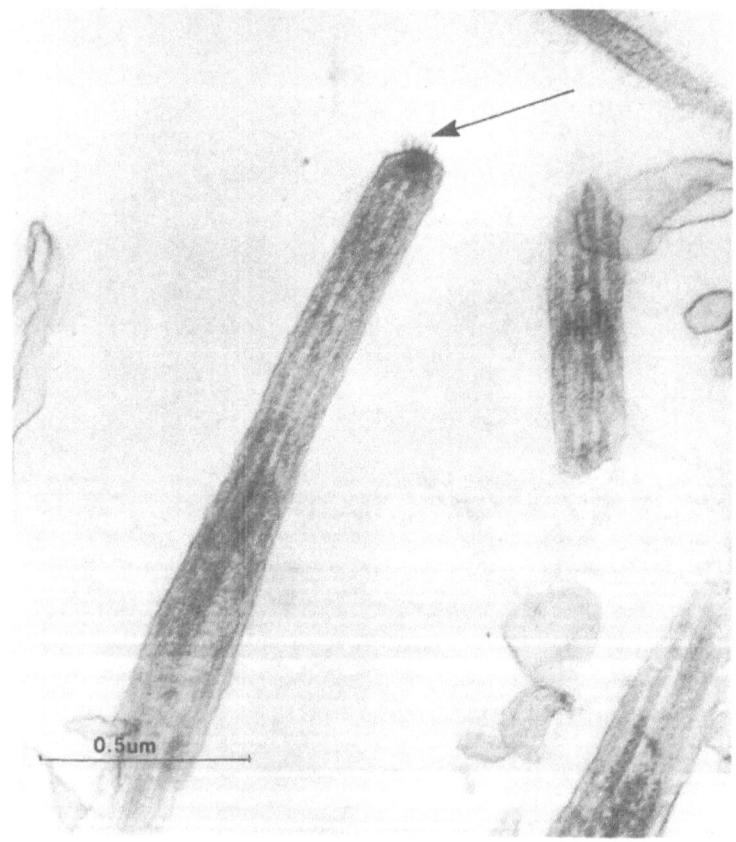

Fig.4. Electron micrograph of the tips of human bronchial cilia to
       show claw-like projections (arrows). Note how the diameter of
       the ciliary shaft reduces towards the tip. Glutaraldehyde +
       osmium tetroxide: uranyl acetate + lead citrate X.

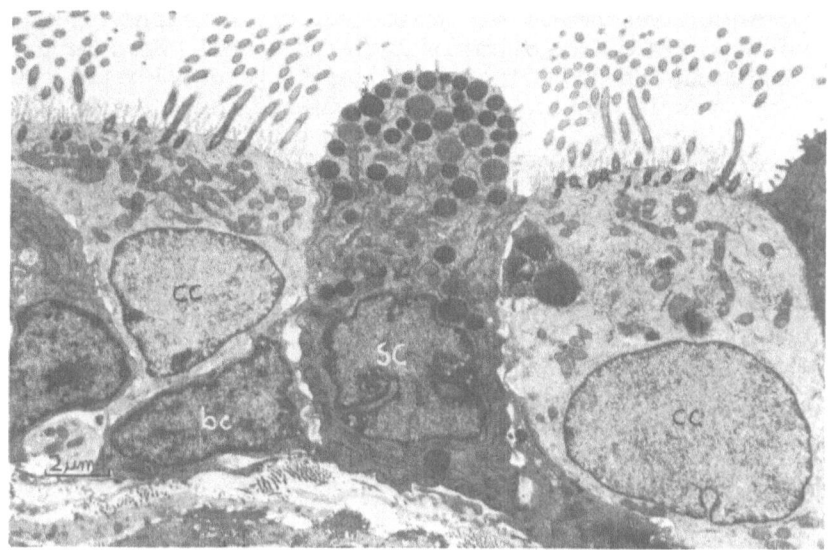

Fig. 5. Electron micrograph of rat main bronchial epithelium showing
an epithelial serous cell (sc) with ciliated (cc)and basal
(bc) cells adjacent. The serous cell has electron-dense
secretory granules and rough endoplasmic reticulum.
Glutaraldehyde + osmium tetroxide: uranyl acetate + lead
citrate X.

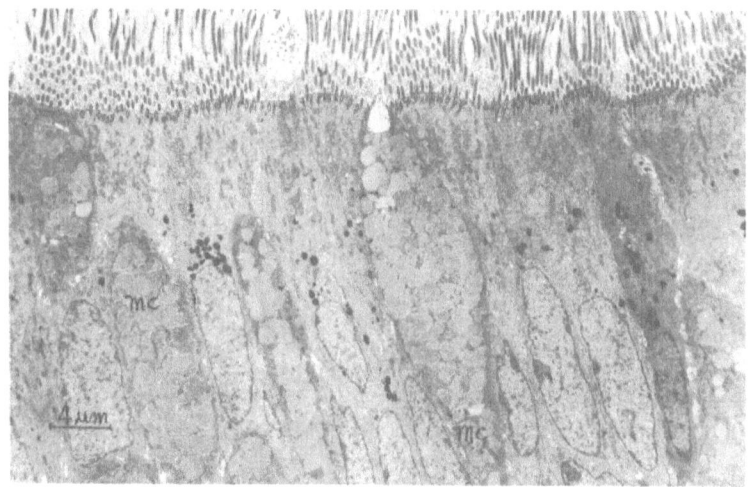

Fig. 6. Human bronchial resection prepared for electron microscopy
showing a typical mucous cell (mc) with a large number of
electron-lucent granules often confluent. Glutaraldehyde +
osmium tetroxide: uranyl acetate + lead citrate X.

Their name derives from their morphology which is similar to that of
the serous cells of the submucosal glands. While descriptions of
these cells in the human foetus (Jeffery & Reid 1976) and cat
(Jeffery & Das in preparation) have appeared they are infrequently
found in these species. In adult human bronchial epithelium,
obtained from resection or biopsy, the most frequently found
secretory cell is the mucous, each cell containing rough endoplasmic
reticulum and a large number of electron-lucent (pale) granules which
are often confluent one with the other (fig.6) and similar to the
mucous cells of the underlying glands. The morphological
characteristics of serous and mucous cells supports the idea that the
discharge of secretion by each is different; merocrine in the serous
and apocrine in the mucous.

There is a species variation in the airways distribution of
Clara cells. In man, rat and cat they are restricted to the intrapulmonary airways and their frequency increases with airway
generation distally. In the original descriptions (Clara 1937) they
were reported to be characteristic of the terminal bronchiolus of man
and rabbit but have now been found as far proximally as the trachea
(in mouse) and nose also (Hansell & Moretti 1969; Mutalionis & Parks
1973). Characteristically the apex of each cell bulges into the
airway lumen and contains an abundance of smooth endoplasmic
reticulum and small angular electron-dense secretory granules (fig.
7). This morphology is similar to that seen in other cells (eg.
testicular) concerned with the elaboration of steroids from
cholesterol-derived precursors. Considerable controversy has
surrounded the function of this cell some claiming it contributes, by
apocrine secretion, to bronchiolar and alveolar surfactant (Niden
1967; Etherton et al 1973). It is fair to say that the evidence to
date favours the type II alveolar cell as the main source of
surfactant and that the function of the Clara cell is still
unresolved.

3)   Kulchitsky cell (syn. enterochromaffin, argyrophil,
fluorescent, granulated and Feyrter cell). Cells resembling the
Kulchitsky or enterochromaffin cells of the gut have now been found
in the airway epithelium of a wide range of species including man
(Gmelich et al 1967, Basset et al 1971, Rosan & Lauweryns 1972,
Hage 1973). They may appear singly or in organized groups as so-
called neuroepithelial bodies. In the former case the cell is
basally located and occurs with greatest frequency in the foetus
and neonate. When grouped as neuroepithelial bodies they usually
reach the airway lumen and are associated with intra-epithelial
nerves. Each cell contains large numbers of dense-cored
("neurosecretory") granules distributed throughout an electron-
lucent cytoplasm (fig.8). It is generally agreed that the bronchial
Kulchitsky cell should be included as part of the diffuse system of
APUD (amine precursor uptake and decarboxylation) cells (Pearse 1969,
Hage 1973) which are all thought to have a common neural crest origin.

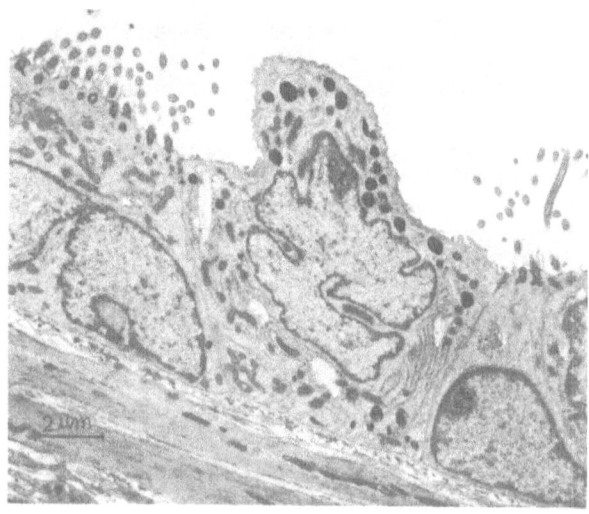

Fig. 7. Rat bronchiolar epithelium showing the ultrastructural
features of the Clara cell: a bulging apex containing an
abundance of smooth endoplasmic reticulum and small angular
electron-dense granules.

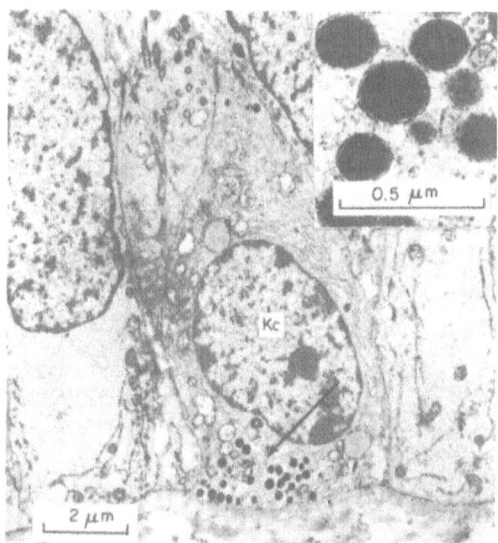

Fig. 8. Bronchiolar epithelium from a 16 week-old human foetus
showing a Kultchitsky cell (Kc) with electron-dense granules
at its base. Inset shows the typical dense-cored structure
of each granule. Glutaraldehyde + osmium tetroxide:
uranyl acetate + lead citrate X

There is evidence of amine uptake and storage of serotonin by such
cells of the trachea (Ericson et al 1972) and of discharge of
granules in response to hypoxia (Moosavi et al, 1973). Recently
neuro-peptides with powerful effects on vasculature and muscle have
been localized to cells with a similar location and distribution by
immunocytochemistry (Polak personal communication). Yet the role
of this cell in normal foetal and adult tissue is still not clear.
In disease there is evidence that these cells form the origin of
hormone-secreting carcinoid/oat cell tumours of the respiratory tract
(Gmelich et al 1967; Bensch et al 1968).

Nerves

     For a long time physiological data has suggested the presence of
receptors within the surface lining epithelium which may respond to
a gentle mechanical stimulus (dust or epithelial deformation) or to
chemical stimuli such as ammonia or sulphur dioxide (Widdicombe 1954).

     Nerve bundles associated with autonomic ganglia can be seen in
the airway wall at all levels of the bronchial tree. With the
electron microscope intraepithelial nerve fibres have been
identified in a wide variety of species including man (Rhodin 1966,
Jeffery & Reid 1973, Walsh and McClelland 1974). In man, rat and
cat nerves penetrate the submucosal gland also (Bensch et al 1965,
Das, Jeffery & Widdicombe 1978, Jeffery unpublished). In each
species the intraepithelial nerve is unmyelinated, has no Schwann
cell covering nor basement membrane and lies in close apposition –
15 nm – to adjacent epithelial cells. They may be recognized by
their electron-lucent cytoplasm, characteristically elongated
mitochondria of small diameter and neurotubules. While most
frequently found at the base of the epithelium they may be extremely
close to the lumen at a place where intraluminal stimulation would
be expected to affect them (fig. 9). In rat and cat they are most
frequent in the extrapulmonary airway and are not found within
intrapulmonary bronchi distal to the hilum (Jeffery & Reid 1973,
Das et al 1978). In our study of 14 cases of human lung resection
(carcinoma) and biopsy material (recurrent infection) intraepithelial
nerves have been found in several of the extrapulmonary bronchi
examined but not yet in segmental bronchi or bronchioli. (Jeffery –
in preparation).

     Whether these intraepithelial fibres are sensory (and represent
the "irritant" and "cough" receptors described by Widdicombe or
motor supplying the secretory or ciliated cells of the epithelium
is still unclear. The presence of neurosecretory vesicles in
terminals or varicosities of some is suggestive of motor function
and this especially if they are of the densecored type. However,
the degeneration studies of Das, Jeffery & Widdicombe (1979) have
shown that in the cat the vast majority of the epithelial nerves of
this species are sensory.

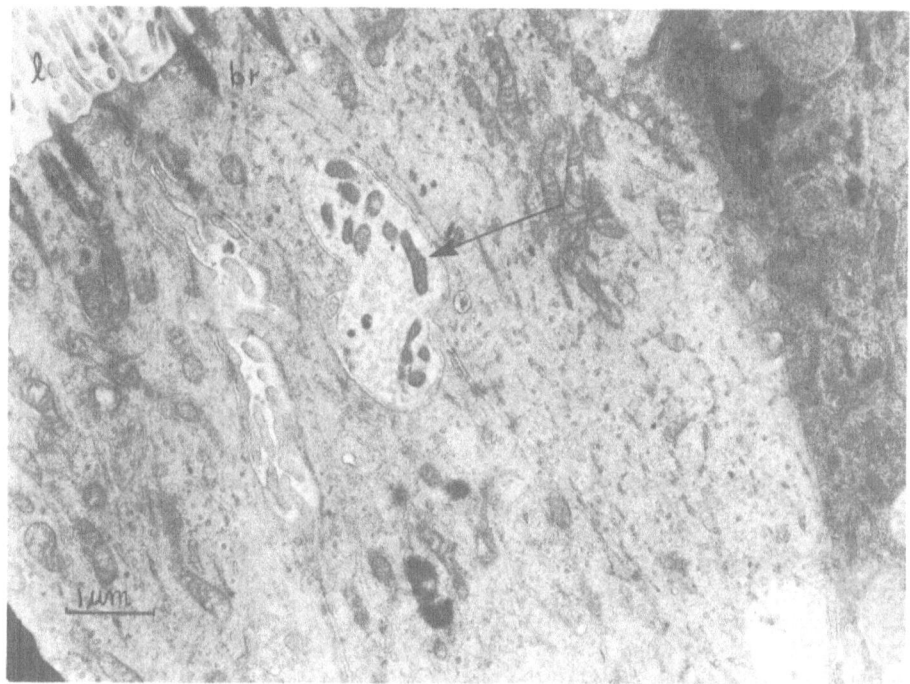

Fig. 9. Electron micrograph of human main bronchial epithelium
        (resection) showing a section through an epithelial nerve
        lying within 2 μm of the airway lumen (l) and surrounded
        by ciliated cells. Note the electron-lucent cytoplasm,
        vesicles and characteristically small mitochondria (arrow).
        Ciliary basal rootlet (br) Glutaraldehyde + osmium
        tetroxide: uranyl acetate and lead citrate X.

Migratory cells

     In the absence of inflammation migratory cells are normally
found within the lining epithelium. Three mononuclear cell types
are of interest: the lymphocyte, globule leucocyte, and mast cell.
The lymphocyte is usually located at the base of the epithelium:
its scant cytoplasm is electron-lucent and contains few organelles.
In contrast the globule leucocyte has large homogeneous electron-
dense granules. The mast cell of various non-primates superficially
resembles the globule leucocyte but the granules are smaller and
show variability in their electron-density. Those of man have
variable granules many of which show a characteristic scroll-
like pattern (fig.10). The distinction between globule leucocyte

and mast cell is further made by light microscopy with differences
in the staining characteristics of their granules emerging.
Ultrastructural changes of mast cells in smokers suggestive of
degranulation have been described (Lumsden & Lamb - personal
communication). It is thought that degranulation may proceed
following nonspecific irritation, the mediators released then
stimulating local irritant receptors and reflexly initiating
bronchoconstriction (Salvato 1976). While of obvious importance in
asthma it may also be a factor in chronic bronchitis.

In the rat trachea the lymphocyte and globule leucocyte
constitute 14% of epithelial cell in the trachea with their numbers
decreasing to less than 1% in peripheral airways. In the cat,
globule leucocytes appear in some but not all animals. The mast cell
and lymhocyte is a frequent feature of all the human resected
extrapulmonary and intrapulmonary airways so far examined but the
mast cell is a rare finding in the epithelium of experimental
animals. While the globule leucocyte has been identified by light
microscopy (Salvato - personal communication) in human material
electron microscopy has failed to demonstrate cells with the
appropriate ultrastructure in man. The functions of the lymphocyte
and globule leucocyte at this site are still unclear although the
latter cell is thought to play a role in the expulsion of nematode
parasites from some species (i.e. self cure phenomenum).

Bronchus-associated lymphoid tissue (BALT)

Subepithelial bronchus-associated lymphoid aggregates (BALT)
have now been described in a wide variety of species and appear to be
analogous with the similar mucosal aggregations of the tonsils and
intestine (GALT). Indeed Bienenstock et al (1976) have shown that
these lymphocytes are not static and that there is a continuous
traffic of lymphocytes between one lymphoid associated organ and
another forming a diffuse mucosal system. Macroscopically these
aggregations can be detected by acetic acid fixation and are
distributed along the whole of the bronchial tree with particular
concentrations at the points of airway branching. Histologically
BALT consists mainly of lymphocytes but plasma cells, mast cells,and
macrophages have also been described. Germinal centres are not
normally a feature of BALT the notable exception being that of the
chicken. In the rat Chamberlain et al (1973) describe associated
postcapillary venules a site typically connected with lymphocyte
emigration. While BALT is mainly subepithelial there is often
extensive migration of lymphocytes into the overlying epithelium
forming a so-called "lymphoepithelium": in these cases the
epithelium is flattened and lacks cilia and often mucous cells also.

The development of BALT is not antigen dependent (as it develops
even in germ free conditions) although it proliferates in infection

or following antigen challenge. Its function is being investigated
with the current hypothesis that BALT may form a first line of
defence against penetration of wet mucosal surfaces by antigen and
that sensitization of such lymphocytes in one organ may result,
following their emigration, in subsequent protection at mucosal
surfaces far removed from the local stimulus. If this is so, this
has interesting implications for the practice of local immunization;
i.e. local immunization would not require local administration of
antigen.

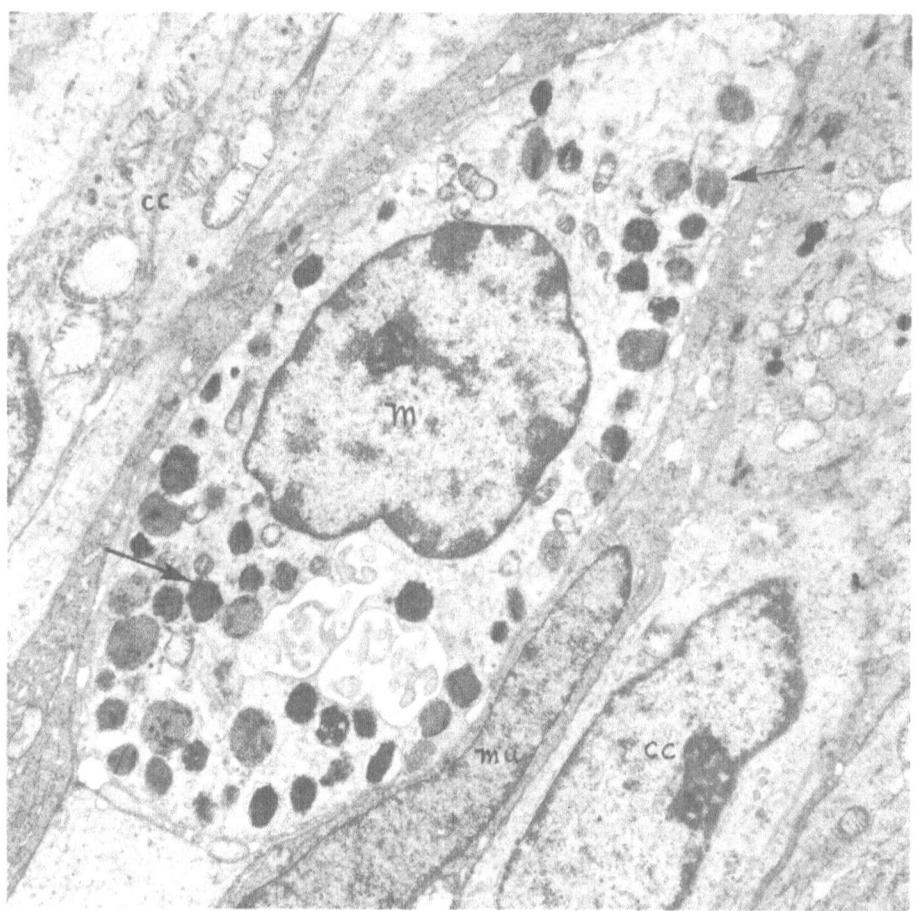

Fig.10. Human bronchial epithelium (resection for Ca) showing an
         intraepithelial mast cell (m) surrounded by mucous (mu)
         and ciliated (cc) cells. The migratory cell has granules
         varying in electron-density some of which contain scroll-
         like inclusion bodies characteristic of the human mast cell
         (arrow). Glutaraldehyde + osmium: uranyl acetate + lead
         citrate X.

In summary, the epithelium is normally a complex of several cell types the presence, distribution and concentration of each depending upon the species examined. Three of the four types of secretory cell have been described and the secretions produced by them impart visco-elasticity to the mucus which along with entangled pollutants, is moved by the tips of cilia to the pharynx where it is normally swallowed. Intraepithelial nerves have been identified and may be associated with specialized granulated (Kultchitsky) cells. These cells are likely to play a part in the lung's response to irritation or hypoxia either directly or following the release of chemical mediators, some of which may derive from nonspecific degranulation of intraepithelial mast cells. Lymphocytes are present both within and below the epithelium, in the absence of infection, and may well form a more diffuse immunological mucosal system linked by a continuous traffic of lymphocytes. The functions of the airways are diverse and are far more than mere conduction, humidification and purification of inspired air.

The author gratefully acknowledges the support of the Cystic Fibrosis Trust and Medical Research Council for many of the studies reviewed here: also to the following publishers for allowing reproduction of figures; Journal of Anatomy for figs. 2 & 7, Wiley Pub. fig. 5 and Marcel Dekker figs. 3 & 8.

References

Basset, F., Poirier, J., le Crom, M., and Turiaf, J., 1971
     Etude ultrastructurale de l'epithelium bronchiolaire humain
     Z. Zellforsch 116:425.
Bensch, K.C., Gordon, A.B., and Miller, L.R. 1965 Studies on the
     bronchial counterpart of the Kultschitsky (argentaffin) cell
     and innervation of bronchial glands.
     J.Ultrastruct. Res. 12:668.
Bensch, K.G., Corrin, B., Pariente, R. and Spencer, H. 1968
     Oat cell carcinoma of the lung. Its origin and relationship
     to bronchial carcinoid. Cancer (Phil) 22:1163.
Bienenstock, J., Clancy, R.L. and Percy, D.Y.E., 1976
     Bronchus-associated lymphoid tissue (BALT): its relationship
     to mucosal immunity, in: "Immunologic and infectious reactions
     in the lungs. (ed. C.H.Kirkpatrick and H.Y.Reynolds) Vol.1.
     of Lung Biology in Health and Disease (ed.C.Lenfant) Dekker
     pp.29-58.
Breeze, R.G., and Wheeldon, E.B., 1977, The cells of the pulmonary
     airways. Am. Rev. resp. Dis. 116:705.
Chamberlain, D.W., Nopjaroonsri, C., and Simon, G.T., 1973,
     Ultrastructure of the pulmonary lymphoid tissue.
     Am. Rev. resp. Dis. 103:621.
Clara, M., 1937, Zur histobiologie des bronchalepithels
     Z. Mikrosk .Anat. Forsch, 41:321.

Das, R.M., Jeffery, P.K., and Widdicombe, J.G, 1978
    The epithelial innervation of the lower respiratory tract
    of the cat. J. Anat. 126:123.

Das, R.M., Jeffery, P.K., and Widdicombe, J.G 1979
    Experimental degeneration of intra-epithelial nerve fibres
    in cat airways. J. Anat. 128:259.

Ericson, L.E., Hakanson, R., Larson, B., Owman Ch., and Sundler, F.
    1972 Fluorescence and electron microscopy of amine-storing
    enterochromoffin-like cells in the tracheal epithelium of mouse.
    Z. Zellforsch 124:532.

Etherton, J.E., Conning, D.M., and Corrin, B. 1973
    Autoradiographical and morphological evidence for apocrine
    secretion of dipalmitoyl lecithin in the terminal bronchiole
    of mouse lung. Am. J. Anat. 128:11.

Gmelich, J.T., Bensch, K.G., and Liebow, A.A. 1967
    Cells of Kultschitsky type in bronchioles and their relation
    to carcinoid tumour. Lab. Invest. 17:88.

Hage, E. 1972 Endocrine cells in the bronchial mucosa of human
    foetuses. Acta. path. microbiol. scand. 80:225.

Hansell, M.M., and Moretti, R.L. 1969 Ultrastructure of the mouse
    tracheal epithelium. J. Morphol.128:159.

Jeffery, P.K., and Reid,L. 1975. New features of rat airway
    epithelium: a quantitative and electron microscopic study.
    J. Anat. 120:295.

Jeffery, P.K., and Reid, L. 1977 Ultrastructure of airway
    epithelium and submucosal gland during development .
    In:"Development of the Lung" (ed. W.A.Hodson) Vol.6 of Lung
    Biology in Health and Disease (ed. C. Lenfant) Dekker pp 87-134.

Jeffery, P.K. and Reid, L. 1977 The respiratory mucous membrane.
    In:"Respiratory Defense Mechanisms Part I (ed. J.D. Brain et al)
    Vol.5 Lung Biology in Health and Disease (ed. C. Lenfant).
    Dekker pp 193-246.

Jeffery, P.K. 1978 Structure and function of mucus-secreting cells
    of cat and goose airway epithelium. In:"Respiratory Tract
    Mucus"(ed. R. Porter) Ciba Foundation Symposium 54 (new series)
    Elsevier pp.5-19.

McCarthy, C. and Ried, L. 1964 Intracellular mucopolysaccharides
    in the normal human bronchial tree. J. Exp. Physiol. 49:86.

Moosavi, H., Smith, P., and Heath, D. 1973 The Feyrter cell in
    hypoxia. Thorax 28:729.

Mutalionis, D.H., and Parks, H.F. 1973. Ultrastructural morphology
    of the normal nasal respiratory epithelium of the mouse. Anat.
    Rec. 176:65.

Niden, A.H. 1967 Bronchiolar and large alveolar cell in pulmonary
    phospholipid metabolism. Science 158:1323.

Pearse, A.G.E., 1969 The cytochemistry and ultrastructure of
    polypeptide hormone-producing cells of the APUD series and
    the embryonic, physiologic, and pathological implications of
    the concept. J. Histochem. Cytochem. 17:303.

Reid, L. 1954 The pathology of chronic bronchitis. Lancet 1:275

Rhodin, J. 1966 Ultrastructure and function of the human tracheal mucosa. Am. Rev. resp. Dis. 93:1

Rosan, R.C. & Lauweryns, J. 1972. Mucosal cells of the small bronchiles of prematurely born human infants (600-17009) Beitr. Pathol. 147:145.

Salvato, G. 1976. Mechanism of asthma attack. B.M.J. 2:179.

Sleigh, M.A. 1977 The nature and action of respiratory tract cilia. In: "Respiratory Defense Mechanisms" Part I (ed. J.D.Brain et al) Vol5 of Lung Biology in Health and Disease (ed. C. Lenfant) Dekker pp 247-288.

Sturgess, J.M. 1977 The mucous lining of major bronchi in the rabbit lung. Am. Rev. resp. Dis. 115:819.

Van As, A. & Webster, I. 1974 The morphology of mucus in mammalian airways. Environ. Res. 7:1.

Walsh, C.,and McClelland, J. 1974 Intra-epithelial axons in avian trachea. Z. Zellforsch, Mikro. Anat. 147:209.

Widdicombe, J.G. 1954 Respiratory reflexes from the trachea and bronchi of the cat. J. Physiol. 123:55.

D I S C U S S I O N

LECTURER: Jeffery   CHAIRMAN: Cumming

DENISON: How do cilia know where the head is? Remembering
                that many of the airways are pointing downwards.

JEFFERY: The question of why do cilia beat towards the head is
                of course a very interesting one and one which it is
                easy to answer teleologically, which I won't do. It
                appears that attachment and arrangement within the
                epithelium is very specific. The direction of the
                cilium is determined by its orientation within the
                epithelium. It has a lateral spur or a basal foot
                which is orientated towards the head. The structure
                of the cilium with its characteristic arrangement of
                neurotubules is so organised that it also has a
                direction and will be towards the head. But on this
                point may I also say, and I think that my second or
                third slide showed this, the cilia are arranged in
                fields. The consequences of this are that the cilia
                in any given field beat in a similar direction and
                are coordinated with each other. The cilia of
                another field, while beating in coordination with
                themselves, are out of step with another field and
                beat in a slightly different direction. The net
                resulting movement of all the ciliary fields moves
                mucus towards the head, but each ciliary field has
                slightly differing patterns of movement and
                direction.

DENISON: The Chairman, like me, is aware that you have not
                answered my question. You have made some
                observation, but you have not told me yet in which
                direction they beat. Do they beat in one direction?
                I accept that they have a little foot which indicates
                in which direction they are beating, but do we know
                in which way they are pointing their foot?

CUMMING: Peter, how will the cilia in the bronchus to the
                upper lobe, which have to beat in the direction of
                the toes, distinguish them from the lower lobe, which
                has to beat in the direction of the head?

JEFFERY: I sympathise with your question because I have asked
                it many times myself. But the fact of the matter is
                that the insertion of the cilium in the cell is
                genetically determined and already precludes the
                beating of the cilia in other directions. The cilia

of the mammalian respiratory tract are perhaps rather different than those of the water propelling cilia of paramecium, where it has been well shown that if you alter the potassium concentrations you will alter secondarily the calcium flux and then the direction of beat. This does not happen in the mammalian system. Experiments have been performed where isolated segments of mammalian cilia have been excised and reversed, but they continue beating as before and do not return to the correct direction. So it would suggest a genetic determination by one greater than ourselves for the direction of ciliary beat.

PRODI: How do cilia behave at a cellular bifurcation?

CUMMING: The question is quite clear, I think the answer is less clear.

PRODI: It could be the reason of formation of some hot spots in dust deposition perhaps because of some slower clearance mechanism.

JEFFERY: I think what I would like to do is pass the question over to one who is more experienced in that particular field, and I would ask Sanderson to answer on my behalf.

SANDERSON: Towards the bifurcation of an airway the effective stroke becomes more and more oblique to the axial direction if we are moving up towards the head. The ciliary beat towards an orifice will swing around so that the mucus comes up and is transported around the orifice. On the other side of the orifice the separated mucous stream will remain separated, draining to the top half which is then left virtually uncovered by mucus. So at bifurcations the direction of the ciliary beat changes to compensate.

JEFFERY: Again, presumably, this is genetically determined.

BELFA: I missed the point of the function of the crown-like apparatus at the upper ending of the cilium. Probably you told me, but I did not understand.

JEFFERY: This finding is interesting in as much as why we looked so carefully at the tip; the work of Professor Sleigh at Southampton describes the cilia pushing the mucus forward, so in response to this statement we decided to look at the ciliary tip to see if there was any specialised structure concerned with moving the mucus and interacting specifically with mucus. To our surprise we found what we described as small claws at the tips of the cilia, which interact specifically, we think, with the mucus and move it forwards.

CUMMING: How do the claws know when to let go?

JEFFERY: They need not know when to let go, because it is the method of motion and recovery stroke of the cilia which determines whether or not these claws project into the mucus or are removed.

MOLINA: I would like to ask two questions to Jeffrey. What do you think of the role of Clara cells in the metabolism of lipids? And, second question, what do you think of the hypothesis that the so-called broncho-alveolar carcinoma is derived from Clara cells?

JEFFERY: In terms of the metabolism of Clara cells, and this cell type has been under much dispute for many years, the presence of smooth endoplasmic reticulum suggests that the cell is handling cholesterol and manufacturing steroids inasmuch as you can draw the analogy of morphology with other steroid-producing cells such as testicular cells of Leydig, for example. So it is thought that these cells may handle cholesterol and be involved with lipid biosynthesis. On the other hand this cell has a multipotential. In our experimental work we have shown that if you inject repeatedly isoprenaline sulphate into animals you can stimulate goblet cell hyperplasia in the distal airways. Under those circumstances the Clara cell can be shown to go through a series of transitional steps and produce mucus. So I think that these cells, although normally handling lipid biosynthesis, may under experimental conditions produce mucus. The relationship with broncho-alveolar carcinoma I would like to pass to my colleague, Corrin.

CORRIN: There have been a number of electron microscopic studies of this tumour and I think the name bronchiolar or alveolar tumor is very appropriate because they have shown that this pattern of neoplasia can involve both alveolar cells and bronchiolar. And of the bronchiolar cells the tumor may be composed of mucus, goblet cells, or of Clara cells, sometimes the tumour consists of a multiplicity of these cell types.

BIGNON: There was a meeting on the biochemistry of emphysema in Sardinia, the Lund group showed us the bronchial inhibitor inside the airways by the immunoperoxidase method using light microscopy. They showed a very marked location at the level of the mucous gland, but only at the level of the serous cells in the mucous gland of the large airways. At the level of the smaller airways they showed several cells filled with cilia. And my question is: this cell did not look like a Clara cell, but in your data you say that there are few such cells in the lung, in human airways. In my experience I found the same. Where is the location of this bronchial inhibitor in the airway cells?

JEFFERY: Are we talking about lysozyme?

BIGNON: No, the specific bronchial inhibitor. It is an antiprotease specific for elastase, and is a very important protection for the smaller airway and perhaps also for the larger airways to protect the lung structure at this level.

JEFFERY: The answer to the question as to the specific localisation of this enzyme inhibitor is: I do not know. I think we have to be careful when talking about serous cells to distinguish between those serous cells in the gland and the serous cells of the epithelium which we have described. It is true to say that the serous cells in the epithelium of the human airway are few indeed. They appear to be more frequent in the neonates or in the prematurely born than they are in the adult. When you come to look at the smaller airways you find of course that the proportion of cells changes; for example the basal cell is not present in the small airway, but there are more than two cell types; there are Clara cells and ciliated cells, but there are also intermediate cells. Perhaps these latter, which are flattened at their apices, are the cells involved. I cannot say.

CANDURA: If I may, I would like to go back to the question of ciliary movement. Let us imagine a patient who has lobectomy, with the remaining lung expanding to occupy the vacated space, or a retraction, or any other cause determining a totally different position of a bronchus, as compared to its initial position.

Do you think that the cilia, whose beat is genetically determined, will continue to beat in the same way or not? And referring to some of the findings you showed about the interaction between nerve endings and cells, don't you think that there might be a nervous regulation, besides the genetic determination?

JEFFERY: I think the consensus of opinions is that the direction of beat in the mammalian system is genetically determined. So in the situation you described with the lobectomy one would anticipate that the cilia would continue to beat in the direction they had been previously beating. With regard to the neural involvement, yes, we think that nerves are involved in the regulation of the epithelium, but this would not be concerned with the direction of the beat, but rather in either the speed of beat or deciding whether a group of cilia would cease beating for a short period and then continue. For example, neurotransmitters such as acetylcholine have been shown to stimulate the beat, while 5HT inhibits the beat.

HEATH: At Liverpool we have been looking for the best part of ten years to find type III pneumocytes and we have never come across them. Do you still think they exist, and, if so, what do they do?

JEFFERY: This work I must immediately ascribe to Barbara Meyrick who worked with Lynne Reed and indeed was a colleague of mine. She described them in the rat, they have now been described in the small airways of pigs and a number of other species, but not yet adequately in man. In terms of the so-called brush cell in the alveolus, or type III pneumonocyte. I myself have found these cells with some difficulty, but not with great difficulty. I cannot understand why you have not found them. The only thing I will say is that Weibel also for some years doubted the existance of the cells, but certainly when he made his finding he sent us several pictures of the cell as confirmation.

CUMMING: Are you dealing with the same species, Donald, the rat. Your answer is yes.

JEFFERY: The strain of the rat we used was the Sprague-Dowley, SPF.

CUMMING: The same, Donald, that you used?

RICHARDSON: You mentioned that the Kultschitzsky cells might
                    release biogenic amines in response to hypoxia. Is
                    this purely a speculation or is tnere any
                    experimental evidence for it? Has anyone for
                    instance collected the effluent blood coming from
                    hypoxic lungs and looked for biogenic amines in it?

JEFFERY: There are two lines of evidence for making that
                    statement. One is the histochemical evidence of the
                    presence of biogenic amines within these cells. The
                    second is the ultrastructural studies of Heath and
                    Smith and his group, who have looked at the
                    morphology of these cells following exposure of rats
                    to hypoxia. In these studies they have very
                    elegantly shown that these cells do degranulate. If
                    one puts both of these bits of evidence together I
                    think one forms the conclusion which I have put
                    forward. In answer to your specific question: has
                    anyone cannulated and recovered biogenic amines in
                    pulmonary circulation – I do not know. I have not
                    come across this in the literature, have you?

RICHARDSON: No, I have not. But if the role of these cells is to
                    influence the circulation, don't you think it would
                    teleologically be a better place to have them,
                    somewhere nearer the circulation than they are?

                    They are in fact in the airways, some of them a long
                    way from the pulmonary arterioles, so, don't you
                    think it is rather unlikely that even if they do
                    release their granules during hypoxia, the chief role
                    of them is to regulate the pulmonary circulation?

JEFFERY: I have in fact considered that myself and raised that
                    issue. But as I am constantly reminded these cells
                    are present in their greatest frequencies in the
                    small bronchioles, which are closed to the pulmonary
                    circulation.

CUMMING: It would depend very much on the way you thought the
                    signal was generated to produce the motor effect that
                    you observe. the hypoxia may be determined within
                    the airways, and a local reflex then shunts the blood
                    away. That may be the mechanism, in which case you
                    would find the detectors in the airways.

JEFFERY: The other point to make is of course that it is a cannulated, you would have to do it in rather large vessels, and I would not imagine one could pick out the small local variations by cannulation of a large pulmonary vein.

WIDDICOME: So far we have heard two pieces of evidence that nerves in the epithelium may be motor nerves. In the first place some of them contain small vesicles that look like secretory vesicles and other autonomic motor nerves, dense core vesicles or clear vesicles, but those of us who are not histologists will not be very convinced by that evidence. The second observation is that if you apply mediators, such as acetylocholine, or alpha-agonists to the epithelium, you may produce changes - either secretory changes or histological changes - but of course this tells us nothing about the innervation of the epithelium. I wonder if there is any direct evidence, experimental evidence, that nerves can change the epithelial activities. To answer my question partly, the only experimental experimental evidence that I know of is that in the bird, nerves may cause goblet cell secretion from the epithelium. Is there any equivalent experiment in the mammal?

JEFFERY: By saying in the mammal you have cut out my answer.

WIDDICOMBE: In species other than bird?

JEFFERY: Well, in amphibia, looking at the frog palate, changes in ciliary motility have been described in response to neural stimulation. I cannot say, in that species, with regard to goblet cells secretion.

CHEVALIER: With reference to your last slide, is there any evidence that the Clara cells would be involved in hydro-mineral metabolism or Ionic balance?

JEFFERY: There is no evidence that the Clara cell is involved in ionic balance. There is much more evidence, and indeed a lot more is known with regard to the transport of ions across epithelial cells other than Clara cells. There is not specific evidence in terms of the Clara cell. It is indeed more likely that the ciliated cell on its microvillous border is more concerned with that.

CORRIN: To return to the Kultschitzsky cell, I just said in passing in response to Richardson's remark about what mediating a purely local response between ventilation and perfusion; I would have thought that the situation in the smaller airways immediately nextdoor to the arterioles would be an excellent one. what I wanted to bring out is that we have several references that these cells contain, and are perhaps secreting, biogenic amines, but you, Jeffrey and I have recently enjoyed listening to the work from the Hammersmith Hospital's department of cytochemistry. These cells are mentioned as members of the APUD series, many members of the APUD series have developed the APUD characteristics a little bit further to put these aminoacids together, which produce a whole series of peptide hormones, and Dr. Polak has looked for a variety of peptide hormones in many APUD cells. Perhaps Jeffrey would correct me if I am wrong, but I remember, I think, that she located the peptide hormone with vasoactive properties first isolated from an Italian frog, I think its name is Bombesina bombesina. So no doubt there is no necessity for me to remind you whether it is vasoconstricting or vasodilating.

JEFFERY: It is bombesin which has been in fact localised in these cells. The other peptide localised in the respiratory tract is VIP, vasoactive intestinal peptide. Just for the benefit of our audience the cells called APUD system are cells with a common origin, thought to be the neural crest, which then by migration become distributed to various parts of the body. They all have the ability to take up amines and to decarboxylation cells, from which the name APUD derives.

CUMMING: With that little bit of semantics I think, ladies and gentlemen, we will terminate the session on the structure of the normal epithelium. Can we now procees to a further study of ciliary function, by Michael Sanderson.

THE FUNCTION OF RESPIRATOR TRACT CILIA

M.J. Sanderson and M.A. Sleigh

Department of Biology
University of Southampton
Southampton, Hants, England

INTRODUCTION

The function of the respiratory tract is gas exchange; the respiratory mucosa is continually exposed to the inhaled air and its contaminants, in the form of dust particles, bacteria and toxic gases. To function normally under most conditions, the respiratory tract has become specialized, developing the necessary defence mechanisms to prevent the otherwise inevitable dehydration, damage or clogging of the delicate pulmonary tissue. The defence mechanisms of the lungs can be arbitrarily divided into two categories; the specific mechanisms, often mediated by the immunological responses of the lung, and the nonspecific mechanisms. The latter include the aerodynamic filtration by the nasal chambers and the removal from the lungs of respiratory tract fluids with the deposited contaminants, by mucociliary transport or coughing. (Brain et al., 1977; Wanner, 1977).

The specific defence mechanisms and the protection provided by nasal breathing are the subject of other reports in this book; the function of this article is to outline the present understanding of the mechanism of mucociliary transport and to provide an awareness of its importance in lung defence.

CHARACTERISTICS OF MUCOCILIARY CLEARANCE

Mucociliary transport occurs throughout the respiratory tract, however, the direction and velocity of the mucus transport varies. The respiratory tract can be divided at the level of the larynx into upper and lower respiratory tracts. The upper tract includes the airways from the nares to the oropharynx. The lower respiratory

tract extends from the oropharynx to the lung periphery. The lung
airways are classified on the basis of structure, rather than size;
the trachea branches into two major bronchi which lead via several
generations of branching to the bronchioles, followed by terminal
bronchioles and finally respiratory bronchioles before reaching the
alveolar tissue (Fig. 1.).

The appearance of the respiratory mucosa varies within the lung
according to the location, constituent cell types and animal species.
Generally, the lung is lined with a pseudostratified ciliated
epithelium (Breeze and Wheeldon, 1977), containing ciliated cells,
basal cells and mucus secreting cells e.g. goblet cells; in all
there are eleven different recognized cell types.

The density of the ciliated cells becomes progressively less
towards the lung periphery, similarly, the length of the cilia
decreases; in the upper tract and trachea the cilia are approximately
6 μm in length, decreasing to 4.5 μm in the bronchioles. The structure and appearance of the cilia will be considered later.

Although the importance of mucociliary transport in lung defence
has been recognized and many studies completed, the activity and
interactions of the components of the system are not fully understood. However, many studies have described the movements and
appearance of the mucus within the lung. Therefore, before considering the components of the mechanism further, summaries of the
mucus transport, ciliary organization and methods of study are
provided.

Methods of Study

Our limited understanding of mucociliary clearance mainly
results from the technical difficulties involved with studying these
ciliated epithelia. Many observations have been made using reflected
light, assuming that the close association between ciliary activity
and mucus transport allows the activity of one component to reflect
the activity of the other. Until a full understanding of the system
is acheived, observations of this nature should be avoided. The
shortcomings of these techniques are often emphasised by the qualitative nature of many results and variability of quantitative data,
(Dalhamn, 1956; Toremalm et al., 1975) in vitro.

Mucus transport has been readily studied in vitro by diffuse
reflected light and often particles were added to mark the mucus.
The study of mucociliary transport in vivo is obviously more difficult; radioactive tracers have been inhaled into the lung and
their clearance rates monitored. Care must be taken to ensure that
the deposition of particles between trials is similar if the clearance rates are estimated as a proportion of the radioactivity

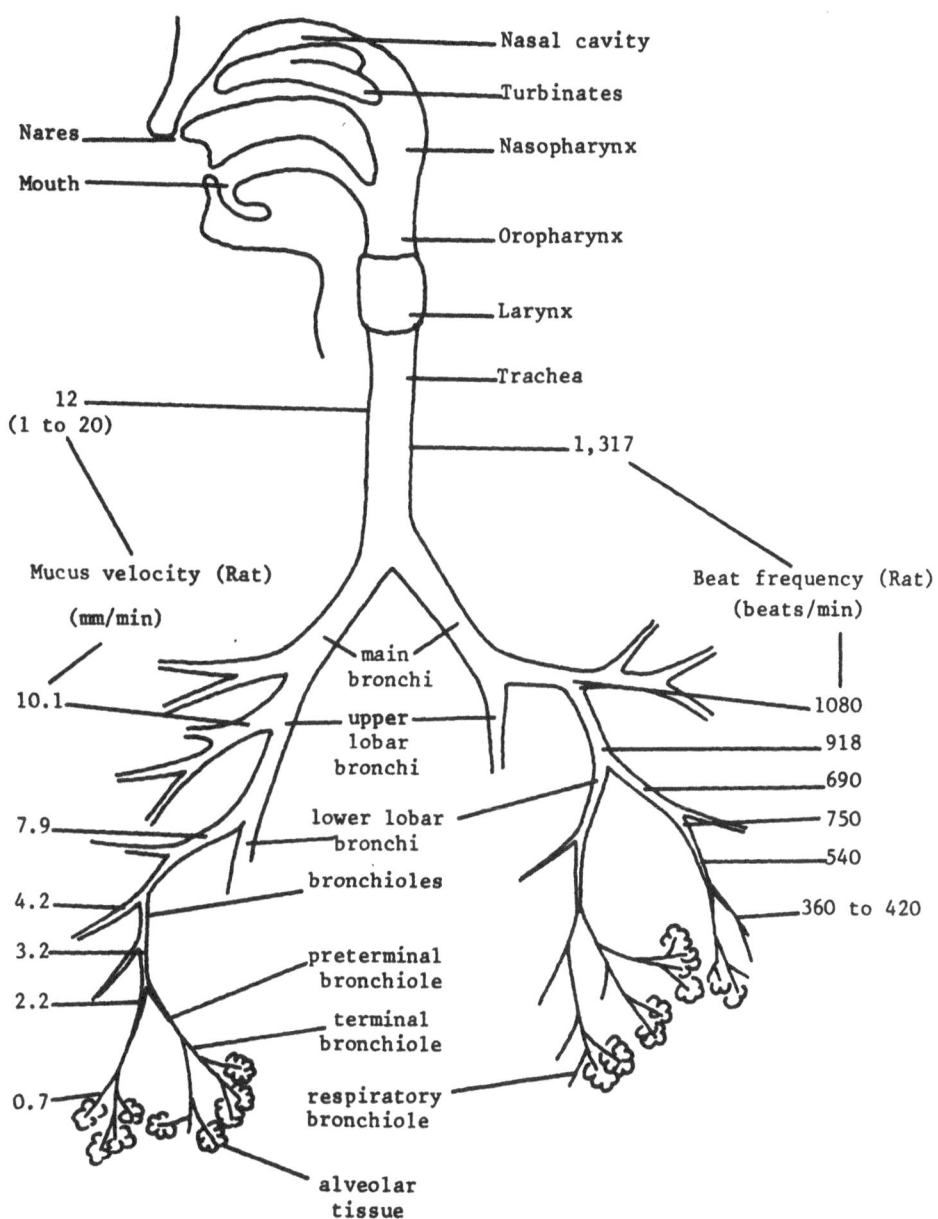

Fig. 1.   The general organization of the mammalian respiratory tract,
          showing the mucus transport velocities and ciliary beat
          frequencies at different levels within the rat lung reported
          by Iravani and Melville (1976).

remaining in the lung. False readings are readily induced by the
proximal deposition of particles. Greater accuracy is achieved if
the actual movement of a bolus of radioactive particles is followed;
the direct observation of a teflon disc in the trachea by bronchoscopy clearly reflects mucus transport; however this technique is
invasive, which in itself may induce unpredicted changes.

Most of these techniques have been restricted to the upper
airways, due to their size and accessibility. Iravani and Van As
(1972) have described a technique to study mucociliary function in
the lower airways. The lungs and trachea of rats are quickly removed
and dissected, perfused with saline and mounted for viewing. The
mucociliary clearance is examined through the transparent bronchiole
wall. More recently the use of Scanning Electron Microscopy (SEM)
has contributed significantly to knowledge of the lung ciliary
apparatus (Sturgess, 1977).

Mucus Transport

The continually secreted mucus provides a partial or complete
lining to the respiratory airways and traps particulate matter. The
mucus is transported towards the oropharynx from all parts of the
respiratory tract; the nasal cilia predominatly moving the mucus
backwards and caudually whilst the cilia of the trachea and lower
airways transport the mucus cranially. At the oropharynx, the small
quantity of mucus (10 to 100 ml per day) produced by healthy lungs
is removed by swallowing; in diseased lungs coughing aids the
removal of sputum.

An important observation pertinent to mucociliary clearance is
the reduction of total airway circumference at different levels towards the oropharynx from approximately 30 metres at the lung
periphery to 50 millimetres in the upper trachea (Hilding, 1957).
Obviously the mucociliary transport mechanism can compensate for
this convergence, otherwise lung failure would be common. There are
several possible ways by which the reduction of surface area can be
offset; the mucus transport can increase proximally, the depth of
the mucus blanket could increase or the respiratory epithelium may
absorb some of the secretions.

The depth of the mucus layer has not been observed to change
significantly throughout the bronchial tree; Dalhamn (1956),
recorded a mucus depth of 5 μm in the trachea, whilst Alder et al.
(1973) have observed a mucus lining of up to 20 μm, although in most
cases the thickness was less than 10 μm. However, a change in the
spatial distribution of the mucus would equally compensate for the
reduction in area. Iravani and Van As (1972) described the mucus
lining as discontinuous, existing in the lower bronchioles as discrete droplets approximately 4 μm in diameter. Moving towards the

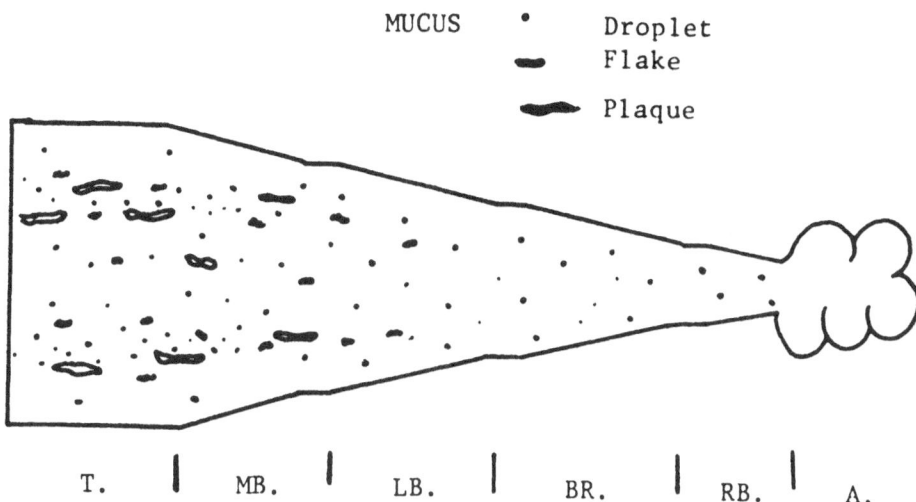

Fig. 2.   The distribution and appearance of mucus throughout the
          bronchial tree as described by Iranani and Van As (1972);
          small 'droplets' of mucus are present in the lower airways
          which aggregate into mucus 'flakes' in the upper bronchi or
          into larger mucus plaques' in the trachea. T., trachea;
          MB., main bronchi; LB., lobar bronchi; BR., Bronchioles;
          RB., respiratory bronchiole; A., alveoli.

trachea the droplets were observed to coalesce forming flakes of
mucus, whilst larger aggregates or plaques were formed higher in the
airways, (Fig. 2.).

    The accuracy of this report has been contested by Sturgess (1977)
who has presented evidence that the mucus blanket is more extensive
in the rabbit trachea. However, the preservation of the life like
appearance of mucus is difficult, owing to its weak association with
the ciliated surface. It would seem that both of these views are
partially correct, depending on the level of observation and animal
species.

    The absorbtion of lung secretions has not been fully investigated.
Whilst it is possible that the epithelium could absorb the sol component of the mucus lining (Nadel and Davis, 1978), it would seem
unreasonable for the mucosa to absorb the mucus containing the
particles destined for removal.

    The simpler explanation of proximally increased mucus flow is an
attractive hypothesis to counter the effects of surface reduction by
convergence. Increased flow rates have been observed towards the
oropharynx in the nasal cavities, and between the lower and upper

trachea. Iravani and Melville (1976) have reported a gradient of
mucociliary transport throughout the bronchial tree (Fig. 1.),
measuring from 1.7 mm/min in the preterminal bronchioles to
10.1 mm/min in the lobar bronchi.

The route followed by mucus within the airways is varied;
Iravani and Van As (1972) reported streams of mucus 'zig zagging' up
the trachea as the mucus encountered numerous metachronal fields of
various orientations. Asmundsson and Kilburn (1970) noted a helical
path of mucus transport in dog trachea, whilst axial streaming
appeared to be present within cow trachea (Hilding, 1957). The
mucosal surface is commonly interrupted by the apertures of joining
airways; to prevent the build up of mucus around each orifice, the
mucus stream diverges into two streams approximately 300 to 500 μm
ahead of the opening. The two streams flow around the orifice and
remain separated on the downstream side, the mucus of the bronchial
tributary draining into the downstream edge (Hilding, 1957). A
similar division of the mucus stream must occur in the bronchiolar
branch; the mucus on the surface of the lower half of the airway
must diverge towards the top half to be transferred to the upward
stream of the major airway.

The Organisation of Ciliary Activity

By contrast to the prominent metachronal waves formed by water-
propelling cilia of protozoa or such larger animals as Mytilus,
respiratory cilia do not show a pronounced coordination. The lack
of prominent regular metachronal waves appears to be characteristic
of mucus transporting cilia; only short, erratic waves being observed
on frog palate epithelium, mouse oviduct and Mytilus frontal cilia.

Dalhamn (1956) reported that small patches of activity
(0.16 mm x 0.02 mm) were apparent on rat tracheal epithelium observed
by reflected light. Similar patches of coordinated activity, or
metachronal fields, were observed in the major bronchi of rats
(Van As and Webster, 1972). Each field varied in shape and size,
ranging from a few cells to several hundred cells. The direction of
beat and consequently the direction of travel of the metachronal wave
was not consistent between adjacent fields; often neighbouring fields
could have the effective strokes diverging by as much as 180°.
However, most activity was generally in a cranial direction. In
accordance with the bifurcation of the mucus streams to avoid
bronchial obstruction at airway junctions, the direction of the
effective stroke became progressively more oblique approaching the
orifice until the stroke direction parallelled the sides of the
opening (Hilding, 1957).

The ciliary beat frequency has been estimated by many workers
and a considerable range has been recorded, probably as a result of

the different techniques used. Dalhamm (1956) observed a frequency
of rat tracheal cilia of 1,317 beats/min. at 37°C. Studies by
Iravani and Melville (1976) have documented an increase in beat
frequency from the rat lung periphery towards the trachea; the
bronchiole frequency of 360 to 420 beats/min. increased to 1200 to
1300 beats/min. in the lobar bronchi (Fig. 1.). This gradient of
beat frequency, coupled with the increased ciliary length towards
the trachea, will generate greater ciliary tip velocities; an effect
reflected by the increasing gradient of mucus transport velocity in
a cranial direction (Iravani and Van As, 1972).

THE COMPONENTS OF THE MUCOCILIARY CLEARANCE SYSTEM

The mechanism of mucociliary clearance was first proposed by
Lucas and Douglas (1934), whilst studying the nasal ciliary activity
of various mammals. This has remained the preferred hypothesis in
view of the additional evidence discovered by numerous subsequent
studies. Although many of the exact details of the system still
await documentation, it is generally accepted that the mucociliary
interface consists of a double fluid layer; a 'gel' or mucus layer,
which possesses the necessary visco-elastic properties essential for
particle transport, supported at the ciliary tips by an underlying
'sol' or watery periciliary layer which bathes the cilia. The mucus
is transported by a clawing action of the cilia; the tip of each
cilium penetrating the underside of the mucus during each effective
or propulsive stroke (Fig. 3a, b).

Sol Layer

In comparison to the mucus, far less is known about the origin
or constituents of the sol layer; its watery nature prevents the
preservation of this layer by conventional fixation techniques. The
spaces between the cilia in many sections prepared for histology
and transmission electron microscopy appear to be empty, the
presence in vivo of an actual sol layer being assumed rather than
proved by the tissue appearance. Biochemical analysis has been
attempted with little success, as the quantities of secretions produced
by healthy lungs are small (10 to 100 mls/day) and not easily sampled.
Increased amounts of secretions can be collected after the administration of stimulants such as prostaglandin F2α. Although the resulting
sputum is readily separated into two fractions, with the gel fraction
corresponding to the mucus secretions, it is unknown to what extent
the sol phase represents the normal constituents of the periciliary
fluid (Lopez-Vidriero et al., 1977).

The presence and influence of the periciliary layer has been
more readily recognized by studies investigating mucociliary clearance on live tissue. Several studies have demonstrated that

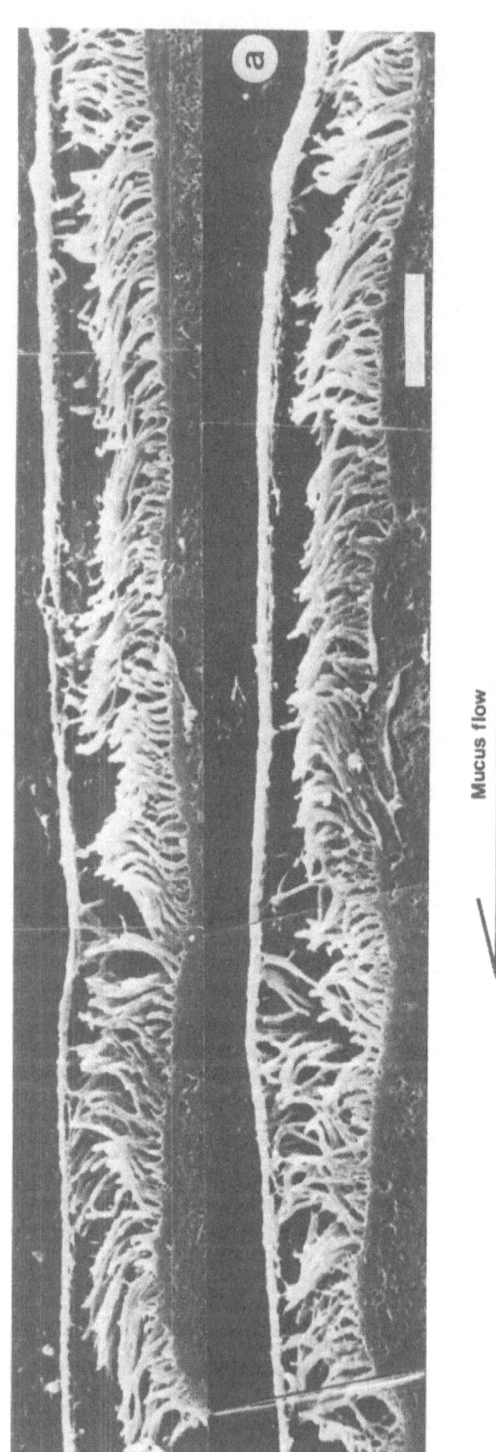

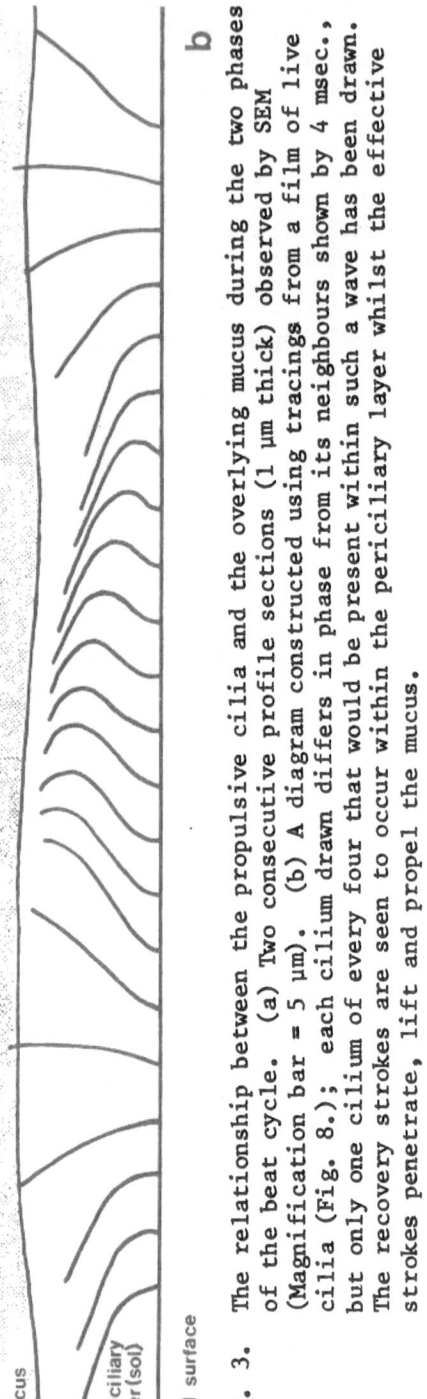

Fig. 3.   The relationship between the propulsive cilia and the overlying mucus during the two phases of the beat cycle. (a) Two consecutive profile sections (1 μm thick) observed by SEM (Magnification bar = 5 μm). (b) A diagram constructed using tracings from a film of live cilia (Fig. 8.); each cilium drawn differs in phase from its neighbours shown by 4 msec., but only one cilium of every four that would be present within such a wave has been drawn. The recovery strokes are seen to occur within the periciliary layer whilst the effective strokes penetrate, lift and propel the mucus.

mucociliary transport could come to rest, the mucus remaining station-
ary whilst the cilia beneath continued to beat. Dyes or small particles subsequently added were often observed to be transported,
although no further transport of the mucus had occurred (Lucas and
Douglas, 1934; Proctor et al., 1978). These observations have been
explained by a change in the periciliary layer; an increased depth
would tend to float the mucus above the cilia, preventing ciliary
contact and leading to impaired mucociliary transport. The movement
of the dyes and small particles occurs below the mucus but above the
cilia, carried by the weak fluid currents generated in the excess
periciliary fluid beyond the ciliary tips; appreciable penetration
of particles between the cilia does not appear to occur. However,
it should be emphasised that under normal conditions, when the sol
layer reaches only as far as the ciliary tips, the transport of the
periciliary fluid is probably very slow and its movement may be
merely oscillatory. This assumption, in the absence of data, is
based on simple hydrodynamic principles which relate to the propulsion
of fluids by cilia. It will be an advantage to the reader to be
familiar with these basic principles if the significance of the
different beat patterns and associated coordination of cilia in
systems specialized for the transport of water or mucus are to be
appreciated. A brief discussion of these principles pertinent to
this report is provided.

Fundamental hydrodynamic principles. The behaviour of moving
fluids can be influenced by the inertia of the propulsive body or
the viscosity of the fluid. A non-dimensional relationship, the
Reynolds number, which provides an estimate of the predominant forces
within a system, has been formulated as follows:-

Reynolds Number (Re) = fluid density x size x velocity
                            fluid viscosity

Large Reynolds numbers (often exceeding 10^3) are recorded in
fluid propulsive systems where inertial forces predominate, a familiar
example being the propellers of a ship. However, in the case of
cilia, having only a small size and low velocity, a low Reynolds
number is recorded (often less than 10^{-3}). In this situation the
forces of inertia are negligible and viscous forced predominate.

At low Reynolds numbers, the phenomenon of 'Non Slip' fluid
layers becomes progressively more important. The fluid immediately
surrounding a body remains in contact with its surface as the body
moves through the fluid. The influence of the moving object on the
surrounding fluid layer decreases with distance until the limits of
influence are reached, at which point the fluid is no longer affected
by the object. The limits of influence are related to the size and
velocity of the moving body and the viscosity of the fluid.

These simple principles apply to a beating cilium, the fluid

near its surface remaining in contact throughout the beat. They also
apply to the cell surface, fluid being progressively easier to move
as the distance from the cell surface increases; there is therefore
competition between the cilium attempting to move fluid and the cell
surface tending to prevent movement, the ciliary influence dominating
towards the tip. Thus the zone of influence of cilium on the fluid
changes with alterations in ciliary shape and ciliary tip velocity
that take place during various phases of the beat cycle.

During the effective stroke (see below), when the cilium maximises its extension from the cell and its velocity, the zone of
fluid influence is greatest and is approximately the shape of a cone
with an elliptical cross section (Fig. 4). The zone of influence is
reduced during the recovery stroke, a result of the slower tip
velocity and lower profile during this phase   (Fig. 4.). Because
more fluid is moved in one direction (by the effective stroke) than
in the other (by the recovery stroke), the ciliary beat results in
a net flow of fluid.

Although the same hydrodynamic principles apply to mucus-propel-
ing cilia and water-propelling cilia, a comparison of these cilia
helps to explain some of the differences in function and propulsive
capability. One of the most prominent differences is length; rabbit
cilia are very short (6μm), when compared to Mytilus lateral cilia
(16μm). Obviously the zone of influence will be reduced. More
important, however, are the relative heights of beat envelopes of
recovery and effective phases. In water-propelling cilia the effective stroke projects well above the recovery stroke giving appreciable
net propulsion; in mucus-propelling cilia the envelopes of the
effective and recovery phases substantially coincide (Fig. 4), and
most of the fluid influenced to move forwards by the effective stroke
will be moved back by the recovery stroke.

The movement of the sol layer is hindered further by the
phenomenon of 'Non Slip', as this applies to all surfaces; both
the epithelial surface and the surrounding cilia will resist the
movement of the fluid in their immediate vicinity. In addition,
the tips of mucus-propelling cilia commonly make contact with the
mucus, resulting in a marked reduction in the velocity of ciliary
movement and in coordinated activity, both of which will reduce
the fluid movement. Consequently, the movement of the fluid within
the sol layer is very complicated but the net result would be an
oscillation of the fluid rather than a net flow towards the
oropharynx.

The Mucus Layer

Mucus consists mainly of a mixture of glycoprotiens, each
molecule having a peptide core rich in the amino acids threonine

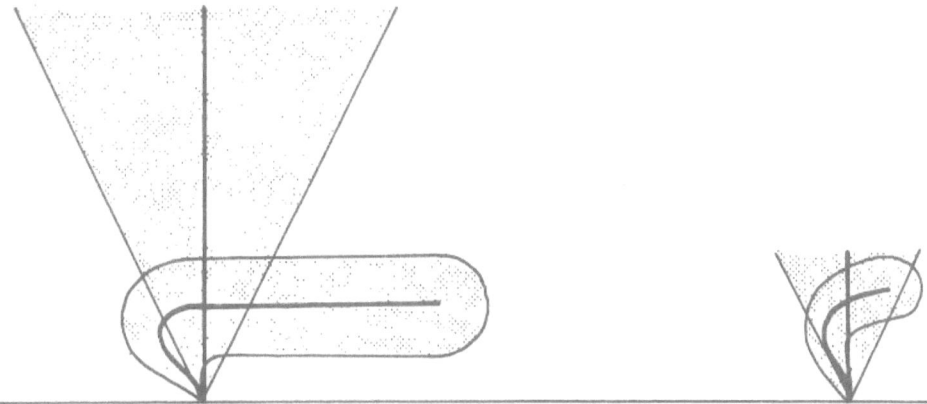

Fig. 4.   A comparison of the extent of hydrodynamic influence
          (dotted area) of the effective and recovery phase during
          each beat of longer (16 µm) water-propelling cilia (left)
          and shorter (6 µm) mucus transporting cilia (right).

              The effective stroke of the long cilia moves more fluid
          forward at greater distances from the cell surface then
          is moved back by the recovering cilia, whilst most of the
          fluid moved forward by the effective stroke is returned by
          the recovery stroke of short cilia.

and serine, to which are attached a variety of polysaccharide side
chains containing neuramic acid, fucose, galactose, N-acetyl-
galactosamine and N-acetylglucosamine.   These molecules are secreted
from the epithelial goblet cells and from the serous and mucous cells
of the subepithelial glands   (Lopez-Vidiero et al., 1977).

    Mucus is very important to mucociliary transport, its visco-
elastic nature provides the essential rheological properties for
particle transport.   Particles of various shapes and sizes are
entrapped within the mucus and are transported at similar velocities;
in the absence of mucus the same particles can not be transported.
The partial transport of a variety of mucus simulants which also
possess a macroscopic structure demonstrates the importance of the
physical form of the mucus.   Indeed changes in viscosity or
elasticity of the mucus have been shown to alter mucociliary transport rates   (Dulfano and Alder, 1975).   The presence of the mucus
layer at the ciliary tips is well established, but the form and
extent of the mucus covering in different parts of the respiratory
tract is believed to vary.

THE RESPIRATORY CILIA

Ciliary Structure and Mechanism

The mucus is propelled by the cyclical movement of cilia on the lining epithelium of the respiratory airways (reviewed by Sleigh, 1977). Cilia can be regarded as little motors using chemical energy to perform the work of fluid propulsion, the chemical energy driving mechanochemical cycles of activity of protein cross links within the cilium. It is known that activities within the cilium itself, and not at its base or within the cell to which it is attached, are responsible for the bending activity of the cilium, because cilia detached from cells can perform essentially normal movement as long as they are provided with a suitable energy source and ions.

The cilia of all types of animals and plants (but not bacterial flagella) are built on a common plan. All cilia have a similar diameter of about 250 nm, but vary greatly in length; mucus propelling cilia are generally only 5 to 7 μm long and in at least some cases the tips of mucus propelling cilia bear a crown of short projections. A cylindrical projection of the cell membrane encloses a bundle of protein fibres called the axoneme that has its origin from a basal body just below the cell surface (Fig. 5). The protein fibres of the axoneme are called microtubules, each being a cylinder about 25 nm diameter whose wall is formed of 13 rows of globular molecules of the protein tubulin. Two of these microtubules run the length of the centre of the axoneme, but do not extend into the basal body, and nine double microtubules form a cylinder about 200 nm in outside diameter around the two central ones. These outer doublets continue into the cell beneath to form the wall of a basal body about 0.5 μm long, each doublet becoming a triplet by the addition of a third microtubule near the level of the cell surface.

These microtubules form the backbone of the axoneme and carry a range of other components with a predominantly transverse orienta- tion (Fig. 5). Radial connections between the outer doublets and the central microtubules are formed by radial spokes that project inward from each doublet and make connections with curved projections from the central microtubules. Two types of links can connect adjacent doublet microtubules; pairs of arms project from each doublet towards the next doublet in a clockwise direction (when seen looking from the base towards the tip), and occasional 'nexin' links connect adjacent doublets at their inner side.

Our understanding of the functions of these components owe much to two types of experiments, fractionation of the components, and reactivation in vitro of ciliary axonemes from which the membranes have been removed and which have been treated in various ways - the latter are referred to as ciliary 'models'. The arms of the doublet

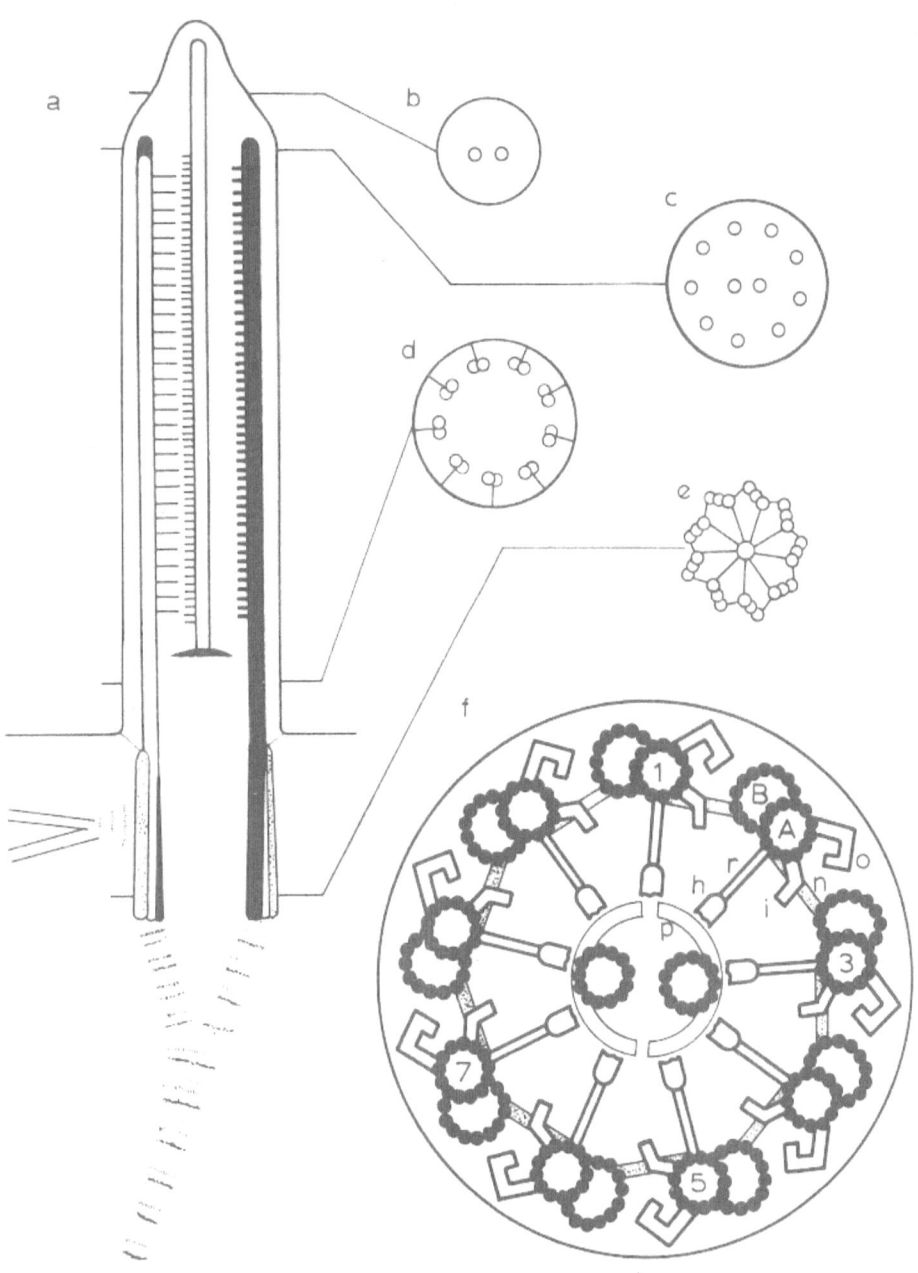

Fig. 5.   Structural features of a cilium. A longitudinal section is
          shown at (a) and a series of transverse sections at various
          levels at (b) - (f), each shown as seen from the basal end
          of the cilium. In the transverse section of the main shaft

(f) some of the peripheral doublets are numbered in the
conventional way; each doublet consists of an A and a B
microtubule, the A microtubule carries outer (o) and inner
(i) arms and nexin links (n) which project towards the
next B microtubule, and radial spokes (r) with dilated heads
(h) that project towards the central complex. The central
complex consists of 2 microtubules with their projections
(p). A pattern of radial links and central projections
is shown at the left of the longitudinal section (a) and
at the right the dynein arms are figured.

The ciliary shaft is enclosed by the cell membrane,
which therefore surrounds cross sections of the shaft.
The basal body of the cilium lies in the cell cytoplasm
and gives rise to striated roots beneath and microtubular
fibrils diverging from the striated basal foot at the left
side of the basal body. Reprinted from Sleigh (1977) p. 256,
by courtesy of Marcel Dekker Inc.

microtubules may be removed by treatment of demembranated axonemes
with 0.5 M NaCl; the resulting supernatant contained large
molecules of a protein with ATPase properties which were able to
reattach to the doublets after removal of NaCl and addition of low
concentrations of Mg^{++}. Since the arm protein was the principal
ATPase protein of the axoneme and occupied a suitable site for
causing movement, this protein was given the name dynein (force
protein).

Ciliary models prepared by demembranation of detached cilia in
an appropriate medium can be reactivated to produce propagated
bending waves and will swim in the presence of ATP and suitable
ions (particularly Mg^{++}), the ATP being dephosphorylated in
proportion to the number and activity of the cilia. The frequency
of bending is reduced if some of the dynein arms are dissolved by
treatment with KCl, or if the model axoneme is treated with an
antibody to dynein. If the model axoneme is briefly treated with
trypsin before reactivation with ATP, the resulting movement takes
the form of an extension of the fibril bundle of the axoneme by an
active longditudinal sliding of the doublets relative to one another
rather than the normal bending. This experiment indicates both that
the result of ATPase activity is a relative sliding of the doublets,
and that the components attacked by trypsin were responsible for
the restriction of sliding, converting relative sliding movements
of the doublets into bending of the axoneme. The trypsin removes
radial spokes and nexin links, so that these are thought to be
concerned in regulating the extent of sliding between the doublets;
the mechanism of their action is not yet well understood, but it
appears that nexin is an elastic component and that the radial
connections can be broken and reformed during the more limited

sliding that accompanies ciliary bending. If a reactivated axoneme
is suddenly deprived of ATP, the dynein arms are found to be attached
distally to the adjacent doublet. The dynein arms of model axonemes
normally lie with their free ends directed towards the ciliary base
at an angle of about 30°; during reactivation the dynein arms of
each doublet undergo a mechanochemical cycle in which they are
assumed to attach to the adjacent doublet and propel it unidirection-
ally towards the ciliary tip, before detaching and returning to
their initial position. Because of the arrangement of the arms on
the doublets, the doublets at one side of the axoneme must be
responsible for the sliding that bends the cilium towards one side
in the effective stroke, and the doublets at the other side of the
axoneme cause the bending that produces the recovery stroke.
Variants of the bending pattern found in different types of cilia
could be explained by regulation of the rate and extent of sliding
of the microtubule doublets.

Although the ciliary axoneme is capable of independent movement in these in vitro experiments, it is normally dependent upon
the cell body to which it is attached for both anchorage and a supply
of energy. The ATP that powers the mechanochemical cycles of
activity of the dynein arms is synthesised in the mitochondria of
the cell body and enters the axoneme by diffusion. In the basal
body the peripheral triplets are cross-linked by various fibrous
components that bind them firmly together. Different types of
rootlet structure are attached to the basal body and serve the
function of restricting the movement of the ciliary base during
the ciliary beat cycle. Striated rootlets often diverge downwards
into the cell cytoplasm and microtubules which radiate outwards just
beneath the cell surface are often principally attached to a projecting knob, the basal foot, at the side of the basal body towards
which the cilium bends in its effective stroke. The orientation
of this movement is normally perpendicular to the plane occupied
by the two central microtubules of the axoneme.

Basic Principles of Beat Patterns and Metachronism

A full appreciation of the activity, function and specializa-
tion of rabbit tracheal cilia for mucus transport, requires an
understanding of ciliary beat patterns and of the basic principles
related to ciliary coordination and metachronism.

Beat pattern. The basic beat pattern of a cilium is divisible
into a recovery or preparative stroke and an effective stroke.
During the effective phase the cilium is generally upright and rigid,
moving with a relatively high velocity, bending only near to its
base. Before a second beat can be performed, the cilium must
return to its starting position by progressing through the recovery
phase of the beat cycle. The cilium maintains a low profile and

moves with a slower velocity to reduce its influence on the surrounding fluid, being drawn backwards by an unrolling action produced by a bending of the cilium progressing from the base to the ciliary tip.

A simple beat pattern contains the effective and recovery strokes within one plane - the beat is said to be planar; however, there are relatively few examples of planar activity, in life, the comb plates of ctenophores being the most familiar. The majority of cilia possess a beat pattern where the effective and recovery strokes are not in the same plane. Often the effective phase remains upright and almost planar with the recovery stroke displaced to one side or the other.

Metachronal coordination. As previously mentioned, each beating cilium influences its surrounding fluid, consequently within a field or band of cilia, each beating cilium will influence neighbouring cilia through the intervening fluid. This hydrodynamic coupling induces the cilia to beat in succession; the movement of one cilium stimulates the neighbouring cilium to beat, which in turn will activate further cilia to beat. Consequently the adjacent cilia in a ciliary row beat slightly out of phase with each other and the field of cilia is said to possess metachrony.

In association with this coordination, the cilia within rows at right-angles to the rows of metachronous cilia beat in synchrony. (see later Fig. 9). The observation of cilia beating with metachronal coordination often shows the apparent movement of waves across the field of cilia. Each wave travels in the direction of metachrony and is constructed from cilia at different stages of their beat cycle; the cilia parallel to the wave front beat in synchrony.

The characteristics of the metachronal wave are related to a number of factors, the beat pattern and fluid viscosity being important. A classification of the wave forms has been proposed, which relates the wave travel to the direction of the effective stroke. The waves generally travel in the direction of the recovery stroke. When an observer looks in the direction of the wave travel, if the movement of the effective stroke is towards the left, the waves are laeoplectic, or towards the right - dexioplectic, in the same direction - symplectic, in the opposite direction - antiplectic (Fig. 6).

Metachronal activity has been extensively investigated in water-propelling cilia, where the waves are very prominent and generally diaplectic (either laeoplectic or dexioplectic). The function of metachrony is to reduce the amount of mutual interference between the cilia and to maximize their efficiency for water transport. The nature of the ciliary beat makes it impossible

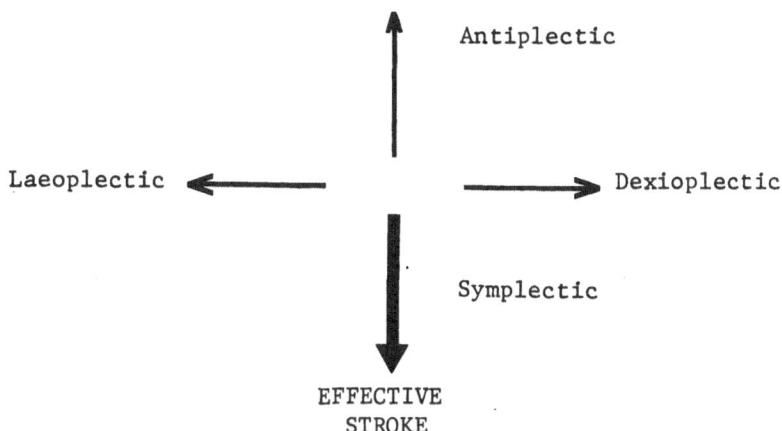

Fig. 6. Terminology to describe the directions of wave travel
         in relation to the orientation of the effective stroke
         that are found in different ciliated systems.

for a single cilium to generate a continuous water current, the
viscous forces in the fluid arrest the fluid immediately the cilium
stops moving. As a result, the fluid current would stop after each
effective stroke whilst the cilium recovers, and consequently energy
would be wasted in overcoming the initial viscous forces to produce
flow in each cycle. However, a metachronal sequence aids fluid
transport since the movement of fluid initiated by one cilium is
maintained by successive beating of the adjacent cilia.

The Beat Pattern and associated Metachronism of Rabbit Tracheal Cilia

     In comparison to water-propelling cilia, the activity of mucus
propelling cilia is not fully understood in terms of beat pattern,
metachrony and its relationship to mucus transport.

     Aiello and Sleigh (1977) have investigated the mucociliary
transport of the frog palate epithelium and described a beat pattern
for these cilia which·is similar to those patterns found to be
associated with water transporting cilia. Although the cilia were
observed to penetrate the mucus, little further information was
provided to account for the coordination and propagation of the
metachronal waves or how their activity may relate to mucus transport.

     The reason for this neglect of the mucociliary systems is
undoubtedly related to the difficulties of studying these cilia.
Their small size and high density on an opaque mucosa greatly hinder
microscopical observation; their inaccessibility and the apparent
lack of extensive metachronism have similarly contributed to

ignorance about these cilia.

Limited descriptions of weak metachronism in tracheal cilia
have been presented by workers observing these cilia by reflected
light or through the bronchial wall. Proetz (1933) and Lucas and
Douglas (1934) observed short antiplectic metachronal waves which
travelled only short distances across the epithelium; by contrast
Iravani (personal communication) reported that the wave direction
was symplectic. However, it must be remembered that the reflections
can originate from several sources, the mucus blanket, the cilia
or the epithelial surface (Toremalm et al., 1975) and may
misrepresent the true ciliary events; similarly, the observation
of the cilia through the bronchial wall is difficult, therefore few
reliable conclusions can be drawn from these studies.

The situation is further complicated by the observations of
Cheung and Jahn (1976) who, having recognized the difficulties of
studying these cilia, reported that the ciliary beat takes the form
of a 'nodding action'. According to these authors, the cilium
maintains a rigid hook-like shape throughout its beat, the ciliary
tip penetrating the mucus for an arc of $5 - 8^{\circ}$ near the mid point
of the effective stroke. Whilst it is feasible that the functional
needs of mucus-transporting cilia are different from water-propelling
cilia, it seems unlikely that this unique beat form is ideal for
the transport of mucus, especially since other mucus-transporting
cilia appear to have a conventional beat pattern. Cheung and Jahn
(1976) concede that the limited contact made with mucus by these
cilia is inadequate for transportation of the mucus and suggest that
the major propulsive force is transmitted from the cilia to the
mucus by the rapid movement of the periciliary layer, frictional
interaction between the two layers producing mucus transport.
However, as previously described, the hydrodynamic forces acting on
the periciliary layer strongly suggest that the movement of the
periciliary layer is very limited. Therefore it would seem unlikely
that this explanation is correct. It is possible that the beat
pattern observed is the result of the preparative techniques, since
cilia beating with this form have been observed by the present
authors on preparations subjected to prolonged study or partial
compression, both causes of unhealthy tissue.

In many lung diseases, mucociliary clearance often appears to
be impaired; the activity of the cilia is an obvious possible source
of failure. Similarly, the disease cystic fibrosis has been
associated with a serum factor which can disorganise tracheal ciliary
activity (Conover et al., 1973). However, without an adequate
understanding of the functioning of mucus transporting cilia, their
role in lung disease or their use in its diagnosis cannot be fully
evaluated. Inadequacies in this area of ciliary research provided
the impetus for further studies concerning the activity of rabbit
tracheal cilia which have only recently been completed (Sanderson

and Sleigh, 1980).

    Beat pattern of rabbit tracheal cilia. At any instant, a sub-
stantial proportion of tracheal cilia lie at rest with their tips
pointing in the direction of mucus transport, that is towards the
oropharynx, in the position reached at the end of the effective
stroke (Fig. 7). This contrasts with water-propelling cilia which
do not normally rest between cycles, each beat running directly into
the next cycle. Further more, should these cilia ever come to rest,
they generally lie in the position at the beginning of the effective
stroke (eg. Mytilus lateral cilia, ctenophore comb plates). The
beat sequence of the two forms of cilia is slightly different; the
mucus-transporting cilia must first perform a recovery or preparative
stroke before the effective stroke can contribute to mucus transport;
water-moving cilia begin the beat cycle with the effective stroke.
This point is of significance to mucus transport and will be discussed later.

    The areas of ciliary activity (metachronal waves) on cultured
rabbit tracheal epithelium are generally small, but each is clearly
visible amongst the regularly orientated resting cilia. Each wave
or active patch has a characteristic appearance in the SEM - a
hollow surrounded by cilia with different orientations showing most
of the successive stages of the beat cycle (Fig. 7). Within each
wave a cilium performs a single beat before coming to rest.

    The shape of the recovery stroke was difficult to establish
by light microscopy as these cilia move even closer together
during this phase of the beat, preventing the resolution of a single
cilium. However, by using plan and profile views in SEM, a description of the recovery phase was possible. The initial movement of
the recovery stroke takes place at some distance from the hollow,
the cilium being drawn backwards and sidewards by an enlarging
bend progressing from the ciliary base to the tip. As the cilium
'unrolls', the tip describes a clockwise arc when viewed from
above. These changes can be seen to occur in the plan SEM views
(Fig. 7.) by the alteration of the tip orientation (the only part
of these cilia visible in such views) in cilia lying nearer to the
hollow. The curved profiles of the cilia are clearly demonstrated
in SEM profiles (Fig. 3a), which correspond to tracings of live
beating cilia (Fig. 8).

    The effective stroke is easier to observe by light microscopy,
as such cilia stand a little above the other cilia and remain in
focus throughout the stroke as they move in an almost planar manner
around the back edge of the hollow (Fig. 7). The effective
stroke begins at an inclination of about 40° to the cell surface
and moves through an arc of about 110°, the cilia coming to rest
at about 30° to the cell surface (Fig. 8).

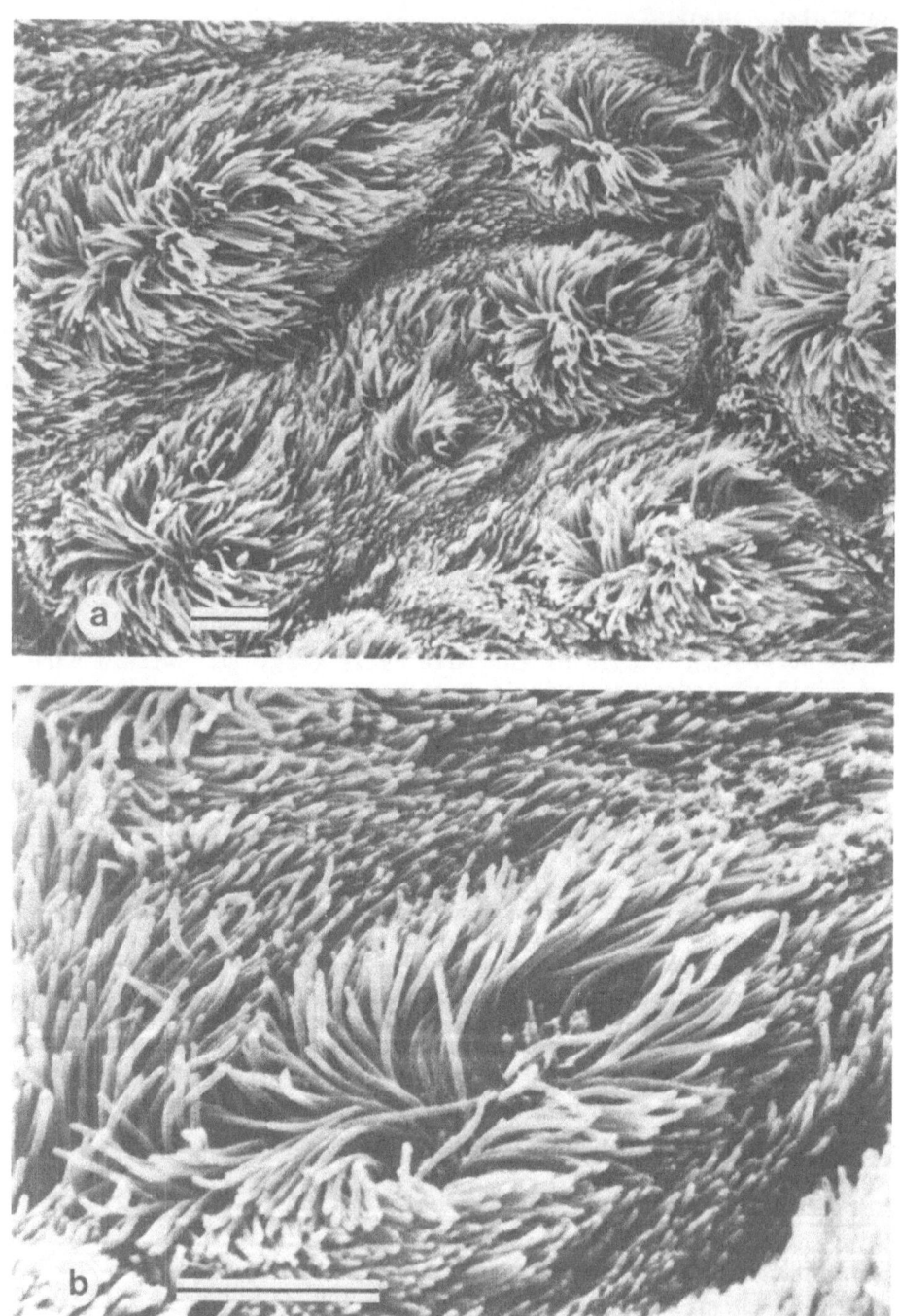

Fig. 7 Scanning electron micrographs of the ciliated epithelium
        of rabbit trachea. (a) A lower power view (Magnification
        bar = 5 μm) clearly shows the numerous patches of active
        cilia amongst a background of regularly orientated resting
        cilia, whose tips point in the direction of mucus transport.
        Each area is surrounded by a surface cleft where the cilia are
        tightly packed together. (b) A higher power view (Magnifica-
        tion bar = 5 μm) of one active area. The cilia at the lower
        side of the active area are progressing through the recovery
        phase of the beat cycle, with the effective stroke directed
        diagonally from bottom left to top right.

    Metachronal coordination. Each of the previously mentioned
active areas displays antilaeoplectic metachronal coordination.
that is, if the reader faces in the direction of wave travel, the
effective stroke occurs backwards and to the left.

    In contrast to water-moving cilia, the metachronal coordination
of tracheal cilia relies only on the hydrodynamic coupling generated
by the recovery stroke, rather than on both effective and recovery
strokes. Similarly, the resting position of the cilia strongly
influences the appearance of the metachronal activity. As previously
mentioned, the cilia lie at rest, with their tips directed towards
the oropharynx.   Before each cilium can contribute to mucus transport, the cilium must first complete a recovery or preparative stroke.
The wave of activity originates from a single pacemaker cilium amongst
the resting cilia. This cilium begins to beat, moving backwards and
sideways in a clockwise direction as described.

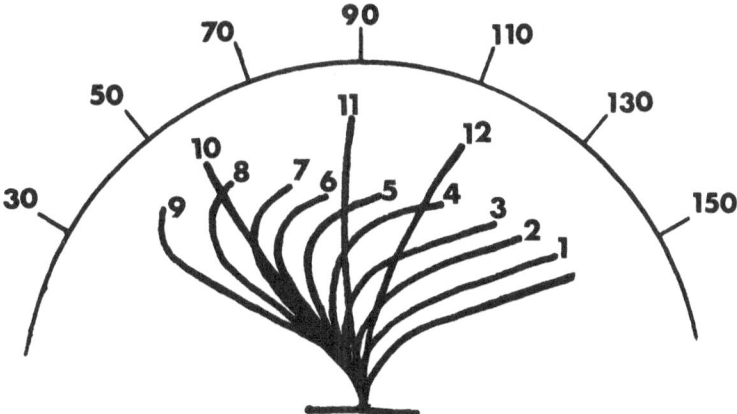

Fig. 8. A sequence of ciliary profiles traced from high speed cine
         film, showing the ciliary shape at intervals of 4 msec.
         The angular position of the cilium is indicated on an arc
         drawn around the profiles.

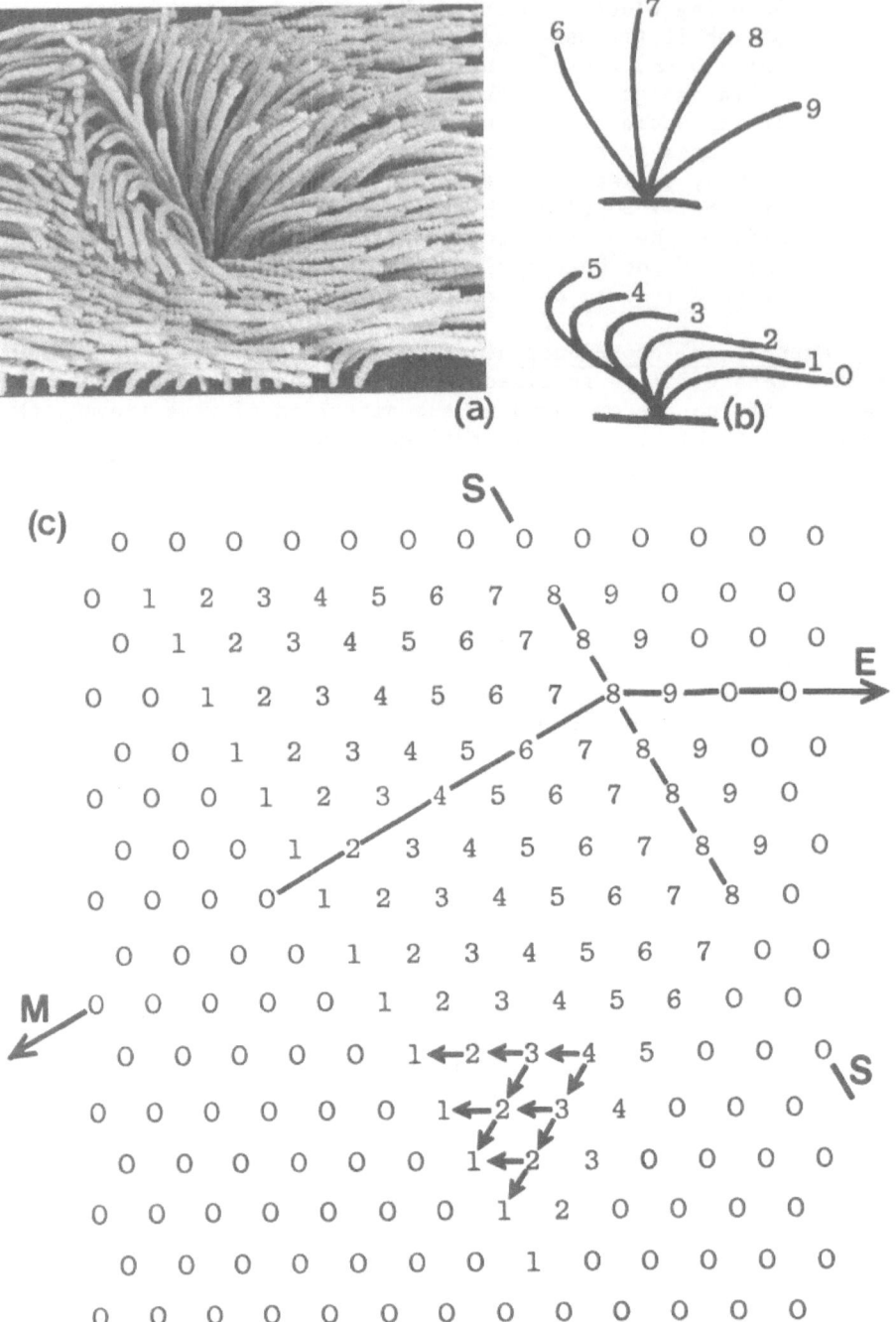

(a)

(b)

(c)

Fig. 9 Interpretation of an area of ciliary activity on cultured
←——— rabbit tracheal epithelium represented by a model (a) which
        is based upon a reconstructed three dimensional beat cycle
        (b) and upon a plan which indicated the spatial relation-
        ships of cilia on the epithelial surface (c).

        All cilia in the plan (c) are equally spaced and each one
        is represented by a single numeral, whose value denotes the
        phase of the cilium in its beat cycle (b) . The pattern of
        ciliary activity represented in (c) indicates the metachronal
        relationships of the component cilia at an instant during the
        extension and propagation of the wave across the epithelium.
        Resting cilia (phase 0) have their tips directed towards the
        right in all figures. The metachronal wave has originated
        from a single cilium. Each cilium commences its movement with
        a clockwise recovery stroke that has two components of hydro-
        dynamic coupling which induce the movement of neighbouring
        cilia. With respect to the mucus propulsion (E), this cou-
        pling by sideways and backward forces, acting respectively
        downwards and to the left, as indicated by the short arrows,
        causes the propagation of the metachronal wave in the direc-
        tion indicated by the arrow M. The effective stroke returns
        the cilia to the resting position by moving in the direction
        of the arrow E and has little influence on the metachronism.
        The metachronal wave therefore travels at an obtuse angle
        ($\approx 135°$) to the effective stroke. The line of synchrony (S)
        is at $90°$ to the direction of the main line of metachrony.

        A model of this pattern using cilia shaped on the basis
        of the observed profile appearances, has been constructed and
        photographed (a), from a similar angle ($\approx 30°$ to the hori-
        zontal) to that used in the SEMs (Fig.7).

The strong hydrodynamic coupling stimulates the adjacent cilia into
activity and these in turn activate more cilia. In this way the
metachronal wave is extended and propagated. The beat cycle is
completed by the effective stroke which propels any mucus it
penetrates and returns the cilium to rest.

The activity within such an area has been mapped (Fig. 9.);
the cilia are equally spaced and each is represented by a number,
the value indicating the phase of the beat.

This form of activity would be expected to produce waves moving
long distances across the epithelium. However, such long waves
of activity are seldom observed. This is not because the areas of
activity and the direction in which the waves move are randomly
changing. Generally, the activity is repetitive and restricted to
small patches of regularly beating cilia. These areas or patches

range in size from appriximately 4 cells long and two cells wide
to the commoner smaller size covering only two cells. The activity
within each field is not normally influenced by neighbouring areas;
the direction of the effective stroke, and consequently the
metachronal wave travel, varies between areas, although a general
oropharyngeal direction is maintained throughout.

The metachronal coordination was not observed to change
significantly upon an increase in beat frequency. Stimuli increasing
ciliary activity, such as increases in temperature or mechanical
contact, served only to raise the beat frequency of active areas
and bring previously resting areas into activity; the metachronal
waves did not increase in length or distance travelled.

The restriction of the metachronal waves contrasts with the
activity observed on the superficially similar frog palate
epithelium. Whilst there are likely to be some physiological
differences, the cilia density, cell size.and beat pattern are
similar in both epithelia. However, the metachronal waves of frog
palate are more prominent and travel for some distance across the
epithelium. This suggests that some discontinuities are present
within the rabbit tissue which would not be present in the frog
tissue. A clue to the difference between the functioning of these
tissues comes from the appearance of the epithelia. In life the
frog palate maintains a regular fixed shape whilst the trachea may
undergo changes in both diameter and length. In order to study
these tissues it is necessary to dissect them from the animal. In
the case of frog palate this procedure is relatively simple, the
epithelia can be streched back to its life-like appearance and
studied fresh. However, it is necessary to subject rabbit trachea to
a period of organ culture before the epithelium can be conveniently
studied. This often results in the formation of many surface
convolutions. These irregularities may isolate groups of cells,
each surrounded by a surface cleft in which the ciliated surfaces
are pressed together. The SEM often shows the activity to be
isolated to the tips of mounds (Fig. 7), the size of which compares
favourably with the discrete patches of activity observed with the
light microscope. It is therefore likely that these surface
irregularities provide the discontinuities of the ciliated surface
which cause functional independance of the patches of activity by
limiting the propagation of the metachronal waves.

Although it is likely that these surface irregularities have
been exaggerated by the preparative techniques, there is evidence
that the undulations exist in life. The trachea must be flexible
and capable of changes in length and diameter. The mucosa may
be without convolutions in the fully extended trachea; however,
any constriction or reduction in length would be expected to fold
the mucosal lining. Whilst it is difficult to avoid any damage or
disturbance of the ciliated tissues, studies by SEM and light

microscopy of the whole trachea have demonstrated that surface
folding is present. Prominent transverse clefts were present in
the excised trachea as expected, because of the shortening of the
tracheal length.

The extent of surface irregularities within the trachea appears
to be less than on cultured epithelia; however, it is possible
that a similar restriction of metachrony applies in life.
Iravani and Van As (1972) described independently coordinated
fields of cilia similar to those described here, the size of these
areas differs however, since the tracheal fields ranged from areas
covering a few cells to several hundred cells. If surface clefts
limit these fields, it is more likely that larger functional areas
will exist in those epithelia with less folding.

Mucus transport. The primary function of the cilia is to
transport the mucus towards the oropharynx. This action is achieved
by a clawing action of the cilia. The cilium first recovers within
the sol layer beneath the mucus and this movement has little influence
on the mucus. During the effective stroke in which the cilium stands
a little above the resting and recovering cilia, the tip penetrates
less than 0.5 μm into the lower surface of the mucus, irrespective
of the depth of the mucus blanket (Fig. 3). This contact is maintained through an arc of approximately 80°, from the initial penetration at about 50° to the cell surface to the disengagement at 130°.

The mucus is lifted by the cilia during its transport, increasing
the dept of the sol layer (Fig. 3). This action would be expected if
the cilia do not continue to penetrate further into the mucus during
their effective stroke; such lifting and pushing may facilitate the
mucus transport. The 'claws' or crown of ciliary hairs present at
the ciliary tips would serve to increase the frictional grip of the
cilia.

Again, the resting position of the cilia is related to mucus
transport; the tips point downstream and produce a non-return surface
helping to prevent the recoil of the mucus. Similarly, the tips of
the recovering cilia will not interfere with the transport of the
overlying mucus.

Although the beat pattern of tracheal cilia reflects the activity
of many other cilia, having on upright effective stroke with a
laterally inclined recovery stroke, the metachronal coordination
differs from that found in water-transporting cilia, having become
specialized for mucus transport. The transport of a more solid
visco-elastic layer of mucus is achieved readily by a clawing action,
without the need for strong coordination. An analogy can be drawn
between the movement of water by an archimedian screw and the movement of loosely woven cloth by a 'spiked' roller.

The metachronal activity of water transporting cilia is strongly influenced by the high hydrodynamic coupling generated by the fast-moving effective stroke. However the tracheal cilia do not rely on the effective stroke to generate and maintain the wave propagation. The recovery stroke generates the wave, the effective stroke serving only to propel the mucus and return the cilium to rest. This has a great advantage in connection with mucus transport. Often the effective stroke will become entangled with the overlying mucus, reducing its angular velocity and consequently its hydrodynamic coupling, with the result that the coordination of the cilia at this phase will be reduced. In other ciliated systems this would probably result in the loss of the wave propagation. However, in tracheal cilia, the wave propagation has been secured by the action of the recovery stroke stimulating the cilia before any contact is made with the mucus.

FACTORS INFLUENCING MUCOCILIARY TRANSPORT

The efficiency of the mucociliary clearance mechanism relies on the interaction of the cilia, the mucus and the periciliary layer. Any factor producing a change in any one or more of these components will induce a change in mucociliary transport.

Humidity

Proetz (1933), first noted the detrimental effects of drying on nasal ciliary activity in vitro; humidities above 70 % had little effect on the activity, whilst relative humidities of 30% or 50% impaired ciliary activity within 5 and 10 minutes respectively. Normally, in life, the relative humidity (RH) of the ambient air has little chance to influence the mucociliary transport of the airways. During quiet breathing, and to an extent during forced nasal breathing, the nasal chambers condition the inspired air; the air reaching the oropharynx has been warmed almost to body temperature and humidified to 100% saturation. The efficiency of the human nose has been demonstrated by Anderson and co workers (reviewed by Proctor et al., 1977); subjects breathing dry air (9% R H at 23°C) for 78 hours suffered no impairment of their mucociliary clearance rates. Similarly, the nose can protect the lungs against excessive moisture.

The effects of humidity are readily observed on in vitro preparations; experiments of this kind are particularly relevant to the surgical procedure of tracheotomy, after which the patient must breath the ambient air without the protection of the nasal air conditioning.

The effects of drying on the mucociliary transport mechanism are severe, transport is readily reduced to a standstill. Humidities

above 75% had no effect on mucociliary transport, whilst 50% and
25% impaired the transport rates. The effects of drying are
reversible if not imposed for to long, but cell damage is often
induced. (Iravani and Melville, 1976).

Whilst the effects of temperature (see below) are readily
explainable in terms of increasing cell metabolism, the effects of
humidity cannot be so easily identified. The low humidities will
dehydrate both the mucus, resulting in an increase in the viscosity
of mucus, and the periciliary layer reducing its depth, either of
which will lead to impaired mucus transport. It is interesting to
find that dehydration of the animal's body can also impair mucociliary function.

Temperature Effects

Many biological activities are temperature dependant, and
mucociliary transport does not appear to be an exception.Dalhamm
(1956) demonstrated an increase in beat frequency from 1,472 beats
per minute to 1,727 beats per minute over a temperature increase
from 37° to 41°C (body temperature). Other studies have estimated
the effects of temperature in terms of wave frequency and have
reported a linear increase from 420 to 1,000 waves per minute over
a temperature range of 21 to 40°C (Toremalm et al., 1975).
Temperatures above 45°C are generally lethal to respiratory cilia.

THE CONTROL OF MUCOCILIARY CLEARANCE

It has been suggested that the control of mucociliary clearance
is mediated by neural control. At the present time the exact
mechanism of control is unknown, however, ultrastructural studies
have revealed the presence of epithelial nerves and pharmacological
studies have demonstrated the sensitivity of the respiratory
epithelial cells to neuropharmacological agents.

Neural Control

The ciliary activity of the frog palate epithelium and the
lateral cilia of Mytilus edulus are under neural control.
Acetylcholine acts as a neurotransmitter for the frog cilia,
5-hydroxytryptanine and dopamine are released for the control of the
Mytilus cilia. The need for a control system in these organisms
is understandable; the frog palate system is activated by feeding
to facilitate the transport of the food particles down the oesophagus;
the Mytilus cilia create the feeding currents passing through the
filter apparatus. However, mucociliary clearance is constantly
active and does not require to be switched on and off.   The neural

control of mucociliary transport is likely to be limited to the
control of the mucus velocity.

Whilst the presence of nervous tissue within the subepithelial
tissue, innervating the smooth muscle, vasculature and submucosal
glands, is established, the extent of intraepithelial innervation
remains undecided. The presence of intraepithelial nerves has been
demonstrated in many species including man. Each fibre has a
similar ultrastructure, being unmyelinated and often devoid of any
secretory granules; an appearance consistent with sensory function.
Jeffery and Reid (1973) have provided quantitative evidence for the
innervation of the epithelium; it was estimated that 33% were
adrenergic motor neurones and 17% were cholinergic fibres, the
remaining 50% being sensory. This evidence provides the basic
anatomical evidence for neural control of mucociliary transport.

Pharmacological Agents

An extensive range of pharmacological agents have been shown to
influence mucociliary clearance. Of these compounds, the adrenergic
and cholinergic compounds are of particular interest, as their action
mimics the activity of sympathetic and parasympathetic nerve endings,
supporting the contention of neural control. Similarly the action
of these drugs have been of interest to clinicians with respect to
the treatment of various lung disorders.

Adrenergic compounds. Adrenergic agents or catecholamines
including adrenaline and noradrenaline are associated with the
sympathetic nervous system. In addition, a range of sympathomimetic
drugs are available, a powerful agonist is Isoprenaline. These drugs
influence the cell by combining with specific receptor sites within
the cell membrane. Adrenaline, isoprenaline and a range of
sympathomimetic drugs acting on β_2 receptors, eg. terbutaline,
salbutamol and fenoterol have been clearly demonstrated to stimulate
mucociliary clearance and ciliary activity in humans and other
mammals both in vitro and in vivo. Noradrenaline, having predominantly an α activity, does not appear to stimulate transport rates. This
capability of catecholamines to increase mucociliary clearance has
led to their therapeutic use in obstructive lung disease. Increased
clearance rates have been observed in bronchial asthma, chronic
bronchitis and cystic fibrosis after the administration of these
agents by inhalation, sublingual absorption or subcutaneous injection.

The exact mode of activation is not fully understood, but these
compounds do not increase the levels of secretion; an increase in
ciliary beat frequency remains the preferred hypothesis.

Cholinergic agents. Mucociliary transport also appears to be
influenced by the parasympathetic system; the cholinergic drugs

acetylcholine, pilocarpine and methacholine have been demonstrated
to increase mucus secretion, cilia beat frequencies and mucociliary
clearance rates. Conversely the antagonists atropine or hyoscine
have been shown to reduce the stimulating effects of the cholinergic
drugs.

The therapeutic use of these kinds of drugs is limited because
their action increases secretions and produces a constriction of the
airways. Again, the exact mode of influence on mucociliary clearance is unclear, especially since these drugs appear to stimulate
several components of the system.

Other drugs. A range of miscellaneous drugs have been examined,
for example biological amines such as 5HT and histamine, central
nervous system stimulants, and anesthetics. The physiological involvement of these drugs is not envisaged, the investigation into their
effects being useful to determine their side effects when used for
treatment of other conditions. Neither 5HT or histamine have much
influence on mucociliary clearance, although anaesthetics, whether
general or local, impair transport velocities.

Water Transport across the Epithelium

Hitherto the mechanism of mucociliary control is not understood;
the action of the neurotransmitter drugs is commonly thought to be
mediated via a change in cell metabolism, leading to greater energy
production and a consequent rise in the frequency of ciliary beat.
Recently, the movement of water across the epithelial surface has
been considered as a method of influencing mucociliary transport.

It has been previously pointed out that the rheological
properties of the mucus alters with its hydration, drying tending to
increase the viscosity, which can lead to impaired mucus transport.
Similarly the depth of the sol or periciliary layer, consisting
mainly of water, is critical to the efficiency of the system.
Consequently mucus transport may be influenced by the water content
of the epithelial surface.

The movement of water across an epithelium is generally achieved
in association with the active transport of electrolytes; the water
moving down the local osmotic gradient. The tracheal epithelium of
dogs has been shown to possess the ability to pump Cl^- ions towards
the lumen and Na^+ ions towards the submucosal tissue, providing
this epithelium with the capacity to control water movements. The
contention that ionic pumping may influence mucus transport is
supported by the evidence that adrenergic compounds, such as terbutaline and adrenaline, or cholinergic compounds such as acetylcholine,
all of which increase mucus transport, also increase the net flux of
ions towards the tracheal lumen, suggesting that a net flow of water

towards the lumen may occur.

DEFECTS OF THE MUCOCILIARY CLEARANCE MECHANISM

The cause of obstructive lung diseases such as chronic bronchitis or cystic fibrosis is often attributed to a failure of the lung defence mechanisms particularly mucociliary clearance. Although evidence is available that mucociliary transport is impaired in these diseases, the aetiology of the conditions is not fully understood.

Chronic Bronchitis

Chronic bronchitis is clinically recognized as a lung condition with excessive mucus secretion within the bronchial tree, often accompanied by a productive cough and respiratory infections. It is commonly associated with increasing age, smoking and industrialized environments. A predisposition to the disease may result from the decreased mucociliary clearance in elderly pateints.

The characteristic changes of gland hypertrophy and goblet cell metaplasia account for the increased mucus production. Mucociliary transport is impaired, although there is some discrepency between studies concerning the extent of impairment, resulting from the different techniques used to measure the transport rates.

Iravani and Van As (1972) have noted changes in the mucus flow of bronchitic rats, a greater diversity of beat direction was observed with many of the pluricellular metachronal fields demonstrating caudal transport. In other areas, whirlpools of activity were seen, whilst the total area of inactive cilia had increased.

More recently, Iravani et al (1978) have reported that the sol layer is reduced in bronchitic rats.

Cystic Fibrosis

Severe pulmonary disease is characteristic of the genetic disease cystic fibrosis; however, this lung condition is not always evident at birth, but generally develops later in life. The impairment of the mucociliary clearance mechanism has frequently been suspected as the cause of the pulmonary disease; the presence of a serum factor thought to disorganise ciliary activity would tend to support this hypothesis (Conover et al., 1973). However, the relationship of defective mucociliary clearance to cystic fibrosis is unclear.

Similarly, the viscous nature of CF sputum has been thought specific for the disease, its tenacious nature leading to impaired

mucus transport. Whilst it has been established that purulent mucus
is different from mucoid or normal secretions, a significant
difference between the rheological properties of mucus from CF and
other bronchial diseases has not been demonstrated. (Lopez-Vidriero
et al., 1977). The viscosity of mucoid sputum also seems to be
similar between diseases (Lopez-Vidriero and Ried, 1978). It appears
that the changes in lung tissue and mucus secretion are secondary to
the basic cystic fibrosis defect.

An interesting idea has been recently proposed by Nadel and Davis
(1978) and Proctor et al. (1978), concerning the decreased mucus flow.
CF is a disease that affects the exocrine glands, and often the
secretions of these glands are abnormal; the sweat gland is a good
example, consistently having high levels of Na^+ and Cl^- in the exudute.
As mentioned previously, the mucus flow rate may be influenced by the
water transport of the epithelium, normally controlled by the ionic
pumping of the epithelial cells. If in CF the capacity of these
cells to regulate the ionic movements across the respiratory surface
is impaired, in a manner similar to the abnormalities of the sweat
gland, the mucociliary transport mechanism may be impaired by the
abnormal hydration of the respiratory surface.

Immotile Cilia Syndrome

Whilst the cilia of the previous lung diseases appear to be
normal, evidence has been recently presented that immotile cilia are
responsible for a congenital human syndrome; the patients suffering
from recurrent chronic sinutisis, bronchitis and otitis media, all
resulting from an impairment of mucociliary clearance (Afzelius, 1979).
A clue to the cause of the condition was presented by infertile males
whose spermatozoa were inactive, the sperm tail lacking motility.
Similarly, the respiratory cilia of these subjects did not demonstrate
any beating activity. Ultrastructural studies revealed that the
dynein arms, essential to the sliding mechanism of the axoneme, were
missing in both sperm flagella and respiratory cilia. Such an
omission readily explains their inactivity.

A number of other ciliary defects are also known, Sturgess (1979)
has described cilia which lack the radial spokes, whilst unusual
axoneme configurations of 10 + 2 or 9 + 4 have also been discovered
(Howell et al., 1980). In addition, immotile cilia have been
reported in which the defective structure is not visible by
transmission electron microscopy.

REFERENCES

Afzelius, B.A., 1979, The immotile-cilia syndrome and other ciliary
        diseases, Int. Rev. Exp. Pathol., 19 : 1.

Aiello, E., and Sleigh, M.A., 1977, Ciliary function of the frog oropharyngeal epithelium, Cell Tiss. Res., 178 : 267.

Alder, K., Wooten, O.J., and Dulfano, M.J., 1973, Mammalian respiratory clearance, Arch. Environ. Health , 27 : 364.

Asmundsson, T., and Kilburn, K.H., 1970, Mucociliary clearance rates at various levels in dog lungs, Am. Rev. resp. Dis., 102 : 388.

Brain, J.D., Proctor, D.F., and Reid, L., (Eds), 1977, "Respiratory defense mechanisms", (Part I and Part II), Marcel Dekker, New York.

Breeze, R.G., and Wheeldon, E.B., 1977, The cells of the pulmonary airways, Am. Rev. resp. Dis., 116 : 705.

Cheung, A.T.W., and Jahn, T.L., 1976, High speed cinemicrographic studies on rabbit tracheal (ciliated) epithelia : Determination of the beat pattern of tracheal cilia, Pediat. Res., 10 : 140.

Conover, J.H., Bonforte, R.J., Hathaway, P., Paciuc, S., Conod, E.J., Hirschhorn, F.B., and Kopel, F.B., 1973, Studies on ciliary dyskinesia factor in cystic fibrosis I. Bioassay and heterozygote detection in serum, Pediat. Res., 7 : 220.

Dalhamn, T., 1956, Mucus flow and ciliary activity in the trachea of healthy rats and rats exposed to respiratory irratant gases, Acta. physiol. scanda., 36 (Suppl.123) : 1.

Dulfano, M.J., and Adler, K.B., 1975, Physical properties of sputum, VII. Rheologic properties and mucociliary transport, Am. Rev. resp. Dis., 112 : 341.

Hilding, A.C., 1957, Ciliary streaming in the lower respiratory tract, Am. J. Physiol., 191 : 404.

Howell, J.J., Schochet, S.S., and Goldman, A.S., 1980, Ultra-structural defects of respiratory tract cilia associated with chronic infections, Arch. Pathol. Lab. Med., 104 : 52.

Iravani, J., and Van As, A., 1972, Mucus transport in the tracheo-bronchial tree of normal and bronchitic rats, J. Pathol., 106 : 81.

Iravani, J., and Melville, G.N., 1976, Mucociliary function in the respiratory tract influenced by physicochemical factors, Pharmac. Ther., 2 : 471.

Iravani, J., Melville, G.N., and Horstmann, G., 1978, Tracheobronchial clearance in health and disease with special reference to interciliary fluid, in : "Respiratory tract Mucus", Ciba foundation symposium 54, Elsevier. Excerpta Medica. North Holland, Amsterdam.

Jeffery, P.K., and Reid, L.M., 1973, Intra-epithelial nerves in normal rat airways: A quantitative electron microscopic study, J. Anat., 114 : 35.

Lopez-Vidriero, M.T., Das, I., and Reid, L., 1977, Airway secretion: Source, biochemical and rheological properties, in: "Respiratory defense mechanisms, Part I", J.D. Brain, D.F. Proctor, and L.M. Reid, eds., Marcel Dekker, New York.

Lopez-Vidriero, M.T., and Reid, L., 1978, Chemical markers of mucus and serum glycoproteins and their relation to viscosity in

mucoid and purulent sputum from various hypersecretory
diseases, Am. Rev. resp. Dis., 117 : 465.

Lucas, A.M., and Douglas, L.C., 1934, Principles underlying ciliary
activity in the respiratory tract. II. A comparison of nasal
clearance in man, monkey and other mammals, Arch.
otolaryngol., 20 : 518.

Nadel, J.A., and Davis, B., 1978, Regulation of Na^+ and Cl^- transport
and mucous gland secretion in airway epithelium, in:
"Respiratory tract mucus", Ciba foundation Symposium 54,
Elsevier. Excerpta Medica. North Holland, Amsterdam.

Proctor, D.F., Andersen, I., and Lundquist, G., 1977, in:
"Respiratory defense mechanisms 'Part I; J.D. Brain,
D.F. Proctor, and L.M. Reid, eds., Marcel Dekker, New York.

Proctor, D.F., Adams, G.K., Andersen, I.B., and Man, S.F.P., 1978,
in: "Respiratory tract mucus", Ciba foundation Symposium
54, Elsevier. Excerpta Medica. North Holland, Amsterdam.

Proetz, A.R., 1933, Studies of nasal cilia in the living mammal,
Ann. Otol. Rhinol. Lar., 42 : 778.

Sanderson, M.J. and Sleigh, M.A. 1980, Ciliary activity of cultured
rabbit tracheal epithelium; beat pattern and metachrony,
J. Cell Sci., In press.

Sleigh, M.A., 1977, The nature and action of respiratory tract cilia,
in: "Respiratory defense mechanisms, part I", J.D. Brain,
D.F. Proctor, and L.M. Reid, eds., Marcel Dekker, New York.

Sleigh, M.A., and Barlow, D.I., 1980, Metachronism and control of
locomotion in animals with many propulsive structures, in:
"Aspects of Animal Movement", H.Y. Elder, and E.R. Trueman,
eds., University Press, Cambridge.

Sturgess, J.M., 1977, The mucus lining of major bronchi in the rabbit
lung, Am. Rev. resp. Dis., 115 : 819.

Sturgess, J.M., Chao, J., Wong, J., Aspin, N., and Turner, J.A., 1979,
A cause of human respiratory disease. Cilia with defective
radial spokes, New Eng. J. Med., 300 : 53.

Toremalm, N.G., Mercke, U., and Reimer, A., 1975, The mucociliary
activity of the upper respiratory tract, Rhinology, 13 : 113.

Wanner, A., 1977, Clinical aspects of mucociliary transport, Am. Rev.
resp. Dis., 116 : 73.

Van As, A., and Webster, I., 1972, The organization of ciliary
activity and mucus transport in pulmonary airways,
S. Afr. med. J. 46 : 347.

DISCUSSION

LECTURER: Sanderson   CHAIRMAN: Cumming

CHEVALIER: I have three questions. First, in your beautiful
                model, where is DNA located?

SANDERSON: Are you referring to the DNA at the base of the
                cilia?

CUMMING: No, he is asking where it is.

SNDERSON: There is no DNA.

CHEVALIER: Isn't there any proof of DNA involved in the ciliary
                action in mammals?

SANDERSON: No, not that I am aware of. There have been some
                reports that there is some DNA like material at the
                base of the cilium, but they were some time ago and
                have not been fully repeated. There is no actual
                use of DNA in the motility of the cilium.

CHEVALIER: I thought there was some genetic basis in the
                direction of action of the ciliary movement.

CUMMING: Perhaps the question should be where is the RNA,
                which contains the genetic message to tell them
                where they beat.

SANDERSON: To tell you which way you beat is an orientation of
                the growth of the cilia, not of the activity of the
                cilia.

CHEVALIER: Because DNA has been found in protozoa, very
                clearly.

SANDERSON: Not in the cilium.

CHEVALIER: So, there have been no mutants which have given
                evidence in relation to this possible genetic
                autonomy in the properties of the cilium?

SANDERSON: Mutant protozoa have been used whereby they are
                missing parts of the axones, as we mentioned in the
                disease case, and those cilia or flagella have been
                found to be abnormal or to lack activity. But the

genetics is related to the production of a component of the axone and not related to the motility of the axone.

CHEVALIER: Is there any evidence that following agression, supposedly. by a toxic agent, that the cilia can easily repair and grow again?

SANDERSON: If cilia are removed, they can grow quite quickly again, within six hours. Certainly this is the case in protozoa. If you remove all the cilia from a tetrahymena, a small water ciliate, then it regenerates all its cilia within six hours. I am not sure whether tracheal or mammalian cilia are capable of doing the same thing.

NESCI: I would like to know whether in the axones of nervous cells where it has been shown that the neurotubules consist of a protein which is called neurotubulin, these globular subunits also consist of tubulin. Again, in the axones, to justify axonic flow, another neuroprotein, called neurostenin has been shown at the level of these neurotubules I would like to know whether a similar structure has been revealed in the cilium.

SANDERSON: The microtubules are made of tubulin and the thirteen units if you look at the microtubules in longitudinal section, appear like that. Each one is a globular unit. On closer examination it appears that the dimers are made of a figure of eight. Two tubulin molecules, but te microtubules, but I do not think they have clearly shown the presence of the dynin ATPase activity to make that link.

SMITH: As I recall, in protozoan ciliary action, such as paramecium, the metachronal rhythm is synchronized by a rather complicated system of a lattice-like interconnection of the basal bodies of all the cilia which keeps this metachronal rhythm in synchrony. Does a similar sort of mechanism obtain in the human bronchial ciliated cell, or is this metachronal rhythm synchronised by some other means?

SANDERSON: The metachronism used to be thought to be synchronised by connections between the cilia, as you suggested. But it is now commonly believed to be formed by a hydrodynamic coupling, that is, one cilium influences the fluid around it and its influence is transmitted to the next cilium by the

intervening fluid. And that is what produces the
metachronal wave. It is not a neurotransmission
between the cilia.

TURNER-WARWICK: Could you tell us a little about the anatomical
variations of cilia? I ask this question
particularly because we studied a bizarre family who
had a severe progressive pulmonary disease, all of
them from a young age. The examination of their
cilia appeared to show very bizarre features. For
instance: three sets of doublets surrounded by a
single membrane; then, outer doublets but no
central ones. There was much discussion as to
whether these were acquired as a consequence of
their serious respiratory infections or whether
these had to be some important genetic derivation.

SANDERSON: Other diseases have been found in which there is a
lack of the components of a cilium, and this in turn
often produces respiratory disease. Whether or not
respiratory disaese can induce the abnormalities you
described I am not sure. But certainly an
inflammatory process, I would have thought, may
affect the development of the ciliated cells, so
that the cilia which are developed are abnormal.
But it would be more likely that if it was occurring
at an early age, that it would be a congenital
effect of controlling the configuration of the axone
rather than being induced by a disease. A good
example is perhaps cystic fibrosis, in which there
is severe lung disease for a long time, but yet the
ciliary structure often appears to be normal.
Perhaps Peter Jeffrey would like to expand this, as
he has done some work on this.

JEFFREY: With reference to whether or not a disease can cause
structural abnormalities in cilia, the only
experimental evidence we have in favour of that is
work with sulphur dioxide. I will be showing some
pictures later in the programme and we can produce
by such an irritant ciliary abnormalities, which
would no doubt affect their function. If I could
add a further comment with regard to the
regeneration of cilia in the respiratory tract, it
has certainly been shown that if you stroke the
epithelium, (I am not talking of just the cilia but
the epithelial cells) that you will have once again
a ciliated and mucus producing epithelium within 14
days. That is after mechanical trauma. After viral
infection it has been shown that the recovery period

is somewhat longer, in terms of months, rather than days.

CORRIN: The sort of abnormalities Turner Warwick described, of multiple axony within one sheath, have been observed by Fox in normal nasal biopsies and in normal bronchial biopsies. In the immotile cilial syndrome there is a lack of dynin arms and Fox and his colleagues are the only ones to have quantitated these abnormalities; they have only studied normals, they have not studied any Kartagener's syndrome. A serious lack in our knowledge is the quantitation in Kartagener's syndrome. It may be necessary in future to quantitate this lack of dynin arms and that will be quite a task to undertake. My own question refers to the continuity or otherwise of the mucous sheet. You showed a diagram of spot secretion of mucus, this coalescing into flakes and then into plaques but later you showed some very nice photographs with the scanning electron micrograph of cross sections of embedded tissue showing a continuous sheet. How do you really see this, is this a continuous sheet or is it discontinuous?

SANDERSON: We have seen both sorts of mucus. In some preparations we have the extensive sheet, whereas in other preparations there may be very little mucus at all. This often depends on the preparative method. The continuous sheet can often be preserved by a freeze substitution method, in which the tissue is fixed very quickly and all the mucus is retained. With conventional fluid fixatives, although you may fix the mucus, it is very easy to wash it off, so that you can get both forms from the same tissue. In addition, tissue culture tends to make the epithelium devoid of mucus. Again, depending on the age of the tissue culture or how long the tissue has been before it is fixed after dissection will alter the results of how much mucus covers the cilia. I think it would be likely that you have a situation similar to that which Irivani proposed in that there are smaller droplets in the lower airways, and because of the convergence of the airways the mucus will coalesce. It is not however essential for mucus transport to have a continuous mucus blanket.

CUMMING: Perhaps I could be forgiven a comment from the Chair. I have seen an excellent film made in Sweden in which there is high speed cinematography of the human trachea, onto which they have instilled a drop of methylene blue, and thus seen the streaming of the mucus quite clearly. That appears to be a continuous sheet, and the lines of streaming round a foreign body diverge, reform and then proceed again in a rectilinear fashion. So it seems likely that in humans, in the trachea at least, the sheet is continuous. If you extend this hypothesis to assume that there is a continuous sheet all the way down the bronchi (very dubious but let us accept this for the moment) then since we know the lateral surface area of the airways and since we know of the velocity at various points, it might be possible to answer the question: is there resorption on the way up? Or could this be explained by fusion into a continuous sheet? And I commend this for those who are looking at ciliary action, to look at it in a total lung quantitative way, rather than in a small sampling way.

JEFFERY: Perhaps, Gordon, I could just include into your equation the known fact that when you are looking at the normal human airway the mucus producing cells stop as you proceed more distally. In terms of the submucosal gland, there is only gland where there is cartilage, so by definition all bronchioles will be devoid of mucus producing glands, and the surface goblet cells stop in airways of less than 2 mm diameter. So that you do not need to postulate, or it would be highly unlikely, that you would have a continuous sheet or even droplets in airways less than 2 mm of diameter.

CUMMING: Yes, it is really in the proximal airway only where this could be carried out.

RICHARDSON: If the point you are making, Jeffrey, is that the mucus can only be made where there are classical mucus making cells, how do you think the cilia work in the distal airways? In the proximal airways they appear to need mucus to function.

CUMMING: Well, perhaps not how, why?

JEFFERY: For myself, I presume that there is no muco-ciliary transport in the airways where there is no mucus. It has been very elegantly shown that without mucus there is no transport and I believe it. And so in the small airways I do not think there is any transport of particulate matter. On the other hand, in disease where we know that with irritation the goblet cells extend more distally and therefore mucus is produced then cilia at this distant site will become more functional and important.

CUMMING: You regard them as being redundant for the moment.

JEFFERY: I regard them as being a good back-up defensive mechanism, ready to take any onslaught.

RICHARDSON: May I come back on that? How do you think that the alveolar macrophages get from the respiratory regions of the lung to the large airways where everyone admits that mucus is present? Do they run?

JEFFERY: Somebody suggested that they swim! But I think the comment of Sanderson with regard to the inefficiency of a short cilium of 6 mu in length for a fluid of low viscosity would indicate that the macrophage would have come to an area of periciliary layer. Without mucus I can only presume that they swim.

SANDERSON: The cell surface may itself be able to act in a similar way to the mucus, because of its flexibility.

CORRIN: If the movement of the alveolar macrophage then is not passive with fluid being carried by fluid, we then must envisage some form of active locomotion, and I like to ask Denison how he thinks the macrophage knows which way to go.

CUMMING: Hoist by his own petard!

DENISON: Denison certainly does not know. If the purpose of the mucus is to trap, or one of its purposes is to trap particles, if it is patchy then it is probable that some particles will fall into ciliary beds where there is no mucus. What mechanism exists to retrieve these particles that fall to the bottom of the field of corn?

SANDERSON: Although the particles may fall in a gap at one instant, it is likely that another plug of mucus will come up at some later stage and collect the particle.

DENISON: My image is that it will fall in the serous layer down to the bottom, down to soil level, and then the mucous plug will float above it and do no good at all. Am I right or wrong?

SANDERSON: The density of the cilia in rabbit trachea would
                strongly suggest that it would be difficult for a
                particle to go down between the cilia. It is only
                the couplings of the cilia that tend to squeeze fluid
                from between cilia or upwards, because you get this
                sort of movement. And I would envisage it as being
                very difficult for a particle to slip down. If it
                did, then how it would be recovered I do not know.
                Maybe the macrophage would then come into effect.

NESCI: I would like to know whether this beautiful activity
                of the cilia exists in the foetus in utero and also
                in the neonate.

SANDERSON: I have looked at foetuses 24 weeks old and they had
                ciliary activity. I have also looked at the trachea
                of 12 week foetuses and I was unable to observe any
                ciliary activity within their trachea.

WIDDICOMBE: The question of the role of cilia in human
                bronchioles, where there are no submucosal glands and
                virtually no goblet cells and the ciliated cells are
                rather widely separated, so they do not form a
                continuous sheet, might be answered by looking at the
                mouse trachea which is very similar in structure to
                the human bronchiole in that it has virtually no
                submucosal glands, very few goblet cells, I think no
                serous cells, but about 50% Clara-like cells and 50%
                ciliated cells, the latter being rather widely
                separated, so they do not form into sheets. The
                experiment to test this would be a simple one, apart
                from the size of the animal, and that is to see
                whether the mouse can clear particles which are blown
                down into its lung. As far as I know this has never
                been done and it would be fairly straightforward to
                do. If it can clear particles, then I think one has
                to assume that the ciliated cells are able to move
                these particles, presumably with the secretion of the
                Clara cells as an adjunct to their motility.

CUMMING: Can I ask, has anyone done this experiment? Or does
                anyone know whether this experiment has been done?
                No? Something you can go back and do. And I think
                on that note of uncertainty, which I am delighted to
                hear, I should bring this session to a close.

DEFENSIVE MECHANISMS OF THE UPPER AIRWAYS

John Widdicombe

Department of Physiology
St. George's Hospital Medical School
Tooting, London, SW17 0RE, U.K.

INTRODUCTION

In healthy conditions, the upper respiratory tract (larynx, pharynx and nose, possibly also the mouth) has a double defensive role: it conditions the inspired air, humidifying and usually warming it, and it contrives to separate ingested material from inhaled gas. The efficiency of the latter process depends on the species concerned: in man the pharynx is a common pathway open, except during swallowing, for food and air; animals such as ungulates, which eat and chew more or less continuously, have functionally separate channels for food and air, and can breathe while they swallow; aquatic mammals such as whales have developed anatomically completely separate channels for air and food, so that the upper respiratory tract does not include the pharynx and nose (Negus, 1958).

In adverse conditions, the upper respiratory tract takes on the role of limiting or preventing invasion of the lungs by harmful materials such as toxic gases, solid or liquid aerosols, liquids such as water and large objects such as foreign bodies and vomitus. If the defensive mechanisms do not keep the invader out, they will aim to exclude it or to minimize any damage that may occur. The patterns of defense are various, and depend on the identity of the invading material and the site where it acts. If there are pathological changes, such as infection or mucosal damage to the airway, the defensive processes may be enhanced or inhibited.

In general four types of defensive mechanism can be activated. In order of speed of response they are (1) nervous reflexes, (2)

121

filtration and deposition of harmful materials to prevent their
entry to the lungs, (3) clearance by mucociliary transport and (4)
cellular and immunological changes in the airway mucous membrane.
The four mechanisms will be described in general next, and in
detail later in the chapter (see also Cohen and Gold, 1975;
Widdicombe, 1977a).

Reflex Responses

These are highly specific to the site from which they are
activated.   They can occur so promptly, within a fraction of a
second, that they can interrupt a normal breath and prevent further
entry and penetration of an irritant gas or foreign body.   They
act also to expel any harmful material, as with coughing and
sneezing.   On a slower time scale, but still within seconds, they
will bring into play a variety of responses which will prepare the
whole body against potential harm – changes in breathing, in the
cardiovascular system, bronchomotor tone, mucus secretion and even
spinal reflex sensitivity.   The reflexes include some of the most
sensitive sensory mechanisms and vigorous motor responses seen in
physiology (Widdicombe, 1977b; Korpas and Tomori, 1979).

Deposition and Filtration

Large objects can be impacted and block repiratory passages.
The narrowest passages are in the nose and the bronchioles; if
the former is blocked mouth breathing can keep respiratory passages
open, although some species and some babies seem unable to switch
to mouth breathing and therefore asphyxiate.

For smaller particles the site of deposition depends on many
factors: particle size and shape, electric charge, hydroscopy,
gas flow, gas turbulence (which aids deposition) etc.   The pattern
of breathing is also important, quite apart from the difference
between nose and mouth breathing.   Increased air flow will draw
small aerosol particles deeper into the lungs, but will also cause
more turbulence and deposition; little is known about which of
these opposing effects is more important in human airways.
Breath-holding will aid deposition by sedimentation.

Soluble gases are dissolved in the walls of the airways and
the greater the solubility the less gas reaches deep in the lungs.

Deposited objects, particles and gases may cause mechanical,
chemical irritative, inflammatory or immunological changes at the
sites of deposition.

Mucociliary Transport

Mucus has a double defensive function (see Richardson, this

volume): it acts as a physicochemical barrier to penetration of noxious substances, and it transports the substances from the respiratory tract by ciliary movement or by reflexes such as coughing and sneezing. Mucus secretion and ciliary transport have different control systems, but their activities are functionally related. For example, ciliary transport is impossible in the absence of mucus (Silberberg et al., 1977). One can consider separately different aspects of mucociliary transport: mucus secretion, ciliary beat frequency and coordination, mucus velocity or flow, or total clearance rate from a part of the respiratory system (Iravani and Melville, 1976; Phipps, 1980). All or any of these related mechanisms may be changed when the respiratory tract is abnormal, and all must be effective if the lungs are to be adequately defended.

Cellular and Immune Changes

Damage to the respiratory tract can cause immediate pathological changes in the epithelium, for example if an irritant gas is inhaled, or later inflammatory changes with accumulation of neutrophils or other cells. Vascular changes associated with tissue damage may also occur.

If the inhaled substance is an antigen to which the subject is sensitive, the immune responses may include degranulation of mast cells with subsequent action of released mediators, such as histamine, 5-hydroxytryptamine and SRS-A. Secondary cellular infiltration may occur.

Immunoglobulins (e.g. IgA, E, G and M) and lysozome are present in respiratory tract secretions, and appear to be actively secreted since their concentrations may be higher than in plasma (Cohen and Gold, 1975).

Bacterial and viral challenge to the respiratory tract may be defeated by immunological defenses, or by macrophage ingestion of the organisms. Mucus can act as a culture medium for many bacteria, and the balance between the beneficial and harmful effects of mucus secretion in respiratory infection is still not clearly defined.

THE NOSE

The human nasal cavity is lined by mucous membrane, except for the immediate anterior part which is lined by skin. The mucous membrane is squamous in the anterior third of the nose, and the posterior two-thirds has pseudostratified columnar epithelium. The epithelium contains ciliated and non-ciliated cells, the latter including goblet and serous secretory cells. The lamina propria below the epithelium consists first of a layer containing submucosal

glands and cells such as fibroblasts and mast cells, and then a
deeper layer rich in blood sinusoids.   In the posterior olfactory
epithelium are the sensory nervous receptors responsible for smell
(see Negus, 1958; Mygind, 1978).

The secretory tissues of the nose consist of three components.
The three types of secretion have not been separately analysed, nor
is there evidence as to possible separate control of the secretions
(Tos, 1976; Proctor, 1977; Proctor et al., 1977; Phipps, 1980).

(1)   The goblet and serous cells of the epithelium.   Little
is known of the control of their secretions but presumably, like
similar cells in the lower respiratory tract, they secrete mucoglycoproteins in response to local stimulation by irritants and released-
mediators.

(2)   The submucosal glands, found throughout the respiratory
region of the nose, especially in the septum.   The glands contain
both mucous and serous cells, and are richly innervated with parasympathetic (cholinergic) fibres and possibly nerves containing other
transmitters such as VIP (Uddman et al., 1980).

(3)   Anterior serous glands with long ducts discharging
presumed watery secretion into the region of the vestibule at the
front of the nose.   This secretion is thought to be important in
the humidification of inspired air.   The glands have a rich cholinergic innervation.

The rich vasculature of the nose, in both the turbinates and the
septum, has capillary networks in the lamina propria and round the
submucosal glands, which drain into the venous plexuses.   The
capillaries are fenestrated and therefore would allow easy passage of
fluid through their walls.   There are many arteriovenous anastomoses.
The main innervation of the vessels seems to be sympatho-adrenergic.
The structural appearance of the nasal vasculature is that of an
erectile circulation, and there has been much speculation as to its
role.   For transepithelial transport of water and soluble gases and,
in particular, of heat in air conditioning, total mucosal blood flow
must be the most important controlling factor, as in the skin.
Vascular congestion which results from distension of the erectile
tissue would narrow the air passages and thereby change the rate of
deposition of aerosol particles and, by altering aerodynamic behaviour, the exchange of soluble gases, water vapour and heat.   Changes
in the vascular bed affect nasal airflow resistance, and this has led
to one of the main methods of study of the nasal circulation, but
little is known about the mechanisms that control this function and
its modifications in disease (Empey, 1980).

Physiological Air Conditioning

Inspired air is humidified and brought closer to body temperature by the time the larynx is reached (Proctor and Swift, 1977). It is generally assumed that this is an automatic result of the anatomy and rich blood supply to the nose, especially in the turbinates. Recent studies by Schmidt-Nielson et al. (1980) have shown that at night the dehydrated camel expires air from the nose at considerably less than core-body temperature and less than 100% humidity. They conclude that the air-conditioning mechanisms of the nose can be controlled to allow the preservation of body water and heat.

Reflex Responses

The nasal mucosa contains afferent nerves, some of which seem to contain substance P (Cauna et al., 1969; Anggard et al., 1980). The nerves are presumably responsible for the reflexes that arise from the nose. Of these reflexes the most studied is the apnoeic reflex in response to cold air or gaseous irritants. It may be the equivalent to the diving reflex prominent in aquatic mammals and birds, but also present in man (Angell-James and Daly, 1969). Breathing stops, the larynx closes and there is a readjustment of the circulation with vasoconstriction in skin, splanchnic bed and skeletal muscle, so that cardiac output goes primarily to the brain and heart. Catecholamines are released from the adrenal medulla and blood pressure increases. These responses seem in general well adapted to enable the animal to escape from the dangerous environment while avoiding damage to the lungs. Obviously their duration of action must be limited.

The sneeze reflex has been little studied, possibly because it is not alway prominent in anaesthetized animals. Although it is set up by many of the same stimuli that cause the apnoeic reflex, it is not known whether the sensory receptors are the same for the two reflexes. The expulsive role of the sneeze is obvious, and the preliminary deep inspiration through the mouth is an important adjunct.

Mechanical, thermal or chemical irritation of the nasal mucosa also causes other reflex responses: secretion of mucus in the trachea (Richardson and Phipps, 1978), inhibition of spinal reflexes (Anderson, 1954) and changes in bronchomotor tone (Tomori and Widdicombe, 1969). The last are especially puzzling because both constrictor and dilator responses have been described. Possibly, as has been suggested for the respiratory responses, more than one group of nervous receptors are involved. Any physiological advantage of reflex changes in bronchial calibre on nasal stimulation is hard to envisage, unless they change the pulmonary deposition of that quantity of the inhaled substance that may penetrate beyond the nose.

Filtration and Absorption

Particles with diameters over 10 μm are almost entirely retained
in the nose, although a few may penetrate to the lungs where, if they
are allergens such as pollen, they may cause pulmonary or bronchial
reactions; particles with diameters 2-5 μm penetrate to the lungs
far more readily, and are deposited mainly in the tracheobronchial
tree; particles with diameters of 0.2-0.5 μm have minimum deposition,
although guinea-pigs will retain 30% of cigarette smoke in the nose.
The degree of deposition depends on particle shape as well as size,
on rate of airflow and on degree of vascular congestion and mucus
secretion in the nose.

Gases soluble in body fluids are well absorbed in the nose.
Thus sulphur dioxide has over 99% absorption in the nose in man, even
after 6hr. of inhalation (Brain, 1970).   Ozone, which is less soluble,
is less well taken up in the nose and is therefore potentially more
toxic to the lungs.

Mucociliary Clearance

Mucus secretion from the nose is prompted by inhalation of
irritant gases, cold air and a variety of mediators such as histamine,
methacholine and prostaglandins (Phipps, 1980).   To what extent this
secretion is a direct action of the stimulant or a reflex has not
been carefully determined; probably both mechanisms exist.   The
secretion contains mucoglycoproteins, immunoglobulins, lysozome and
serum albumen, as well as smaller molecules.   The secreted mucus is
usually carried backwards to the larynx where it is swallowed.

Extensive recent studies by Proctor et al. (1977) have shown
that the flow rates of human nasal mucus are variable in healthy
subjects (1-20 mm. min^{-1}), so that the total layer of nasal mucus is
cleared in about 10-15 min.   Human nasal mucociliary clearance is
not affected by large changes in inspired air relative humidity, and
little affected by changes in inspired air temperature; increases or
decreases from 23°C cause small decreases in clearance rate.
However toxic gases such as sulphur dioxide greatly decrease mucociliary velocity.

There seem to have been no studies on whether secretion of nasal
mucus aids defense of the lungs from inhaled noxious substances, but
presumably if the secretion is copious enough to cause mouth breathing this would be detrimental to lung defenses.

Cellular and Immune Responses

In allergic rhinitis histamine and IgE antibodies appear in the
promoted mucus secretion (Bang and Bang, 1977).   The effectiveness

of cromoglycate and of atropinic drugs in lessening the secretion may
indicate that both mast cell degranulation and reflex secretion are
involved, but the relative role of the two has not been settled.
Mast cells and eosinophils are present both in nasal mucosa and in
secretions.

Viral nasal infections cause copious mucus secretion. The
virus may penetrate any mucus layer aided by neuramidase which acts
on mucoglycoproteins and lowers their viscosity. The resultant
secretion includes serum proteins, presumably as transudate, interferon and nonspecific IgA. Virus antibody appears in the secretion
1-2 weeks after infection.

NASOPHARYNX

This part of the respiratory tract has squamous cell epithelium,
with submucosal glands which presumably lubricate food during
swallowing. The mucosa contains fine afferent nerves which may be
terminals of sensory receptors (Fillenz and Widdicombe, 1970).
Mechanical stimulation of the nasopharynx causes the "aspiration
reflex" - repeated strong brief inspiratory efforts that presumably
clear the nose of blocking material which is swallowed or expectorated (Korpas and Tomori, 1979). There is also reflex hypertension,
venoconstriction, tachycardia, bronchodilation and tracheal mucus
secretion, and a strong arousal response. The aspiration reflex is
one of the most powerful stimuli to phrenic activity (albeit brief)
yet studied (Nail et al., 1972), and is resistent to anaesthesia and
hypothermia. It is unknown whether it plays a part in chemical or
allergic responses from the nasopharynx.

LARYNX

Since the vocal folds have squamous cell epithelium, mucociliary
transport travelling up the respiratory tract must stop at this point.
Any mucus or luminal substance will be cleared either by coughing or
by airflow. The fact that the larynx constricts in expiration and
dilates in inspiration means that expiratory velocities of gas will
be greater than inspiratory, and mucus in the larynx will tend to
move outwards even in the absence of coughing. The larynx including the vocal folds also has submucosal glands, but the control of
their secretion and its role have not been determined. Many different types of nervous receptor have been described in the laryngeal
mucosa: "free" nerve endings, taste buds, and various corpuscular
structures (Widdicombe, 1977b). Surprisingly, the vocal folds have
few afferent nerve fibres, although they are highly sensitive to
mechanical stimuli. Several groups of receptor have been identified
by recording from their nerve fibres and seeing what chemical and
mechanical stimuli activate them, but it is not clear which receptors
are responsible for the many reflexes which can be elicited from the
larynx.

Reflex Responses

With a minimal effective mechanical or chemical stimulus to the larynx, there is brief reflex laryngeal closure, in animals and man. Such closure should limit penetration of the stimulant. Stronger stimuli cause apnoea, expiratory efforts, the "expiration reflex" or coughing (Korpas and Tomori, 1979). While the defensive role of these responses is obvious, a limitation of penetration or expulsion of the inhaled substance, the appropriate stimuli for each reflex and the particular advantage of each response is not clear. In addition to changes in laryngeal calibre and breathing, there may also be reflex hypertension, bradycardia, tracheal mucus secretion and bronchoconstriction.

The cough reflex can be mimicked voluntarily and, so produced, is as mechanically effective as is a spontaneous cough. The aerodynamics of coughing have been reviewed by Leith (1977).

SUMMARY

The upper respiratory tract is abundantly supplied with mechanisms that limit entry of dangerous invaders; if penetration occurs other mechanisms limit the damage that may be done. Failure of these mechanisms could lead to damage to the lower respiratory tract and lungs.

REFERENCES

Anderson, P., 1954, Inhibitory reflexes elicited from the trigeminal and olfactory nerves in rabbits. Acta. Physiol. Scand., 30: 137-148.

Angell-James, J. and de B. Daly, M., 1969, Nasal reflexes. Proc. R. Soc. Med., 62: 1287-1293.

Anggard, A., Lundberg, J.M., Hokfelt, T., Nilsson, G., Fahrenkrug, J. and Said, S., 1980, Innervation of cat nasal mucosa with special reference to relations between peptidergic and cholinergic neurons. Europ. J. Resp. Dis., in the press.

Bang, B.G. and Bang, F.B., 1977, Nasal mucociliary systems. In Respiratory Defense Mechanisms, ed, Brain, J.D., Proctor, D.F. and Reid, L.M., pp. 405-426. Marcel Dekker, New York.

Brain, J.D., 1970, Uptake of inhaled gases by the nose. Ann. Otol. Rhinol. Laryngol., 79: 529-539.

Cauna, N., Hinderer, K.H. and Wentges, R.T., 1969, Sensory receptor organs of the human nasal respiratory mucosa. Am. J. Anat., 14: 295-300.

Cohen, A.B. and Gold, W.M., 1975, Defense mechanisms of the lungs. Ann. Rev. Physiol., 37: 325-356.

Empey, D.W., 1980, Assessment of the nasal passages. Br. J. Clin. Pharmac., 9: 317-319.

Fillenz, M. and Widdicombe, J.G., 1971, Receptors of the lungs and
    airways.   In Handbook of Sensory Physiology, Neil, E., ed.
    3: pp. 81-112. Heidelberg, Springer-Verlag.
Iravani, J. and Melville, G.N., 1977, Mucociliary function in the
    respiratory tract as influenced by physiochemical factors.
    Pharmac. Ther. B., 2: 471-492.
Korpas, J. and Tomori, Z., 1979, Cough and Other Respiratory Reflexes.
    Karger: Basel.
Leith, D.E., 1977, Cough.   In Respiratory Defense Mechanisms.,
    Brain, J.D., Proctor, D.F. and Reid, L.M., ed. pp 545-592.
    Marcel Dekker: New York.
Mygind, N., 1978, Nasal Allergy.   Blackwell Scientific: Oxford.
Nail, B.S., Sterling, G.M. and Widdicombe, J.G., 1972, Patterns of
    spontaneous and reflexly-induced activity in phrenic and
    intercostal motoneurones.   Exp. Brain Res., 15: 318-332.
Negus, V.E., 1958, The Comparative Anatomy and Physiology of the Nose
    and Paranasal Sinuses.   E. and S. Livingstone: Edinburgh.
Phipps, R.J., 1980, The airway mucociliary system.   In International
    Review of Physiology, Respiratory Physiology III., Widdicombe,
    J.G., ed.   University Park Press: Baltimore, in the press.
Proctor, D.F., 1977, The upper airway.   I.   Nasal physiology and
    defense of the lungs.   Am. Rev. resp. Dis., 115: 97-129.
Proctor, D.F., Andersen, I. and Lundqvist, G., 1977, Nasal mucocil-
    iary functions in humans.   In Respiratory Defense Mechanisms,
    Brain, J.D., Proctor, D.F. and Reid, L.M. eds. pp. 427-452.
    Marcel Dekker: New York.
Proctor, D.F. and Swift, D.L., 1977, Temperature and water vapour
    adjustment.   In Respiratory Defense Mechanisms., Brain, J.D.,
    Proctor, D.F. and Reid, L.M. eds.   pp. 95-124.   Marcel
    Dekker: New York.
Richardson, P.S. and Phipps, R.J., 1978, The anatomy, physiology,
    pharmacology and pathology of tracheobronchial mucus
    secretion and the use of expectorant drugs in human disease.
    Pharmacol. Ther. B., 3: 441-479.
Silberberg, A., Meyer, F.A., Gilboa, A. and Gelman, R.A., 1977,
    Function and properties of epithelial mucus.   Adv. exp. Med.
    Biol., 89: 171-180.
Schmidt-Nielson, K., Schroter, R.C. and Shkolnik, A., 1980,
    Desaturation of the exhaled air in the camel.   J. Physiol.
    in the press.
Tomori, Z. and Widdicombe, J.G., 1969, Muscular, bronchomotor and
    cardiovascular reflexes elicited by mechanical stimulation of
    the respiratory tract.   J. Physiol., 200: 25-50.
Tos, M., 1976, Mucous elements in the nose.   Rhinology 14: 155-185.
Uddman, R., Malm, L. and Sundler, F., 1980, The origin of vasoactive
    intestinal polypeptide (VIP) nerves in the feline nasal mucosa.
    Acta. Otolaryngol., 89: 152-156.

Widdicombe, J.G., 1977a, Defensive mechanisms of the respiratory
        system.   In International Review of Physiology, Respiratory
        Physiology III., Widdicombe, G.J. ed. 14: pp. 291-315.
        University Park Press: Baltimore.
Widdicombe, J.G., 1977b, Respiratory reflexes and defense.   In
        Respiratory Defense Mechanisms., Brain, J.D., Proctor, D.F.
        and Reid, L.M. pp. 593-630.   Marcel Dekker: New York.

DISCUSSION

LECTURER: Widdicombe   CHAIRMAN: Bonsignore

CORRIN: You described the efficiency of the nose at removing
                   soluble gases, like sulphur dioxide. Would you like
                   to say how you think sulphur dioxide at very low
                   concentrations can cause bronchoconstriction? If it
                   does not get to the lung, I mean.

WIDDICOMBE: Yes, I can speculate. I think one has to assume that
                   it does not get to the lungs, or at least if one
                   takes the published figures, the concentration to get
                   there would be far too small to stimulate
                   bronchoconstrcitor reflexes from the larynx or from
                   the respiratory tract. Therefore it has to act on
                   the nose, or maybe the pharynx, certainly the nose,
                   and I suppose what it is doing is stimulating nerve
                   fibres or receptors in the nose that cause
                   bronchoconstriction. Another problem is a twofold
                   one: firstly, is the concentration of SO_2 in the
                   nose enough to stimulate these receptors? That has
                   not been studied, so at least it is possible that the
                   concentration would be sufficient but a more
                   difficult problem is that the literature on nasal
                   bronchial reflexes is divided about 75 to 25, at the
                   moment; 75% of the papers say that nasal irritation
                   causes bronchoconstriction, and 25% after a more
                   careful study suggest that it causes
                   bronchodilatation. So I think that the chances are
                   that the nose has at least two different afferent
                   systems, as of course do most of the parts of the
                   respiratory tract, one of which will cause a
                   dilatation reflexly and one a constriction. There is
                   also good evidence for other constrictor reflexes in
                   the nose, for example cold applied to the front of
                   the nose will cause a bronchoconstriction, and it may
                   well be that the same receptors or receptor systems
                   are involved in the two responses.

CORRIN: I wonder if you could tell us something about
                   hormonal effects on the nasal mucosa, particularly
                   sex hormones. I ask this because nosebleed seems to
                   be common in adolescence, and there is a vascular
                   tumor nasal angiofibroma which is peculiar to the
                   nose, and which interestingly is virtually confined
                   to adolescent boys and lastly because you likened the
                   vasculature to rectal tissue.

WIDDICOMBE: I cannot give the answer that you would like to have
because the work has not been done. One can measure
nasal congestion if you regard air flow resistance as
a good index of it. If you do that, then of course
there are observations, giving for example steroids
to people with nasal congestion relieving the
congestion, but that is a very artificial
pathological condition. I know of only one
physiological observation, which is not an
experiment, and that is that pregnant women are often
said to have an increased nasal congestion and this
goes on until they have their baby, when the nose
opens up again. I think this is well documented in
some pregnant women but not all, but I do not know
anybody who has done the direct experiment, to give
hormones, either sex hormones or others, while
measuring air flow resistance, even if they have one
would still like to have a more objective method of
measuring blood flow to be convinced.

CANDURA: Could you tell us why there are some irritants which
act typically on the upper airways or even on the
conjunctivae and others acting predominantly on the
intermediate parts of the lung, and others still
which act on the lower airways? This is of great
importance, as you know, from the point of view of
occupational disease.

WIDDICOMBE: That is a very interesting and important observation.
I can give an example of it. In a recent work done
by Douglas in London, who has been measuring the
threshold for SO_2 and has found, as other people
have, that if you inhale SO_2 you may bronchoconstrict
with a dose of 1 or 2 ppm; if you expose the eyes to
SO_2 you will need 5 or 10 ppm to get reflex
production of tear fluid, and you only reach the
sensory threshold for the eye when the concentration
is about 200 ppm. So, the sensitivity of three
different exposed mucosae is quite different by a
factor of 200. I think that what you described is
probably a question of threshold and sensitivity,
rather than a qualitative difference between the
effective irritants. I do not, for example, know of
any irritant which will irritate at one site and be
anaesthetic at another, apart from some anaesthetics,
of course. So I take that back straight away, they
have both an irritant and anaesthetic action. We
have in different parts of the respiratory tract
these mucosal receptors, both the nervous ones and

also mast cells that release mediators exerting local actions, and the sensitivity of these will vary according to the site and possibly according to the irritant as well.

PAOLETTI: Going back to SO_2, I would like to remind you that its irritant capacity is mostly linked to transport. Recent studies have shown that sulphur, if it is inhaled by aerosol, can reach the lower airways and irritate them. This is not a question, I just wanted to point this out with regard to SO_2.

JEFFREY: I wonder if you would like to comment on the role of bronchoconstriction, the reflex bronchoconstriction, in the lower respiratory tract.

WIDDICOMBE: This is a very controversial matter, there has been a lot of discussion and various teleological explanations of bronchoconstriction in response to inhaled irritants have been thought up, but I think very few have been tested to see if they really help the subject inhaling irritant. The kind of explanations that have been given are for example that with bronchoconstriction, the deposition will be higher up and there will be less penetration to the alveoli. Experiments by Thompson and Short support this, and that would clearly be an advantage. It has been suggested by Mead that the constriction will cause a denser layer of mucus at the site of constriction, that seems inevitable, and the result of this would either be that that layer of mucus absorbs any irritant better or that it would be cleared faster, and perhaps that it promotes coughing more readily, which would also help clearance. Another possibility is that when the airways constrict, they not only get narrower, but they get more rigid, so that if in addition to the constriction one had coughing, the rigid airways would be better able to extend the vigorous movement of coughing. There may be other possibilities.

DENISON: There is some evidence in marine mammals that the muscles stiffen the airways in the way you describe. Is anything known of its cyclic activity in respiration? In which phase of respiration is it likely to be more active?

WIDDICOMBE: Yes, these have been studied. The problem is to dissociate the passive stretch and dilatation of smooth muscle in breathing with any active

contraction. There is no doubt that during
expiration the lower airways contract, but this is
probably not passive. One can show that there is a
superimposed rhythmical modulation in airway tone in
several ways, including for example by recording from
the motor neurons going down to the smooth muscle, or
by taking the smooth muscle in isolation in the neck,
where it is not exposed to mechanical changes in the
lungs. Under those conditions the smooth muscle does
contract in expiration, usually rather late, and as
far as I know only in experiments where the breathing
has been very slow. The interesting question would
be what happens if breathing gets very fast, so that
the neuromuscular control, in which the smooth muscle
is slow, is not able to follow. You might then reach
a state when the contraction is taking place out of
phase. This is only a possibility, it has not been
tested, but I think the question of the phasic
contraction of smooth muscle really has to be looked
at under these kinds of conditions, where there is a
superimposed either bronchoconstriction or
respiratory change and not in anaesthetised animals
breathing very slowly, a rather artificial condition.

BONSIGNORE: This morning's session is devoted to physiology and
we will now hear Sant'Ambrogio speaking on the
afferent activity and the reflexes in the
tracheo-bronchial tract.

AFFERENT ACTIVITY AND REFLEX EFFECTS EVOKED FROM THE TRACHEO-

BRONCHIAL TREE

Giuseppe Sant'Ambrogio

Dept. of Physiology & Biophysics
University of Texas Medical Branch
Galveston, Texas   77550

INTRODUCTION

There are obvious alterations in the rate and depth of
breathing and in the capability of reacting to noxious stimuli,
as for example in coughing, when the vagal afferent activity
originating from the tracheo-bronchial tree is blocked. This
block can be accomplished by local anesthetization of the lumen
of the airways with involvement of several and possibly all
types of nervous endings (1,2), by sulphur dioxide inhalation,
which presumably paralyzes only the slowly adapting stretch
receptors (3). The changes in the pattern of breathing introduced by these experimental procedures are similar to those
observed with cervical vagotomy thus indicating that in this
latter situation it is just the interruption of the afferent
component which plays the major role in altering the respiratory
pattern. There are several observations which indicate that
the economy of respiratory functions suffers when the vagus
nerves are sectioned. For example, it was shown that the work
of breathing was consistently increased after vagotomy (4,5).
Another example of the respiratory impairment introduced by a
block in the vagal conduction is the marked decrease of the
ventilatory response to carbon dioxide (6). Changes in several
respiratory parameters (as oxygen and carbon dioxide arterial
concentrations, lung compliance and resistance, tidal vlume and
breathing frequency) observed when an experimental pneumothorax
was introduced were much greater after vagotomy (7) indicating
that maintenance of normal respiratory values was more effectively accomplished with intact innervation. Furthermore, it
was found that bilateral section of the vagal afferent fibers
(apparently possible at the level of the nodose ganglion)

135

produces pulmonary congestion and oedema which do not allow a
prolonged survival (8). This study however does not seem to be
supported by the relative good condition observed in cats
subjected to chronic vagotomy performed below the recurrent
laryngeal nerves (9).

All these observations justify our interest in the afferent nervous supply to the tracheo-bronchial tree (in its various component) and in the reflex responses that can be evoked
from it.

AFFERENT NERVE SUPPLY TO THE TRACHEO-BRONCHIAL TREE

The afferent fibers distributed to the tracheo-bronchial
tree are part of the peripheral autonomic nervous system. We
know much more about the parasympathetic (vagal) component of
this afferent innervation than about its sympathetic
counterpart. The functional role of the vagal component is
generally thought to be more important.

Morphological data, based on fibers counting in cross sections of the bronchial branches of the vagus nerves after sup-
ranodose vagotomy*, are available for the bronchial afferent
supply in two species: cat (10) and rabbit (11). In the cat
there are about 3,800 non myelinated fibers and 1,200 myelinated fibers, i.e., the non medullated C-fibers are 3.17 times
more numerous than the medullated A-fibers. Viceversa in the
rabbit the non medullated C-fibers are decisively a minority.
This morphological difference between these two species might
have important functional implications: perhaps we should always
remember the possibility of these, even major, species differences, when extrapolating results from one species to others.

The nature of the fibers innervating the extrapulmonary
airways has not been studied. According to a schema provided
by Nagaishi (12, also for references) the trachea and the main
stem bronchus on the left side are innervated mostly from
branches originating from the recurrent laryngeal and the right
main stem bronchus from branches coming directly from the thoracic vagus. Part of the extrathoracic trachea, in its medullated afferent component, is innervated through the recurrent
laryngeal with the other fibers joining the vagus nerves through

---

* The neurons corresponding to these afferent fibers lie in the
nodose ganglion whereas those corresponding to the efferent
fibers are in the brain stem. A section of the vagus nerve
above the nodose ganglion will determine the degeneration of
the efferent fibers, separated from their cells, but not of the
afferent ones still connected to their cells in the nodose
ganglion.

other routes (13).

There are evidences for an impulse traffic related to respiration in the sympathetic chain (14,15), but no data are
available on the number and size of these fibers.

ACTIVITY WITH RESPIRATORY MODULATION RECORDED FROM AFFERENT FIBERS IN THE VAGUS NERVE AND IN THE SYMPATHETIC CHAIN

Action potentials with some kind of fixed time relationship
with the respiratory movements can be readily recorded from vagal and sympathetic afferent fibers mostly having conduction velocities compatible with a medullated nature. Some of this activity, both vagal and sympathetic, can be traced to structures
outside the respiratory tract: for instance in the mediastinum,
in the great vessels of the thorax, in the oesophagus (16,14).
The close proximity of these structures with those strictly respiratory explains, for simple mechanical reasons, their secondary respiratory modulation. For analogous reasons there are action potentials traceable to respiratory structures which have a
cardiac modulation besides a respiratory one: the close proxim
ity with the heart, or other pulsating vascular structures, determines this superimposed rhythmicity (17).

In most instances the activity with a respiratory modulation can be recognized as originating in respiratory structures
and can be distinguished between two separate patterns. One
with a regular change in its rate of discharge when either
volume or flow vary and another with an essentially irregular
discharge pattern, occurring in bursts, during the respiratory
movements, especially when inspiratory and expiratory flows are
high and when the end inspiration is reached. When a given
transpulmonary pressure is applied and maintained the first
type shows a maintained rate of discharge whereas the second
type has a rapidly fading activity. The slowly adapting receptors (SAR) appear to be far more numerous than the rapidly
adapting receptors (RAR); it is usually proposed, but it has
never been substantiated, a ratio of 10:1 for their respective
occurrence.

SLOWLY ADAPTING RECEPTORS (SAR) OF THE TRACHEO-BRONCHIAL TREE

The SARs have always been found to be associated with the
airways having a diameter of at least 0.3 mm (18). This
diameter was measured in collapsed lungs and should therefore
correspond to airways of about the 19th generation (terminal
bronchioles). These endings are not distributed uniformly
throughout the tracheo-bronchial tree: their concentration
(number of endings per unit lateral surface area of the airways) was found to be greater in the extrapulmonary airways and

to decrease sharply along the intrapulmonary airways (Fig. 1, right panel; 18,19).

In the trachea the SARs have been localized, by direct exposure of the tissues, in the membranous posterior wall within the trachealis muscle (20). Furthermore there are other less direct evidences that support their location within the smooth muscles of the airways (21). Bitensky et al. (22) have proposed a location in the superficial layers of the mucosa on the basis of experiments in which the inflation reflex could be blocked by radiolabelled bubivacaine that was found mostly in the superficial layers above the basement membrane. However the presence of the same local anesthetics, though at very low concentration, within the smooth muscle layer, does not guarantee a unique location of these endings within the epithelium.

Since the work of Adrian (23) it is well established that a considerable proportion of SARs are active at end expiratory volume (FRC) with the rest being recruited only during inspiration. On the basis of this behavior they are classified as "tonic" and "phasic". Most of the "tonic" SARs were found to

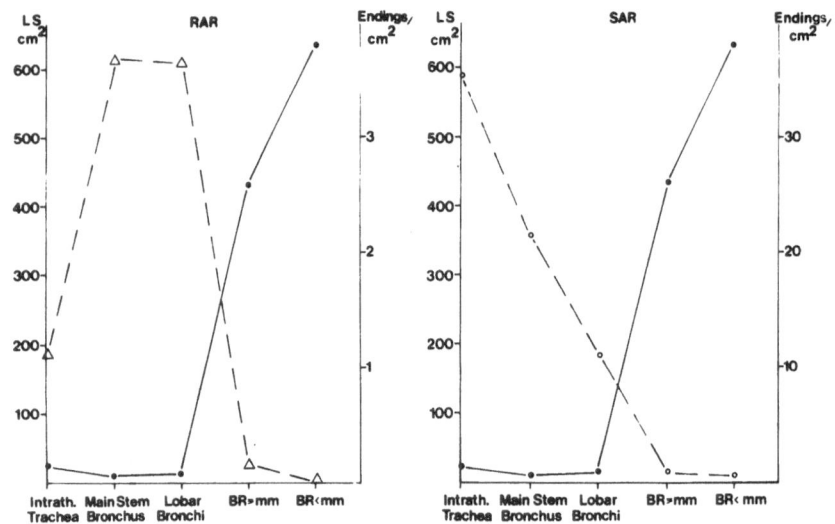

Fig. 1.   In each diagram the continuous line with the closed circles represent the lateral surface area (LS) for airways of different generations (below the lobar bronchi the airways are grouped in two types: greater and smaller than 1 mm). The broken line with the open triangles refer to the concentration of SARs (right panel) and RARs (left panel).

be in the larger extrapulmonary airways (24).

The extrapulmonary airways, in most mammals, have a characteristic structure: a series of U-shaped cartilaginous rings
supporting a membranous posterior wall. Neither one of these
structures (cartilages and posterior wall) is at its resting
position, the cartilages tending to expand and exerting a
transversal stretch on the posterior wall. The mechanical
coupling between these two structures explains most of the
properties of the SARs in the extrapulmonary airways (25).
Most of them are active at zero transmural pressure and respond
to both distending and collapsing pressures showing the least
activity at a low collapsing pressure (Fig. 2).

This asymmetric response curve to positive and negative
pressures has significant physiological implications. Half of
the trachea is outside the thorax and, simply for this situation, the corresponding changes in transmural pressure during
the breathing cycle differ from those in the intrathoracic
trachea and airways. In these latter airways the transmural
pressure at end expiration is +4 cm H_2O and increases to an
higher value at the end inspiration (for instance to +10 in a
quiet breath) leading to a corresponding increase in the SAR
activity which we could predict from the response curve (Fig.
2, arrows at the right). In the extrathoracic trachea the
transmural pressure is zero both at end expiration and at end
inspiration, but it becomes negative (collapsing). when there
is inspiratory airflow and positive (distending) during the
period of expiratory airflow. Since we have an asymmetric
response curve for these tracheal SARs we will have a decrease
in their discharge during inspiration and an increase during
expiration (Fig.3). Both intrathoracic and extrathoracic
SARs send signals related to transmural pressure across their
respective airways, but while the ones inside the thorax can
signal both flow and volume, the one outside, in the extrathoracic trachea, can only send informations related to airflow and
upper airways resistance (26). The activation of the extrathoracic and the intrathoracic airways SARs is thus, during a
quiet breathing cycle, out of phase: in inspiration there is an
increase in activity in the intrathoracic SARs with a decrease
in those in the extrathoracic trachea and viceversa during the
expiration (Fig. 2).

For the same reason, i.e. the different transmural pressure for their different location, the behavior of SARs in the
extrathoracic and in the intrathoracic trachea is markedly different during inspiratory efforts against upper airways occlu-
sion. In this condition the transmural pressure in the intrathoracic airways does not change from the end expiratory value,
accounting for the lack of inspiratory modulation of the cor-

responding SARs, while in the extrathoracic trachea there are
marked changes in transmural pressure which determine considerable modulation of the corresponding SARs (Fig. 4).

Another interesting consideration on the behavior of the
SARs in the larger extrapulmonary airways concerns their possible activation during forced expiratory efforts and/or coughing. In these circumstances SARs of such airways, being downstream from the equal pressure point, should be subjected to
considerable collapsing pressures which could increase their
activity above their FRC value and thus reflexly provide an
inhibitory influence which can delay the next inspiration and
an excitatory influence to the expiratory muscles.

The intrapulmonary airways SARs are distributed among the
lobes of the lung according to their respective volumes (19)

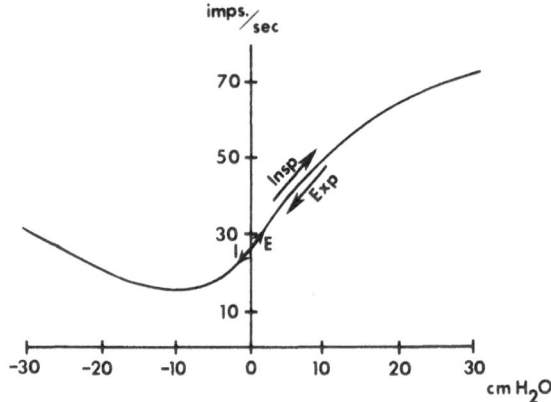

Fig. 2.   Relationship between transmural pressure and dis-
           charge rate of tracheal SARs. The arrows on the
           right side indicate the changes in SARs discharge in
           the intrathoracic trachea during inspiration and ex-
           piration. The shorter arrows originating from the
           zero pressure line indicate the changes in activity
           of extrathoracic tracheal SARs during inspiration (I)
           and expiration (E). In the extrathoracic trachea
           there are changes in transmural pressure and hence in
           receptors discharge only when there is flow.

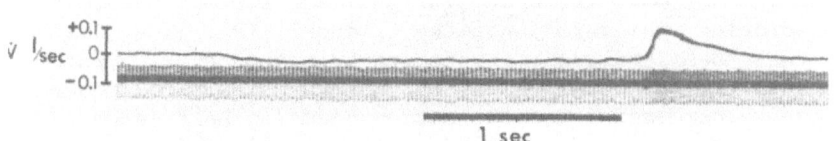

Fig. 3.   Anesthetized dog breathing through a tracheal cannula.
           Upper signal = airflow (V), inspiration downward.
           Lower signal = action potentials recorded from a
           vagal afferent fiber originating from a SAR in the
           lower half of the extrathoracic trachea.

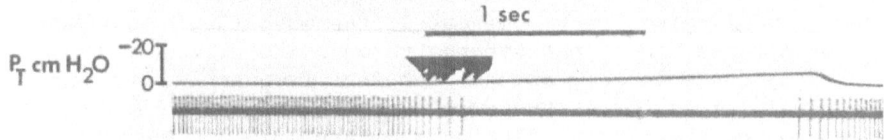

Fig. 4   Anesthetized dog. Inspiratory effort against the
           occluded upper airways. Upper signal = transmural
           pressure (P_T) in the extrathoracic trachea
           (negative, collapsing, pressure upward). Lower
           signal = action potentials from a SAR in the
           extrathoracic trachea.

and, as previously indicated, decrease in concentration toward
the periphery (Fig. 1; 18). An increasing number of the intrapulmonary SARs, as their location becomes more peripheral, is
not active at end expiration being recruited only during inspiration (24).

     Another feature of the more distally located SARs is their
linear relationship with transpulmonary pressure in contrast
with the ones in the larger airways for which the same relationship shows a plateau above 15 cm H_2O (24). This different
behavior could be indicative of an increasing reflex influence
originating from the intrapulmonary SARs at higher transpulmonary pressures.

     The response of the SARs to carbon dioxide, which inhibits
their activity, is a well documented characteristic, but still

very puzzling for its physiological implications. This response
has been shown in rats (27), dogs (28,29), cats (30), rabbit
(31) and a marsupial (32). Bartlett and Sant'Ambrogio (33)
have shown that only the bronchial SARs are susceptible to the
carbon dioxide inhibition, not the tracheal SARs. They also
found that these endings are affected by carbon dioxide introduced into the bronchial lumen, but not when its blood concentration is increased. The CO_2 action seems to be related
with the increase in hydrogen ions concentration. Actually
sulphur dioxide which has stronger acidifying properties can
paralyze these endings, at least in the rabbits (3).

The histological studies of SARs are still very unsatis-
factory and lacking in meaningful details pertaining to either
the endings themselves or to their relationship with the surrounding structures. Using light microscopy Elftman (34) found
within the trachealis muscle endings arising "from a coarse
fiber which branches palmately. The branches show many varicosities and terminate in swellings or reticulations among the
smooth muscle cells." The same author observed similar endings
in the bronchi being "very numerous in the musculature from the
point of bifurcation of the trachea through the hilus of the
lung. From there on they appear much less frequently, but can
be found almost as far as the smooth muscle extends." The close
association of these nerve terminals with the smooth muscle
support their identity with the SARs, but cannot guarantee it.
A more recent study with electron microscopy concerning small
airways of the rat shows an ending within the smooth muscle
reaching toward the epithelium (35).

RAPIDLY ADAPTING RECEPTORS (RAR) OF THE TRACHEO-BRONCHIAL TREE

The RARs, as the SARs, are not uniformly distributed along
the tracheo-bronchial tree: their concentration (number of RARs
per sq. cm of lateral surface area) is greater in the larger
airways with the highest value in the main stem and the lobar
bronchi and then markedly decreases in the more peripheral
intrapulmonary bronchi (Fig. 1; 35). At least in the left lung
the distribution of RARs was not found to be proportional to
the volumes of the two lobes: their concentration appeared to
be higher in the upper lobe (36).

Their distribution along the circumference of the trachea
and the main stem bronchus is definitely different from that of
the SARs: these receptors are found throughout the circumference and not only in the posterior wall as the SARs (37). Even
in the trachea there is only a modest cross innervation between
the two sides (37). Their concentration in the trachea increases toward the carina and appears to be higher at the points
of branching just below the carina on the medial surface of the

main stem bronchus and at the points of branching of the lobar bronchi (37). In the extrathoracic trachea, where this study could be easily performed, the receptive field appears to cover two contiguous cartilages suggesting that the nerve fiber leading to an ending enters the trachea between two cartilages spreading then over them.

These endings respond both to gross deformations of the airways (such as those introduced by large distending and collapsing transmural pressures) and to local probing as can be more easily accomplished in the more proximal airways. The two modalities of activation of RARs could be separated by a superficial lesion of the receptor's field: the response to local probing could then be abolished with retention of the responses to both inflation and deflation (37). This seems to indicate a multibranching structure for these endings which could be distributed to both superficial epithelial structures and to deeper structures.

RARs located in the extrathoracic trachea can be conveniently studied for the characteristics of their responsiveness. The local probing appeared to be a more effective stimulus than gross deformations to inflation and deflation; the response to local probing administered in orthogonal directions did not disclose any particular orientation for its efficacy (Fig. 5).

The respiratory modulation of the RARs is irregular in both its timing with the breathing cycle and in its pattern of discharge: bursts of activity from an ending occurs preferably either during the periods of inspiratory and expiratory flow or/and at peak inspiration (38). As pointed out by Sellick and Widdicombe (7) the activity of RARs is inversely related to lung compliance. A feature also shared by the SARs. This indicates that a preponderant role in their activation is played by the transpulmonary pressure. Presumably for this reason RARs become more active in pulmonary congestion and oedema (38).

Many inhaled substances in the form of gases (ammonia, sulphur dioxide), aerosols and fumes stimulate these endings which therefore are also called "irritant receptors" (38). The action of some of the inhaled irritants is different in different species: for instance ammonia and cigarette smoke are much more effective in the rabbits than in the dog (39).

Very important, for its patho-physiological implications, the stimulation of these endings by histamine, either inhaled or injected intravenously. This activation is partly attributable to the contraction of the smooth muscle (38), but is also contributed by a direct action of histamine on the endings themselves (40).

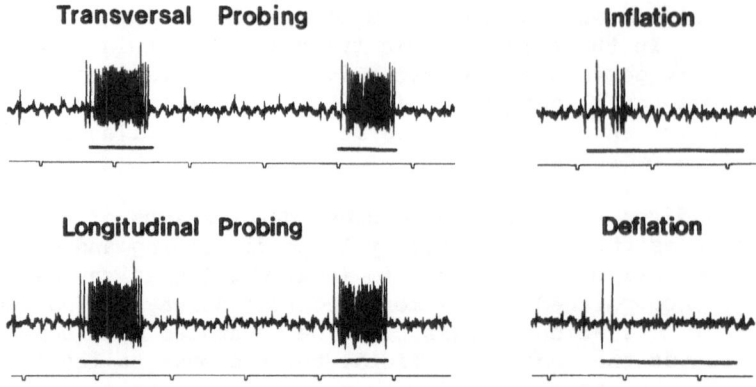

Fig. 5.   Anesthetized dog. Action potentials from an RAR in
           the extrathoracic trachea. Responses to local
           probing having a different direction do not differ,
           they are much greater than those to inflation and
           deflation (given previously to the exposure of the
           luminal surface of the extrathoracic trachea).
           Stretching longitudinally the trachea did not provide
           an effective stimulus. Time Marker = 1 sec.

       Local anesthetics introduced into the lumen of the airways
can block RARs at a lower concentration and/or in a shorter
time than the SARs indicating a more superficial location of
these receptors (41).

       Also in the case of RARs the histological evidence is
scarce. We still do not know how these endings are structured.
We do have evidence both from light microscopy studies (34) and
from electron microscopy (42) on the presence in the airways
epithelium of nerve terminals whose afferent nature seems very
probable. Das el al. (43) working with cats compared the morphology of the nervous supply to the superficial layers of the
airways with intact vagal innervation and after unilateral
vagotomy (below the·nodose ganglion) which should lead to a
degeneration of afferent fibers, but not of the post-ganglionic
parasympathetic efferents. They found that most of the fibers
on the vagotomized side had degenerated and hence established
their afferent nature. The same authors have also found a
greater concentration of these endings in the larger airways of
the cat. Similar results have been found in the rat (44).

The far greater concentration of the RARs in the hilar airways might provide a more efficient defense against various chemical and mechanical irritants to which they are very accessible and hence can readily be activated evoking some protective reflex actions. Their central location should also make these endings capable of monitoring some sort of average lung compliance and possibly provide the reflex actions to correct it.

C-FIBERS RECEPTORS OF THE TRACHEO-BRONCHIAL TREE

At least in the cat the C-fibers entering the lung greatly outnumber the medullated fibers (3.17 to 1) and their total number resulted to be 3,800 (10). These endings are essentially lacking any respiratory modulation, even when the inflating pressure, in dogs with open chest and artificially ventilated, exceeded 10 cm H_2O (45). They are generally studied stimulating their activity by injection or inhalation of drugs such as phenyl-diguanide or capsaicin. Perhaps the greatest difficulty in studying these endings depends on the absence of any definite respiratory pattern. We do not know much, as yet, which structure within the lung, these non medullated fibers innervate. At present the only direct evidence for the location of some of these endings has been provided by Coleridge and Coleridge (45) who established the receptor's field by probing the mucosal surface with a catheter or a bristle in large intrapulmonary bronchi as well as in other smaller airways. Another important contribution of the same authors was that there are important differences among the C-fibers endings which therefore cannot be described as an homogeneous group. They distinguished between "bronchial" and "pulmonary" C-fibers receptors on the basis of their circulatory accessibility through either the pulmonary or the bronchial circulation (injection of an activating substance into the right or the left atrium). This criterion, which undoubtedly differentiates two distinct groups of endings, might be objectionable as a tool for implying their location, since, at least for the SARs, it was found that endings in larger intrapulmonary airways were preferentially reached by an activating substance through the pulmonary circulation, a finding which should imply a very peripheral location (Sant'Ambrogio and Sant'Ambrogio, unpublished results).

Another difference betweent the two groups of endings is the greater mechanosensitivity of the "pulmonary" endings: response to large inflations. The C-fibers endings described by Coleridge and Coleridge as "pulmonary" might be identifiable with the J receptors described by Paintal (46).

Another interesting difference between these two groups of C-fibers endings is their different chemosensitivity to either

estraneous chemicals or naturally occurring substances. For
example "bronchial", but not "pulmonary" endings are activated by bradykinin which can be released in the lung in some
allergic reactions (47). The pathophysiological implications
of these observations appear to be very interesting in consideration of the reflex effects that might originate from these
endings.

AFFERENT ACTIVITY TRAVELLING THROUGH THE SYMPATHETIC NERVES

Action potentials influenced by respiratory maneouvers
(inflation, deflation, probing or moving the respiratory structures) have been recorded in cats from nerves associated with
the stellate ganglion (14). The sites of the corresponding
receptors were indicated as the mediastinum, the lung root,
structures near the trachea and the esophagus at the thoracic
inlet. Their association with the airways was not clearly established and they lacked any definite respiratory modulation.

More recently an activity with a clear respiratory modulation has been recorded in the higher thoracic white rami com-
municantes in dogs (15). Their discharge appeared to be
linearly related to transpulmonary pressure and to have a slow
adaptation. The fibers involved were found to be medullated.

REFLEX RESPONSES ORIGINATING FROM THE TRACHEO-BRONCHIAL TREE

When considering the reflex responses that can be elicited
from the tracheo-bronchial tree it is easier to describe the
alterations which follow the challenges introduced than to
recognize the involvement of any particular receptor. All the
stimuli that we might think of using for "specifically" or
"selectively" stimulating any particular ending is very likely
affecting other endings as well, directly or indirectly, immediately or with some delay. We just need to consider few examples to illustrate this point. Inflation of the lungs stimulates SARs but also RARs, at least transiently; the reflex influences from the stimulation of these two types of receptors
are opposite on both the inspiratory muscles and the smooth
muscles of the tracheo-bronchial tree. Therefore the net outcome of the inflation can be disturbingly different and most
probably depends on the relative balance between the stimulation of the various endings involved.

Another factor that we need considering is the influence
that some of the responses reflexly evoked have on the receptor
itself. For instance RARs cause bronchoconstriction and this
in its turn stimulates the RARs: there is a reverberating self-
sustaining positive feed-back. SARs, viceversa, inhibit the
smooth muscle and this relaxation should lessen the receptor's

discharge: a negative feed-back.

Another condition that we should consider is that most of
the receptors can be influenced by alterations which take place
far from their location. A change in lung compliance can alter
the discharge of a tracheal receptor simply because the transmural pressure of the trachea has changed without any alteration
in the receptor's immediate environment.

The attribution of a particular reflex response to the
activity of a particular receptor is usually done considering
the characteristics of both the response and the receptor's
behavior with the challenge introduced. For instance a long
lasting response to a maintained stimulus is considered to depend more likely on the excitation of a slowly adapting receptor
which can maintain its activity. Viceversa a short lived response is more likely attributable to a receptor with a burst
like discharge. Such sort of circumstantial evidence is not
fool proof, but has nevertheless provided a useful approach.

Reflexes are more often studied in anesthetized animals,
and anesthesia can affect several of these responses in such a
way as to render the observed phenomena very artifactual, without much correlation with real life.

Despite all these limitations, that we should always keep
well exposed in bright light before venturing through unwar-
ranted theories, we have in this last century accumulated
useful indications on the role of airways receptors.

REFLEX INFLUENCES FROM THE SLOWLY ADAPTING RECEPTORS OF THE
TRACHEO-BRONCHIAL TREE

Within the respiratory system, activation of the SARs is
generally considered to exert an inhibitory influence on the
muscles of inspiration (23) and on the smooth muscles of the
tracheo-bronchial tree (48).

Their inhibitory influence on inspiratory activity is man-
ifested both as a prolongation of expiration and as a shortening
of the inspiratory activity. Without the influence from these
receptors the pattern of breathing is considerably affected:
inspiration becomes deeper and the respiratory rate slower.

The best way for studying the influence of these SARs is
perhaps that of blocking their activity with inhalation of sulphur dioxide (200 p.p.m.). This gas determines a complete, and
reversible paralysis of SARs in the rabbit (3). The changes in
the pattern of breathing are similar to those observed after
vagotomy and perhaps we can say that the SARs are the main con-

tributing factor in determining the alterations observed with vagal section.

An important aspect of the breathing pattern after vagotomy is its lower efficiency: the cost of breathing increases, essentially because the elastic component of each breath is much increased. Lim et al. (5) found that the mechanical work per liter total ventilation increased 45% after vagotomy and Zechman et al. (4) have similarly found a 30% increase.

There is good evidence on the inhibitory influence of SARs to the airways smooth musculature (48): since SARs are functionally in series with the smooth muscle fibers and are activated by their contraction this reflex mechanism constitutes a negative feed-back. The possibility that these receptors contributed to the regulation of the tracheo-bronchial smooth muscle tone is of great interest: the size of the airways is serving two conflicting interests, on one side the dead space and on the other airflow resistance (49). A decrease in dead space can be accomplished only with an increase in resistance and viceversa. It is possible that SARs helped in compromising between the two parameters, therefore providing a further mechanism for a more economical alveolar ventilation.

Another aspect of the homeostatic role of SARs is their contribution in reducing the changes in functional residual capacity with postural changes. When SARs had been blocked with sulphur dioxide the shift in FRC observed with head-up tilting in rabbits became considerably larger (50).

The sensitivity of the bronchial (not the tracheal) to carbon dioxide when it is inhaled, but not when it is raised in the blood is still very puzzling in its physiological significance. It might explain the hyperventilation which ensues very rapidly in dogs on cardio-pulmonary by-pass when CO_2 is inhaled (51), but it does not appear to play a significant role in the ventilatory response to hypercapnia in more physiological circumstances. Actually in rabbits with SARs blocked with sulphur dioxide the hyperventilation in response to hypercapnia is increased (3). An involvement of this mechanism in exercise hyperpnea is very unlikely.

REFLEX INFLUENCES FROM THE RAPIDLY ADAPTING RECEPTORS OF THE TRACHEO-BRONCHIAL TREE

The reflex actions attributed to these endings comprise: excitation of the inspiratory discharge with shortening of expiration, tracheo-bronchial constriction and more complex responses as coughing (from the trachea and the extrapulmonary RARs) and the periodic augmented breath (38,52).

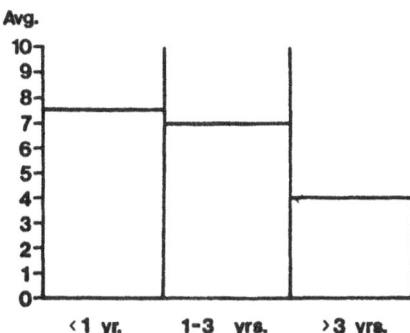

Fig. 6.   Average number of RARs ordinate found in the left
           vagus nerve in dogs of three age groups (abscissa).
           13 dogs   1 yr, 9 dogs = 1-3 yrs and 8 dogs   3 yrs.
           No statistical significance between the first two age
           groups and the 2nd and 3rd, significant difference
           between 1st and 3rd groups.   Dogs of the 1st group
           were older than 5 months.

Cough is attributed to these receptors because the triggering stimulus is known to excite RARs, but not SARs and to
involve only medullated fibers.   Furthermore local anesthetics
are more effective in blocking cough than the Hering-Breuer
inflation apnea (1,2) and they can block RARs before and/or at
lower concentration thatn SARs (41).

In rabbits exposed to sulphur dioxide and with absence of
Hering-Breuer inflation apnea inspiratory time increased and
expiratory time decreased presumably as a result of RARs influences (3).   The overall respiratory activating role of these
receptors is also apparent through the greater responsiveness
to carbon dioxide after sulphur dioxide block of SARs (3).

An important physiological role of RARs is considered that
of triggering periodic augmented breaths which restore the lung
compliance within proper values.   There is good circumstantial
evidence in favor of this action: presence of RARs activity at
the start of an augmented breath (53), occurrence of augmented
breaths when lung compliance is low and RARs activity is high
(54), presence of augmented breaths after sulphur dioxide

inhalation (3).

The marked activation of RARs by histamine is of great pathophysiological interest since this substance is released in anaphylactic and asthmatic reactions. Ensuing bronchoconstriction provides a further activation of RARs which in turn aggravates the bronchoconstriction (a self maintaining positive feed-back; 38,55,56).

With aging there are indications of a progressive loss in the protective responses of the airways (57): part of this decrease in responsiveness might be found in a decrease in the number of RARs in the tracheo-bronchial tree (Fig. 6; Mortola and Sant'Ambrogio, unpublished results).

The role of these endings is both protective (cough, broncho-constriction) and stabilizing of lung mechanics (augmented breath).

REFLEX INFLUENCES FROM THE C-FIBER RECEPTORS OF THE TRACHEO-BRONCHIAL TREE

Coleridge et al (58) have documented the presence of a population of afferent C-fibers corresponding to bronchial endings that can respond to several substances that occur naturally in the lungs where they are produced, released and also catabolized (histamine, some of the prostaglandins of the F and the E series, bradykinin). The activating properties of these substances is unrelated to their effects on lung mechanics, a factor that distinguish markedly these endings from the RARs. For instance bronchial C-fiber endings are stimulated by both bronchoconstricting (PGF_{2a}) and bronchodilating (PGE) prostaglandins whereas RARs are activated only when there is contraction of the bronchial musculature. This indicates a strictly chemosensitive role for the bronchial C-fibers endings which could participate in a regulatory mechanism for both airways and circulatory smooth muscles. In any event the reflex responses from these receptors in both physiological and pathological situations are still mostly obscure and completely open to future studies.

REFLEX INFLUENCES ORIGINATING FROM THE SYMPATHETIC AFFERENT ACTIVITY IN THE LUNGS

The presence of extra-vagal influences having either excitatory or an inhibitory action on respiration and travelling through the sympathetic nerves have been repeatedly reported (59,60,61,62,63). Recently it was reported (64) that distention of an intrathoracic tracheal segment could, after vagotomy, lead to a marked shortening of the expiratory duration. The threshold for this response was very high, the tracheal segment had to be distended at its TLC position. This might indicate a nociceptive nature of the endings involved. Kostreva et al. (63) have, on the other hand, reported inhibitory influences on respiration with sympathetic afferent stimulation involving only medullated fibers, possibly identifiable with those reported to have a respiratory modulation (15). Perhaps the facilitatory influences rely upon non medullated fibers entering the central nervous system through the sympathetic nerves. These latter fibers might subserve an unspecific generalized nociceptive function. The role of the medullated afferents remain to be ascertained, but their functional importance in the respiratory control is not probably prominent.

NIH Grant #5Rol HL20122.

REFERENCES

1. Jain, S.K., Trenchard, D., Reynolds, F., Noble, M.I.M.,
      and Guz, A., 1973, The effect of local anesthesia of
      the airway on respiratory reflexes in the rabbit,
      Clin. Sci., 44:519.
2. Cross, B.A., Guz, A., Jain, S.K., Archer, S., Stevens, J.
      and Reynolds, F., 1976, The effect of anesthesia of
      the airway in dog and man: A study of respiratory
      reflexes, sensation and lung mechanics, Clin. Sci.,
      50:439.
3. Davies, A., Dixon, M., Callanan, D., Huszczuk, A.,
      Widdicombe, J.G. and Wise, J.C.M., 1978, Lung
      reflexes in rabbits during pulmonary stretch
      receptors block by sulphur dioxide, Respir. Physiol.,
      34:83.
4. Zechman, F.W., Jr., Salzano, J., and Hall, F.G., 1958,
      Effect of cooling the cervical vagi on the work of
      breathing, J. Appl. Physiol., 12:301.
5. Lim, T.P.K., Luft, V.C. and Grodins, F.S., 1958, Effects
      of cervical vagotomy on pulmonary ventilation and
      mechanics, J. Appl. Physiol., 13:317.
6. Guz, A., Noble, M.I.M., Widdicombe, J.G., Trenchard, D.
      and Mushin, W.W., 1966, The effect of bilateral block
      of vagus and glossopharyngeal nerves on the
      ventilatory response to CO_2 of conscious man,
      Respir. Physiol., 1:206.
7. Sellick, H. and Widdicombe, J.G., 1970, Vagal deflation
      and inflation reflexes indicated by lung irritant
      receptors, Quart. J. Exp. Physiol., 55:153.
8. Mei, N. and Dussardier, M., 1966, Etudes des lesions
      pulmonaires produites par la section des fibres
      sensitive vagales, J. Physiol. (Paris), 58:427.
9. Remmers, J.E. and Bartlett, Jr., D., 1977, Reflex control
      of expiratory airflow and duration, J. Appl.

Physiol., 42:80.

10. Agostoni, E., Chinnock, J.E., Daly, M. DeBurgh and Murray, J.G., 1957, Functional and histological studies of the vagus nerve and its branches to the heart, lungs and abdominal viscera in the cat, J. Physiol., 135:182.

11. Evans, D.H.L. and Murray, J.G., 1954, Histological and functional studies on the fibre composition of the vagus nerve of the rabbit, J. Anat., 88:320.

12. Nagaishi, C., 1972, "Functional Anatomy and Histology of the Lung", University Park Press, Baltimore and London.

13. Sant'Ambrogio, G., Bartlett, D. and Mortola, J.P., 1977, Innervation of stretch receptors in the extrathoracic trachea, Respir. Physiol. 29:92.

14. Holmes, R. and Torrance, R.W., 1959, Afferent fibres of the stellate ganglion, Quart. J. Exptl. Physiol., 44:271.

15. Kostreva, D.R., Zuperku, E.J., Hess, G.L., Coon, R.L. and Kampine, J.P., 1975, Pulmonary afferent activity recorded from sympathetic nerves, J. Appl. Physiol., 34:37.

16. Widdicombe, J.G., 1954a, Receptors in the trachea and bronchi of the cat, J. Physiol., 123:71.

17. Bianconi, R. and Green, J.H., 1959, Cardio-respiratory afferent fibres in the vagus of the cat, Arch. Sci. Biol., 43:454.

18. Miserocchi, G. and Sant'Ambrogio, G., 1974, Distribution of pulmonary stretch receptors in the intrapulmonary airways of the dog, Respir. Physiol., 21:71.

19. Miserocchi, G., Mortola, J. and Sant'Ambrogio, G., 1973, Localization of pulmonary stretch receptors in the airways of the dog, J. Physiol., 235:775.

20. Bartlett, Jr., D., Jeffery, P., Sant'Ambrogio, G. and Wise, J.C.M., 1976, Location of stretch receptors in the trachea and bronchi of the dog, J. Physiol., 258:409.

21. Widdicombe, J.G., 1954b, The site of pulmonary stretch receptors in the cat, J. Physiol., 125:336.

22. Bitensky, L., Chambers, D.J., Chayen, J., Cross, B.A., Guz, A., Jain, S.K. and Johnstone, J.J., 1975, Evidence concerning the site of receptors mediating the Hering-Breuer inflation reflex, J. Physiol., 249:30.

23. Adrian, E.D., 1933, Afferent impulses in the vagus and their effect on respiration, J. Physiol., 79:332.

24. Miserocchi, G., and Sant'Ambrogio, 1974, Responses of pulmonary stretch receptors to static pressure inflations, Respir. Physiol., 21:77.

25. Mortola, J.P. and Sant'Ambrogio, G., 1979, Mechanics of the trachea and behaviour of its slowly adapting stretch receptors. J. Physiol., 286:577.

26. Sant'Ambrogio, G. and Mortola, J.P., 1977, Behavior of slowly adapting stretch receptors in the extrathoracic trachea of the dog, Respir. Physiol., 31:377.

27. Schoener, E.P. and Frankel, H.M., 1972, Effect of hyperthermia and Pa_{CO_2} on the slowly adapting pulmonary stretch receptors, Am. J. Physiol, 222:68.

28. Sant'Ambrogio, G., Miserocchi, G. and Mortola, J. 1974, Transient responses of pulmonary stretch receptors in the dog to inhalation of carbon dioxide, Respir. Physiol., 22:191.

29. Bradley, G.W., Noble, M.I.M. and Trenchard, D., 1976, The direct effect on pulmonary stretch receptors discharge produced by changing CO_2 concentration in dogs on cardiopulmonary bypass and its action on breathing, J. Physiol., 261:359.

30. Kunz, A.L., Kawashiro, T. and Scheid, P., 1976, Study of CO_2 sensitive vagal afferents in the cat lung, Respir. Physiol., 27:347.

31. Mustafa, M.E.K.Y. and Purves, M.J., 1972, The effect of CO_2 upon discharge from slowly adapting stretch receptors in the lung of rabbits, Respir. Physiol., 16:197.

32. Bystrzycka, E.K. and Nail, B.S., 1980, CO_2 sensitivity of stretch receptors in the marsupial lung, Respir. Physiol., 39:111.

33. Bartlett, D. and Sant'Ambrogio, G., 1976, Effects of local and systemic hypercapnia on the discharge of stretch receptors in the airway of the dog, Respir. Physiol., 26:91.

34. Elftman, A.F., 1943, The afferent and parasympathetic innervation of the lungs and trachea of the dog, Am. J. Anat., 72:2.

35. During, von, M., Andres, K.H. and Iravani, J., 1974, The fine structure of the pulmonary stretch receptor in the rat, 2. Anat. Entwicklungsgesch., 143:215.

36. Mortola, F., Sant'Ambrogio, G. and Clement, M.G., 1975, Localization of irritant receptors in the airways of the dog, Respir. Physiol., 24:107.

37. Sant'Ambrogio, G., Remmers, J.E., deGroot,, W.J., Callas, G. and Mortola, J.P., 1978, Localization of rapidly adapting receptors in the trachea and main stem bronchus of the dog, Respir. Physiol., 33:359.

38. Mills, J.E., Sellick, H. and Widdicombe, J.G., 1970, Epithelial irritant receptors in the lungs. in: "Breathing," Edited by R. Porter, Hering-Breuer Centennial Symposium, Churchill, London.

39. Sampson, S.R. and Vidruk, E.H., 1975, Properties of 'irritant' receptors in canine lung, Respir. Physiol., 25:9.

40. Vidruk, E.H., Hahn, H.L., Nadel, F.A. and Sampson, S.R.,
     1977, Mechanisms by which histamine stimulates
     rapidly adapting receptors in dog lungs, J. Appl.
     Physiol.: Respirat. Environ. Exercise Physiol.,
     43:394.

41. Camporesi, E.M., Mortola, J.P., Sant'Ambrogio, F. and
     Sant'Ambrogio, G., 1979, Topical anesthesia of
     tracheal receptors, J. Appl. Physiol.: Respirat.
     Environ. Exercise Physiol., 47:1123.

42. Das, R.M., Jeffery, P.K. and Widdicombe, J.G., 1978, The
     epithelial innervation of the lower respiratory tract
     of the cat, J. Anat., 126:123.

43. Das, R.M., Jeffery, P.K. and Widdicombe, J.G., 1979, Exper-
     imental degeneration of intra-epithelial nerve fibres
     in cat airways, J. Anat., 128:259.

44. Jeffery, D. and Reid, C., 1973, Intra-epithelial nerves in
     normal rat airways: a quantitative electron
     microscopic
     study, J. Anat., 114:35.

45. Coleridge, H.M. and Coleridge, J.C.G., 1977, Impulse activ-
     ity in afferent vagal C-fibres with endings in the
     intrapulmonary airways of the dogs, Respir. Physiol.,
     29:143.

46. Paintal, A.S., 1969, Mechanism of stimulation of type J
     pulmonary receptors, J. Physiol., 203:511.

47. Kaufman, M.P., Coleridge, H.M., Coleridge, J.C.G. and
     Baker, D.G., 1980, Bradykinin stimulates afferent
     vagal C-fibres in intrapulmonary airways of dogs, J.
     Appl. Physiol.: Respirat. Environ. Exercise, 48:511.

48. Widdicombe, J.G. and Nadel, J.A., 1963, Reflex effects of
     lung inflation on tracheal volume, J. Appl. Physiol.,
     18:681.

49. Widdicombe, J.G., 1966, The regulation of bronchial
     calibre, in: "Respiratory Physiology," C.G. Caro,
     ed., Williams and Wilkins, Baltimore.

50. Davies, A., Sant'Ambrogio, F.B. and Sant'Ambrogio, G.,
     1980, Control of postural changes of end expiratory
     volume (FRC) by airways slowly adapting mechano-
     receptors, Respir. Physiol, in press.

51. Bartoli, A., Cross, B.A., Guz, A., Jain, S.K., Noble,
     M.I.M. and Trenchard, D.W., 1974, The effect of
     carbon dioxide in the airways and alveoli on
     ventilation: A vagal reflex studied in the dog, J.
     Physiol., 240:91.

52. Fillenz, M. and Widdicombe, J.G., 1972, Receptors of the
     lung and airways, in: "Handbook of Sensory
     Physiology," E. Neil, ed., Springer-Verlag, Berlin,
     Heidelberg, New York.

53. Knowlton, G.C. and Larrabee, M.G., 1946, A unitary analy-
     sis of pulmonary volume receptors, A. J. Physiol.,

147:100.

54. Glogowska, M., Richardson, P.S., Widdicombe, J.G. and
    Winning, A.J., 1972, The role of the vagus nerves,
    peripheral chemoreceptors and other afferent pathways
    in the genesis of augmented breaths in cats and
    rabbits, Respir. Physiol, 16:179.

55. Mills, J., Sellick, H. and Widdicombe, J.G., 1969, The
    role of lung irritant receptors in respiratory
    responses to multiple pulmonary embolism,
    anaphylaxis, and histamine induced
    bronchoconstriction, J. Physiol., 203:337.

56. Koller, E.A. and Ferrer, P., 1973, Discharge patterns of
    the lung stretch receptors and activation of
    deflation fibres in anaphylactic bronchial asthma,
    Respir. Physiol., 17:113.

57. Pontoppidan, H. and Beecher, H.K., 1960, Progressive loss
    of protective reflexes in the airway with the advance
    of age, J.A.M.A., 174:2209.

58. Coleridge, H.M., Coleridge, J.C.G., Baker, D.G., Ginzel,
    K.H. and Morison, M.A., 1978, Comparison of the
    effects of histamine and prostaglandin on afferent
    C-fiber endings and irritant receptors in the
    intrapulmonary airways, Adv. Exp. Med. Biol., 99:291.

59. Barry, D.T., 1913, Afferent impressions from the respira-
    tory mechanisms, J. Physiol. 45:473.

60. Banister, J., Fegler, G. and Hebb, C., 1949, Initial res-
    piratory responses to the intratracheal inhalation of
    phosgene or ammonia, Quart. J. Exptl. Physiol.,
    35:233.

61. Cromers, S.P., Young, R.H. and Ivy, A.C., 1933, On the
    existence of afferent respiratory impulses mediated
    by the stellate ganglia, Amer. J. Physiol., 104:463.

62. Widdicombe, J.G., 1954, Respiratory reflexes from the
    trachea and bronchi of the cat, J. Physiol., 123:55.

63. Kostreva, D.R., Hopp, F.A., Zuperku, E.J., Igler, F.O.,
    Coon, R.L. and Kampine, J.P., 1978, Respiratory
    inhibition with sympathetic afferent stimulation in
    the canine and primate, J. Appl. Physiol.: Respira.
    Environ. Exercise Physiol., 44:718.

64. Rao, S.V., Sant'Ambrogio, F.B. and Sant'Ambrogio, G.,
    1979, Respiratory responses to tracheal distention,
    The Physiologist, 22:105.

D I S C U S S I O N

LECTURER: Sant'Ambrogio   CHAIRMAN: Bonsignore

JEFFERY: One interesting observation we made when looking at
                the developing lung in the rat, and we studied the
                rat both prenatally and postnatally at 5 days, 12
                days and 19 days, and found that the nerves supplying
                the epithelium did not appear in any significant
                number until 12 days; they were not there at 5, and
                then they increased in number up to 19 days.   I
                wonder if you could comment on your graph with regard
                to the response of the rapidly adapting receptor
                before the age of one year in the dog.

SANT'AMBROGIO: Our observation in dogs did not refer to new-born
                dogs; all of them were several months old. So, I
                cannot use our data to relate with your observation.
                In terms of general responses, like spinal reflexes
                or reaction times, you find in the literature that
                the response to a shining light or to the ring of a
                bell or something like that, is much greater in very
                young animals, new-born animals, then it goes down
                within very few weeks, and then it goes back up at
                maturity. We have seen that several years ago for
                the spinal reflex and it seems to be true also for
                reaction times, so it is a general property. I do
                not know if that is interpretable in terms of
                peripheral information.

JEFFERY: Yes, that is complex. My second point is that, very
                interestingly, you showed the diminution in terms of
                number of rapidly adapting receptors with age in the
                dog. Is this actually a diminution in the number, or
                is it number per unit area, which might actually be
                due to the increase in surface area as the animal
                grows while the number of fibres remain constant?

SANT'AMBROGIO: These animals were not growing, they were already
                adult animals and I do not think that the surface
                area was increasing; in three year old animals you
                have about the same surface area than in one year
                old. One interesting consideration we can make is
                that if you consider the number of afferent neurons
                in animals of different size you find about the same
                number. In the mammal you always have something like
                30,000 neurons, in rats, cats, dogs, also goats.

That means that the concentration of endings should increase with the decrease in size. It would be interesting to analyse this in terms of development.

FABBRI: I would like to know whether we can think of reflexes coming from the main bronchi and determining a reduction in the vascularisation of the small circulation. This question is of practical interest, as we, among others, have seen patients with a neoplasia involving the main bronchi showing a reduction in the vascularisation in the homolateral lung in the absence of atelectasis and then, after surgery, the neoplasia did not infiltrate or stenose the homolateral pulmonary artery.

SANT'AMBROGIO: So that it could not be a local mechanical factor. I have no data to answer your question. In fact I have some negative data on this point. The experiments where we measure pressure in the systemic circulation and in the pulmonary artery, we distend, not the main bronchus but the trachea near the carina and presumably also the main right bronchus. We do not see pressure variations in the systemic circulation nor in the pulmonary circulation, in anaesthetised dogs. Perhaps someone else may have an idea of any reflex which might affect the pulmonary circulation.

BONSIGNORE: Have you measured the pressure in the pulmonary artery? It was normal. And did you check perfusion by scintigraphy? Yes? and did you find a normal perfusion? A reduction? Then in these cases it might be that there is compression on the pulmonary artery due to a metastasis.

FABBRI: No, the patients have been operated upon and we did not find it. These were young subjects.

BONSIGNORE: So the mechanical factor is excluded.

FABBRI: Yes, including atelectasis of the bed downstream.

SANT'AMBROGIO: And there was no hypoxia.

FABBRI: No

BONSIGNORE: There are many factors, either mechanical ones or reflex ones.
                   I think we have used all the time, so I shall bring this session to a close.

MINERALOGICAL AND BIOCHEMICAL ANALYSIS OF LUNG WASHING

FLUID FROM PATIENTS EXPOSED TO ASBESTOS

Jean Bignon, Marie-Claude Jaurand and Patrick Sebastien

Service de Pneumologie, ERA CNRS et LEPI
40, Ave de Verdun, 94010 Créteil cedex

INTRODUCTION

Recently, it has been shown in the baboon that broncho-alveolar lavage (BAL) is an efficient procedure for the recovery of radioactive particles deposited in the intra-alveolar compartment (1). For this reason, a prospective study was decided in order to investigate the possibility of assessing by means of this technique, the asbestos fiber content in the alveolar spaces of the human lung. The asbestos fibers recovered by BAL were quantitatively and qualitatively studied in order to see if this technique can help in the diagnosis of asbestos related diseases and if the fibers recovered are valuable indicators compared to sputum of past asbestos exposure and/of parenchymal asbestos burden.

In the field of asbestos, in vitro studies have shown that phagocytosis of asbestos fibers by alveolar macrophages (AM) induces extracellular enzyme release (2,3). Moreover, after intratracheal injection of asbestos fibers into the rabbit, the enzyme content of AM has been found modified (4). Since BAL can also be used to recover cells and proteins from the alveolar spaces (5,6,7), it also appeared interesting to determine the enzymatic contents of lung washing fluid (LWF) and alveolar macrophages (AM) from patients with various degrees of asbestos exposure related to the intra-alveolar fiber content as assessed by quantitative mineralogical measurements.

MATERIAL AND METHOD

Subjects

61 cases were studied. Of those, 17 cases had heavy asbestos
exposure (insulation workers, asbestos-cement plant workers,
asbestos-textile plant workers) ; 18 cases had moderate asbestos
exposure (boiler fitter, plumber, glass blower, occasional
asbestos sheet sawer, automobile workers, dental personnel) ;
20 corresponded to blue collar workers (industrial workers
without definite known asbestos exposure) ; and 6 were white
collar workers without known industrial exposure, which included
4 volunteer subjects. Table 1 gives the distribution of cases
on one hand according to respiratory diseases, and on the other
hand according to occupational exposure.

Elsewhere, cytological and biochemical studies have been
carried out on the LWF recovered, on one hand in 14 cases with
asbestos exposure (10 cases with heavy and 4 cases with moderate
exposure) (Table 1) and on the other hand in 9 controls without
any dust exposure or pulmonary pathological feature. The pathological features of exposed subjects were mainly moderate or
light fibrosis. Concerning the smoking habit, the proportion of
smokers was similar in the 2 groups (10 out of 14 in exposed
subjects and 6 out of 9 in controls).

Broncho-alveolar lavage procedure

The fiber optic bronchoscopy was performed on a sitting
subject after local anesthesia of the respiratory tract. The
fiber optic bronchoscope was positionned in a segmental bronchus
of the right lower lobe or the lingula. Fifty milliliters of
filtered saline (0.45 µm nuclepore size) were infused and
recovered by gravity (8). The wash was repeated 3 to 4 times
(a total of 150 to 200 ml of saline was injected). An aliquot
of 10 ml of LWF was submitted to mineralogical analysis.
The remaining was centrifuged for 15 min at 300 g at 4°C in
order to separate the cells from the supernatant for cytological
and biochemical studies.

Mineralogical analysis

Sodium hypochlorite was added to each aliquot of LWF in
order to destroy organic material. Particles were recovered
on a nuclepore filter membrane previously coated with carbon
layer. Then the inorganic particles were transferred directly
on microscopic electronic grids as previously described (9).
The grids were scanned under the light microscope for ferruginous
bodies and under an analytical transmission electron microscope
for TEM size fibers. Asbestos fibers were identified by their

Table 1. Partition of cases according to respiratory diseases and occupational exposure

Occupational Exposure	RESPIRATORY DISEASES					
	Fibrosis	Bronchial Carcinoma	Pleural Plaques	Bronchitis	Miscellaneous	Total
Asbestos Heavy	16 (10)*			1		17
Asbestos Moderate	7 (2)*		5 (1)*	5 (1)*	1	18
Blue Collar	7	6		2	5	20
White Collar		1		1	4+	6
Total	30	7	5	9	10	61

* Number of cases studies both for mineralogical study and for biochemical study

+ Volunteers

morphological features, electron diffraction pattern and elementary chemical analysis. By measuring and counting each identified
fiber, numerical concentrations, mineralogical type, mean length
and diameter were assessed. In 31 subjects, one sputum sample
was collected in the morning for comparative mineralogical analysis according to the same procedure. The quantitative results
refer to the total volume of LWF recovered and to the sputum.

Cytological study

     The cells were counted using an haemtocytometer. 5 ml of LWF
were centrifuged using a cytocentrifuge (Elliot Shandon, London).
The slides were stained by the Maygrunwald Giemsa method.

Biochemical analysis

     After centrifugation of the LWF, biochemical analysis was
performed on one hand, on the supernatant and on the other hand,
on the AM. AM were isolated from other free cells by adherence
on a Falcon flask ; the cells suspended in Hank's solution were
allowed to adhere for 1 hour at 37°C in Falcon flasks (respectively 1.5 and 2.5 x 10^6 cells per flask). Then non adherent cells
were removed and counted, allowing evaluation of the number of
adherent cells. For enzymatic analysis, AM remaining on the
flask were lysed with 0.2 % v/v Triton X-100 (Sigma) in 1 ml
of Hank's solution. For the cells' content, 1 ml of 0.5 N NaOH
was used in order to lyse the cells.

     Ten milliliters of the supernatant were concentrated to 1 ml
by ultrafiltration using Amicon Minicon B concentrator. The
enzyme content of AM was determined as previously described (4).
The activity of the following enzymes was determined : LDH and
acid phosphatases (Boehringer test, Boehringer Laboratories,
Mannheim, Germany), N-acetyl-β-D-glucosaminidase (10),
β-glucuronidase (11) and β-galactosidase (12). The protein content was determined by the Lowry's procedure (13).

     Statistical differences were determined using Wilcoxon's
test (not significant $p_{2\alpha} > 0.05$).

RESULTS

Mineralogical study

     The values from BAL have been compared to the values from
sputum in cases with heavy exposure and in cases with moderate
exposure (Fig. 1). Ferruginous bodies (FB) and TEM size fibers were
found in all LWF samples and in almost all sputum samples from
patients heavily exposed to asbestos. In contrast, FB and/or

TEM size fibers were present in LWF and sputum in about 40 percent of the cases with moderate asbestos exposure. Fig. 1 shows that in general LWF samples were more frequently positive for asbestos material than sputum samples. The fiber yield by BAL has been compared to the fiber yield of the collection of one sputum in the heavy asbestos exposed group. In Fig. 2, the number of fibers per LWF is plotted against the number of fibers per sputum ; fibers recovered by BAL were more numerous than fibers found in one sputum sample. This fact is especially noticeable for FB. Elsewhere, the ratio in percent of FB to TEM size fibers (TSF) was 10 times higher in LWF than in sputum. TSF were shorter in LWF (mean length 3 μm) than in sputum (5 μm). Moreover, for cases with exposure to both chrysotile and amphibole, the proportion of amphibole type fibers was higher in LWF than in sputum (Table 2).

The numbers of FB and TEM size fibers in LWF are graphically shown in Figures 3 and 4 according to the degree of asbestos exposure. There is an overlapping of the number of FB (Fig. 3) between heavy (18 to 2.5×10^6 per lavage) and moderate asbestos exposure groups (0 to 473 per lavage) but no FB were found in LWF from blue and white collar cases without asbestos exposure. In contrast, for TEM size fibers (Fig. 4), there was no overlapping (except for one plot) between heavy (10^5 to 10^8) and moderate group (10^4) while there was some overlapping between the moderate asbestos exposure group and the blue and white collar group.

Cytological study

The data concerning the number of recovered cells indicated great variations from one case to another (Table 3) : nevertheless, the mean value was decreased in subjects exposed to asbestos as compared to control subjects, but the difference was not significant. The Maygrunwald Giemsa staining showed that the formulas for both the heavy and the moderate exposed subjects were not significantly different from the control group. However, there was a trend to a greater proportion of polymorphonuclear leucocytes in the exposed group.

Biochemical study

Protein content and enzymatic activities have been compared in the 3 groups of cases classified according to asbestos exposure : control, moderate and heavy asbestos exposure.

Protein content of alveolar macrophages (Tables 4 and 5). By comparison with control cases, the total cell protein content was increased in cases associated with heavy asbestos exposure, ($p_{2\alpha} < 0.02$) ; the N-acetyl-β-D-glucosaminidase was increased

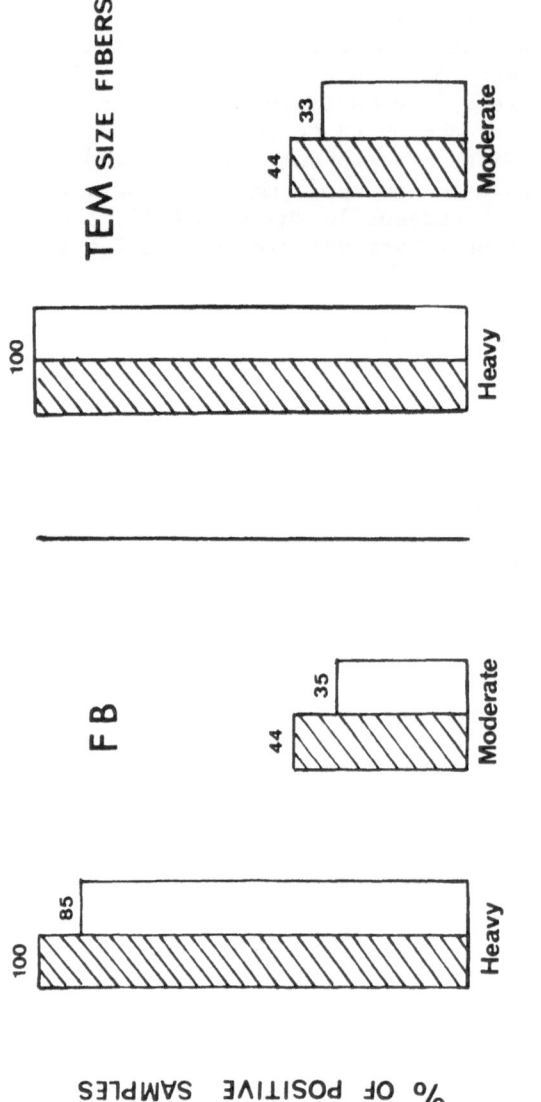

ASBESTOS EXPOSURE GROUPS

Figure 1. Comparisons between LWF ▨ and sputum ☐ of past asbestos exposure : as external indicators of past asbestos exposure : percentage of positive samples for ferruginous bodies (FB) and TEM size fibers in cases with heavy or moderate exposure

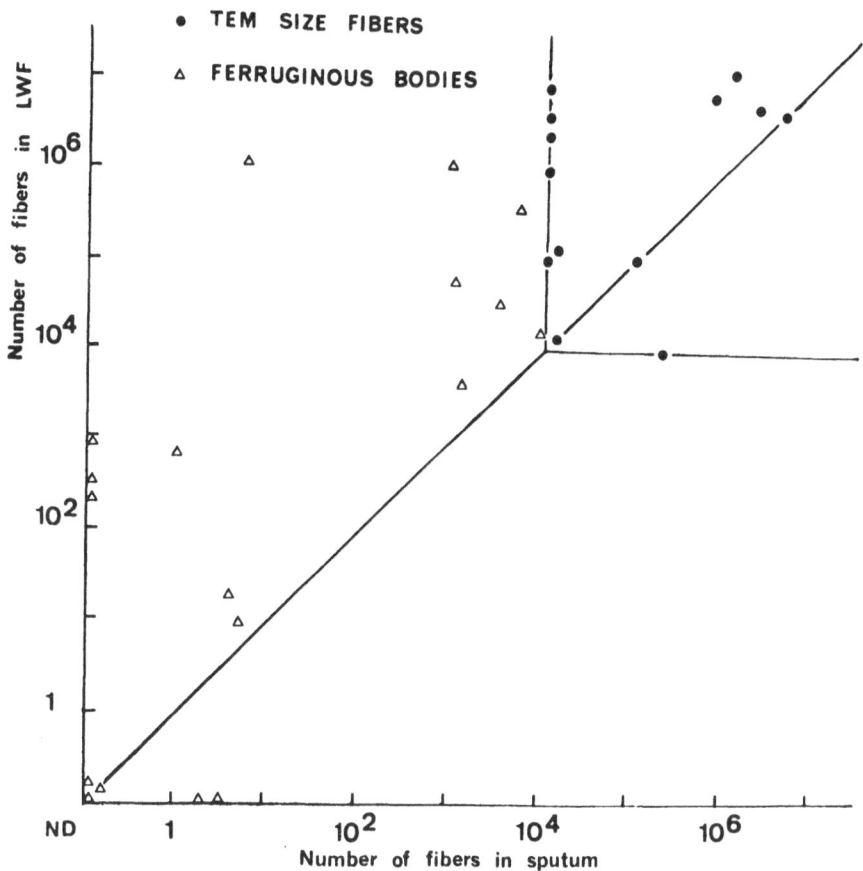

Figure 2. Correlation of the fiber count (transmission electron
          microscopic size fibers and ferruginous bodies) in
          sputum with the counts in lung washing fluids (LWF)
          samples for cases with heavy asbestos exposure.

Table 2. Asbestos material : ferruginous bodies (FB) and TEM size fibers (TSF) in LWF versus sputum in case heavily exposed to asbestos

Samples	FB	TSF	$\frac{FB}{TSF}$ x 100	ASBESTOS FIBERS		
				% amphiboles TSF	Mean Length TSF*	Mean Diameter TSF*
LWF (whole washing recovered)	3.10^4	5.10^6	6	88	3	.13
SPUTUM	7.10^2	1.10^5	.7	65	5	.16

* Expressed in micrometer

Table 3. LWF cytological content

	Controls	Moderate Asbestos Exposure	Heavy Asbestos Exposure
Number of cases	9	4	10
Cells / ml x 10^3	457 (34–850)	360 (37–900)	147 (40–291)
Total of cells x 10^6	35.9 (2–90)	30 (2–90)	11.6 (3–20)
Alveolar macrophages %	82	84	80
Polymorphonuclear Leucocytes %	5	9	11
Lymphocytes %	13	7	9

Table 4. Protein content and enzymatic activities of AM per mg of protein

		Protein*	LDH*	Acid Phosphatase+	β-Glucuronidase**	N-acetyl-β-D-Glucosaminidase**
Control	n++	9	9	9	6	8
	mean	318	445	607	607	10 143
Moderate	n++	3	3	3	2	2
	mean	247	568	1223	526	20 073
Heavy	n++	8	8	8	6	8
	mean	460	680	725	336	13 340

* Expressed as µg per 10^6 cells

+ Expressed as mU per mg of protein

** Expressed as n mole of product formed per hour per mg of protein

++ Numer of cases

Table 5. Enzymatic activities of AM per million of cells

		LDH*	Acid Phosphatase*	β-Glucuronidase*	N-Acetyl-β-D-Glucosaminidase+
Control	n**	9	9	6	8
	mean	141	176	197	3330
Moderate	n**	3	3	2	3
	mean	144	298	143	5047
Heavy	n**	9	8	6	8
	mean	296	309	147	6105

*   Expressed as mU per million of cells

+   Expressed as n moles of product formed by hour per 10^6 cells

** Number of cases

only when expressed as activity per million of cells ($p_{2\alpha}$ = 0.05).
By contrast, the activity of β-glucuronidase was not different
from control cases in the asbestos exposed groups. Concerning
smoking habits, acid phosphatase was the only enzyme showing an
increase in smokers (Table 6).

Supernatant analysis : There was no significant discrepancies in
protein content and enzymatic activities in the LWF supernatant
among the different groups, either relating to asbestos exposure
or to smoking habits. However, there was a slight increase of
N-acetyl-β-D-glucosaminidase activity per mg of proteins in cases
with heavy exposure as compared to the controls (Table 7).

DISCUSSION

The present investigation is the first systematic evaluation
of the value of two external indicators (sputum and LWF) for
evaluating the asbestos pulmonary retention. Thus, our data,
using either light microscopy or transmission electron microscopy
have demonstrated that both sputum and LWF were reliable for
the quantitative estimation of fibers in retention in the lung.
However, the proportion of positive results for FB as well as
for TEM size fibers with both types of samples was higher in
cases with heavy exposure (almost 100 per cent) than in cases
with moderate exposure (about 40 per cent). Nevertheless, the
comparison of both indicators showed that LWF was more sensitive
for detecting a past asbestos exposure, since it yielded a
greater amount of fibers than one sample of sputum. It is possible
that the sensitivity would even have been increased if the mineralogical study had been carried out in a LWF sample having a
larger volume (20 or 50 ml). Thus, BAL which is not a very invasive
technique can help in the diagnosis of asbestos related diseases.

Light microscopy alone can provide, unexpensively and
without consuming time, very useful data for the diagnosis of
asbestos related diseases. Indeed counting only ferruginous bodies
by light microscopy in LWF appeared more reliable to distinguish
those cases with asbestos exposure from those without because of
the absence of overlapping between these two groups. However,
the assessment of TEM size fibers by electron microscopy might
be very helpful for the study of the dose-response relationship
since there was no overlapping between the heavy and moderate
exposed groups. Moreover, electron microscopy is the only method
to obtain data on the fiber types and sizes.

The results obtained here in the heavily exposed groups can
be used to understand some basic points relating to the clearance
and pulmonary retention of asbestos fibers.

Table 6. Acid phosphatase activity of AM

	Control		Moderate Exposure		Heavy Exposure	
	n[*]	mean	n[*]	mean	n[*]	mean
No smokers	3	86[+]	0	–	2	173[+]
	3	230[**]	0	–	2	319[**]
Smokers	6	221[+]	3	298[+]	6	310[+]
	6	796[**]	3	1228[**]	6	725[**]

[*] Number of cases

[+] Expressed as mU per million of cells

[**] Expressed as mU per mg of protein

Table 7. Protein content and enzymatic activities in LWF supernatant

		Protein*	LDH		Acid Phosphatase		β-Glucuronidase		β-Galactosidase		N-Acetyl-β-D-	
			mg +	ml **	mg +	ml ***	mg ++	ml ***	mg ++	ml ***	mg ++	ml ***
Control	n+++	9	9		9		7		9		9	
	mean	124	168	17	13	1.2	5.3	.5	83	8	717	74
Moderate	n+++	4	4		3		4		4		4	
	mean	85	164	13	7.7	.8	9	.9	38	3.5	804	59
Heavy	n+++	10	10		10		7		10		9	
	mean	99	105	14	8.8	.6	3.8	.4	43	3	1015	57

*   μg per ml

+   mU per mg of protein

**   mU per ml

++   n mole of product formed per hour per mg of protein

***   n mole of product formed per hour per ml

+++   number of cases

Table 8. Number and size of asbestos fibers in various pulmonary compartment

FIBER COUNTS		SIZES	
Method for evaluation	Numbers for the whole lung	Mean length (μm)	Mean diameter (μm)
Concentration per cu cm of digested lung tissue x 5000	2.10^{10}	4.9	.13
Total dusts in LWF recovered from one segment x 20 segments	2.10^{8}	3.3	.13

1/ The comparison of fibers in LWF and in sputum of workers heavily exposed to mixed type fibers (amphiboles + chrysotile) allowed
the following comments about the clearance and retention of fibers
according to their type. The larger proportion of chrysotile
fibers in sputum than in LWF suggests that the bronchial clearance
ce might be greater for chrysotile fibers than for amphiboles
type fibers. This hypothesis is in agreement with the findings
of Wagner and co-workers (14) namely that in comparison to
amphibole type fibers, a small amount of chrysotile was retained
in the rat lung after inhalation. A faster bronchial clearance
of chrysotile fibers than amphibole fibers could explain why
a larger proportion of amphibole type fibers has been found
in the parenchyma of human lungs (16, 17) ; in cases with definite
past exposure to a mixture of chrysotile and amphibole type
fibers, that the proportion of amphibole type fibers was always
greater than that of chrysotile type might explain why there were
10 times more FB in LWF than in sputum sample. Indeed, amphibole fibers are more easily coated than chrysotile inside the lung
(18).

2/ Table 8·shows, for the heavily exposed cases, an attempt to
compare the number and size parameters of asbestos fibers recovered from the alveolar spaces by BAL to those found in lung
parenchyma after chemical digestion and microfiltration (the
fibers correspond to intra-tissular + intra-alveolar fibers).
The fiber count is estimated for the whole lung : for the intra-
alveolar fibres by multiplying the number of fibers revovered
from one segment by 20, and for the intra-parenchymal fibers
by integrating the concentration per cubic centimeter of parenchyma samples obtained from comparatively exposed cases (17) to a
standard volume of 5000 ml. From these data, the fraction of
fibers recovered from the alveolar spaces by BAL was found to
represent only 1 per cent of all the fibers retained in lung
parenchyma. If we assume about 300 millions alveoli for an adult
lung (19), that would mean that the BAL recovered about one fiber
from each alveolus in cases with heavy asbestos exposure. This
explains why there were so few fibers found in cases with moderate
asbestos exposure when the LWF recovered from one pulmonary
segment was analysed by electron microscopy. This comparison
showed also a significant mean length difference between the
parenchymal fibers (4.9 μm) and the intra-alveolar fibers
(3.3 μm). These findings agree with the experimental data obtained in rats by several workers where the longest fibers were
found in intra-parenchymal tissue, either after inhalation or
after intra-pleural injection (20, 21, 22).

The cytological studies have shown a decrease of the number
of recovered AM in subjects exposed to asbestos, particularly
significant in cases with heavy asbestos exposure. In previous

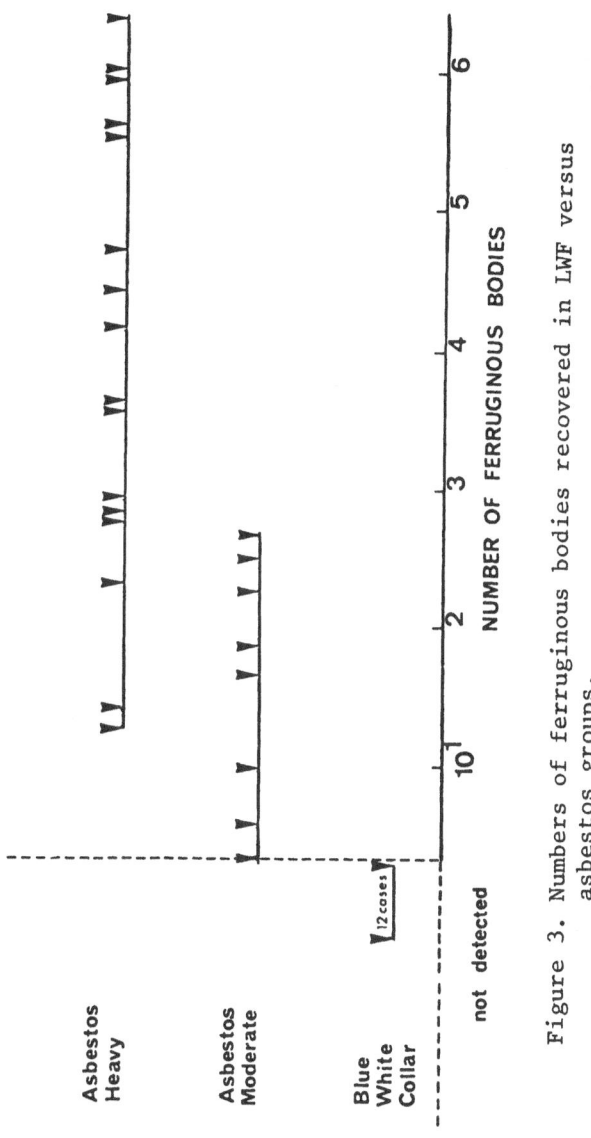

Figure 3. Numbers of ferruginous bodies recovered in LWF versus asbestos groups.

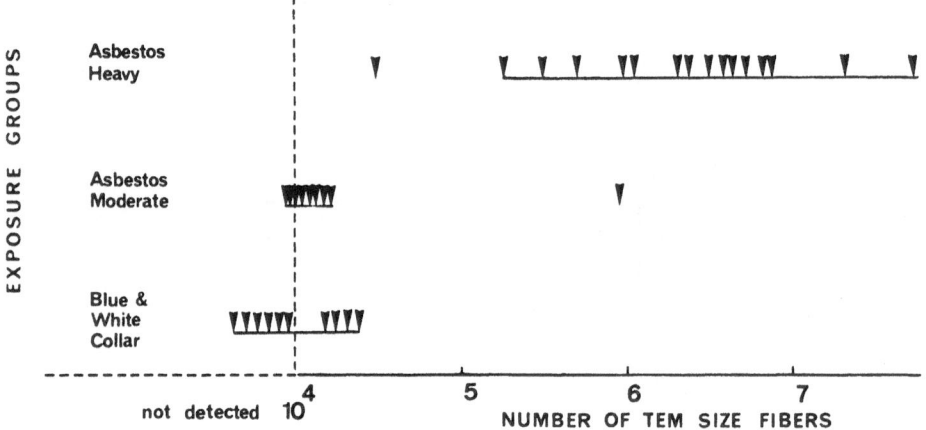

Figure 4. Numbers of TEM size fibers recovered in LWF versus
          asbestos groups

experiment, such a decrease was also observed in rabbits after
intra-tracheal injection of chrysotile fibers (4). This may be due
to the cytolysis of alveolar free cells, to a sequestration of the
AM in the alveoli, or to a reduction of AM renewal. It must be
noted that the cell viability, as measured by the dye exclusion,
was not modified in the different groups. The percentage of the
various types of cells was not different in the three exposure
groups, altough some cases with heavy exposure had lung fibrosis.
Nevertheless, a previous study (23) has shown that in patients
with asbestos related lung fibrosis, there was a cytological
formula close to that of idiopathic lung fibrosis with a decreased proportion of AM compensated by a significant increase in
the proportion of polymorphonuclear leucocytes (6). In this
study, the subject had only moderate or light fibrosis explaining
the discrepancy.

The biochemical study has shown that in heavily exposed
cases, the LDH activity in AM increased when expressed as activity per 10^6 cells or per mg of proteins. Moreover, in every group,
the activity increased with the amount of protein per cell
(23). That suggests an increase of LDH enzyme synthesis
rather than a stimulation. By contrast, the N-acetyl-β-glucosami-
nidase does not show this pattern; there was no relation between enzymatic activity and the total protein cell content. The increase of enzymatic activity of acid phosphatase in
AM according to smoking habits agrees with published observations
in contrast to results obtained on AM, the activity of acid
phosphatase in supernatant fraction was not different in smokers
as compared with non smokers.

In the supernatant fraction of the LWF, enzymatic lysosomal
activities were not found to have increased in exposed patients,
as could have been deduced from in vitro experiments (2, 3)
showing a release of enzymes by AM phagocytosing asbestos fibers.
This failure could be due to an inactivation of the enzyme either
by inhibitors or by the activity of proteolytic enzymes. There
was no longer correlation between the enzymatic activities and
the asbestos fibers content in LWF, except for a slight increase
of N-acetyl-β-D-glucosaminidase. Moreover, there was no correlation between these enzymatic activities and lung fibrosis.
The results have to be taken with caution because of the difficulty in expressing enzymatic activities ; the expression per
ml of supernatant per mg of proteins depends on all protein
content including non cell proteins such as plasma proteins.
However we assume that the amount of fibers used in the in vitro
experiments is much higher than the amount found in human LWF.
Then even if we accept the hypothesis of enzyme release, it is
possible that the amount of enzyme release is lower than the
sensitiveness of the measurement method.

REFERENCES

1. D. Nolibe, H. Metivier, R. Masse and J. Lafuma, Therapeutic
    effect of pulmonary lavage in vivo, after inhalation
    of insoluble radioactive particles. "Inhaled Particles"
    IV, W.H. Walton ed., Pergamon Press (1975)
2. E.G. Beck, P.F. Holt and N. Mandjlovic, Comparison of effects
    on macrophage cultures of glass fibers, glass powder
    and chrysotile asbestos. Br. J. Industr. Med.,
    29 : 280 (1972).
3. P. Davies, A.C. Allison, J. Ackerman, A. Butterfield and
    S. Williams, Asbestos induces selective release of
    lysosomal enzymes from mononuclear phagocytes. Nature,
    251 : 423 (1974).
4. M.C. Jaurand, J. Bignon, A. Gaudichet, L. Magne and A. Oblin,
    Biological effects of chrysotile after SO_2 sorption.
    II. Effects on alveolar macrophages and red blood cells.
    Environ. Res. 17 : 216 (1978).
5. J.O. Harris, G.N. Olsen, J.R. Castle and A.S. Maloney,
    Comparison of proteolytic enzyme activity in pulmonary
    alveolar macrophages and blood leukocytes in smokers
    and nosmokers. Amer. Rev. Resp. Dis., 111 : 579 (1975).
6. H.Y. Reynolds and H.H. Newball, Analysis of proteins and
    respiratory cells obtained from human lungs by bronchial
    lavage. J. Lab. Clin. Med., 84 : 559 (1974).
7. G.A. Warr, R.R. Martin, P.M. Sharp and R.D. Rossen, Normal
    human bronchial immunoglobulins and proteins effects of
    cigarette smoking. Amer. Rev. Resp. Dis., 116 : 25
    (1977).
8. F. Basset, P. Soler, M..C. Jaurand and J. Bignon, Ultrastruc-
    tural examination of broncho-alveolar lavage for diag-
    nosis of pulmonary histiocytosis X. Thorax, 32 : 303
    (1977).
9. P. Sebastien, M.A. Billon, X. Janson, G. Bonnaud and J.
    Bignon, Utilisation du microscope électronique à
    transmission (MET) pour la mesure des contaminations
    par l'amiante. Arch. Mal. Prof., 39 : 229 (1978).
10. J.W. Wollen, R. Heyworth and P.G. Walker, Studies on glyco-
    saminidase. Biochem. J., 78 : 111 (1961).
11. P. Talalay, W.H. Fishman and C. Huggins, Chromogenic subs-
    trates. II. Phenolphtalein glucuronic acid as substrate
    for the assay of glucuronidase activity. J. Biol. Chem.,
    166 : 757 (1946).
12. J. Conchie, J. Findlay and G.A. Lewy, Mammalian glycosidase.
    Distribution in the body. Biochem. J., 78 : 111 (1961).
13. O.H. Lowry, N.J. Rosebrough, A.L. Farr and R. Randall,
    Protein measurement with the Folin phenal reagent. J.
    Biol. Chem. 193 : 265 (1951).

14. J.C. Wagner, G. Berry, J.W. Skidmore and V. Timbrell, The
     effects of the inhalation of asbestos in rats. Br. J.
     Cancer, 29 : 252 (1974).

15. A. Morgan, J.C. Evans and A. Holmes, Deposition and clearance
     of inhaled fibrous minerals in the rat. Studies using
     radioactive tracer techniques. "Inhaled Particles"
     IV. W.H. Walton ed., Pergamon Press, Oxford (1977).

16. L. Le Bouffant, S. Bruyere, J.C. Martin, G. Tichoux and
     C. Normand, Quelques observations sur les fibres
     d'amiante et les formations minérales diverses rencon-
     trées dans les poumons asbestosiques. Rev. Fr. Mal.
     Resp., 4 : 121 (1976).

17. P. Sebastien, A. Fondimare, J. Bignon, G. Monchaux,
     J. Desbordes and G. Bonnaud, Topographic distribution
     of asbestos fibres in human lung in relation to occupa-
     tional and non occupational exposure. "Inhaled Particles"
     IV. W.H. Walton ed., Pergamon Press, Oxford (1977).

18. A. Churg and M.L. Warnock, Analysis of the cores of ferrugi-
     nous (asbestos) bodies from the General Population.
     I. Patients with and without lung cancer. Lab. Invest.,
     3 : 280 (1977).

19. E.R. Weibel, Morphometry of the human lung. Heldeberg
     Springer Verlag (1963).

20. A. Morgan, R.J. Talboj and A. Holmes, Significance of fibre
     length in the clearance of asbestos fibres from the
     lung. Br. J. Industr. Med., 35 : 146 (1978).

21. J. Bignon, G. Monchaux, A. Hirsch, P. Sebastien and J.
     Lafuma, Human and experimental data of translocation
     of asbestos fibers through the respiratory system.
     Ann. N.Y. Acad. Sci., 30 : 745 (1979).

22. M.R. Leadbertter and M. Corn, Particle size distribution of
     rat lung residues after exposure to fibre glass dust
     clouds. Am. Industr. Hyg. Assoc. J., 33 : 511 (1972).

23. M.C. Jaurand, A. Gaudichet, K. Atassi, P. Sebastien and
     J. Bignon, Relationship between the number of asbestos
     fibres and the cellular and enzymatic content of broncho-
     alveolar fluid in asbestos exposed subjects. Bull.
     Europ. Physio-Pathol. Resp. (in press).

24. R.R. Martin, Altered morphology and increased acid hydrolose
     content of pulmonary macrophages from cigarette smokers.
     Amer. Rev. Resp. Dis., 107 : 596 (1973).

D I S C U S S I O N

LECTURER: Bignon   CHAIRMAN: Bonsignore

CANDURA: I would like to thank the speaker for his very
              interesting lacture. Before asking my question I
              would like to point out that the data on synergism,
              the interaction of a number of causes is of practical
              importance in occupational disease. For some years
              the occupational physician has been aware of this
              aspect because the worker is rarely exposed to one
              cause of disease, he is exposed to several, either at
              the same or at different times during his working
              life. My question is: you answered yesterday that
              the macrophage does not recognise silica particles
              because it is covered by a protein blanket; do you
              think that this is or is not in agreement with the
              old data of Pernis, according to which if quartz dust
              is washed with hydrofluoric acid the silicosis
              developed by the animal is much more massive than if
              the animal is exposed to quartz dust as such? This
              has very important practical implications, because we
              are in fact very careful when we expose our workers
              to quartz sand which has previously been leached.

BIGNON: Yes, there is a mechanism of recognition for quartz,
              but it is the same for all particles, as clearly
              shown by Allison. He demonstrated that when quartz
              was embedded by protein, the macrophage could
              recognise the quartz and thereafter there was a
              dysfunction of the lysosomal membrane. I think it is
              a very real model and we can apply it to all fibres
              but I have some difficulty in answering your
              question, because if you wash the fibres before
              inhaling them they will meet with some protein inside
              the alveolar space and then will be covered by the
              protein and phospholipid inside the alveolar space.
              I do not see exactly how we can protect people except
              by treating previously the particle in some way which
              can protect the reaction at the surface of the
              crystal against an attack on the plasma membrane.

HEATH: I would like to return to the question I asked
              yesterday. One can understand that in Turkey, where
              you have these isolated villages, there is a risk of
              pleural mesothelioma because people are born there
              and live all their lives. But I am just beginning to

wonder if volcanoes are as safe from this point of view as you might think. The telescope has just been put up on a volcano in Hawaii and dust samples analysed in Cardiff have been found to contain zeolite in the volcanic dust. Astronomers are going to be working on the summit of this volcano for a period of up to three years. Is that going to constitute a significant risk for the development of mesothelioma, and so, from a practical standpoint, does this mean to say that you are going to ask astronomers to be wearing masks all the time that they are doing their work? Just as a quick extension of this, volcanoes, of course, attract tourists throughout the world, and I suppose many of us in this room, most have been to the Canary Islands and places like this, which is a very dusty place; are tourists absolutely safe? With even a short exposure such as they might get, say a two or three weeks holiday?

BIGNON: You raise the question of the dose-response. This problem of volcanic dust exposure is a new one and will be discovered in many areas in the world. I was in a meeting in Napoli in November and it was said that in Italy there are many locations where zeolite has been found. I do not know if you live three years at the summit of a volcano you are liable to die from a mesothelioma or to die from an eruption of the volcano!

PRODI: Just a comment more than a question. This is going to become a problem because there are fibres also in construction material like gravel for the roads and things like that, so I think this concern for fibres is going to increase, besides the use in brake lining and so on. I think that the emphasis in this course on this problem is just right, especially for the synergistic effect. In addition the example you showed can be of importance for general epidemiology because radon is ubiquitous, especially indoors. With energy conservation problems, this concentration of radon indoors is going to increase. The dose to the lungs from alpha particles from radon is very high, so it can be delivered into hot spots in the respiratory tract, and this might explain some increase in incidence in particular bronchial degenerations.

CORRIN: I was very interested in the biodegradation of the chrysotyle and in particular the leaching of magnesium from the fibres in the first 24 hours, and this is presumably why in interpreting Wagner's data you emphasised his analysis, that chrysotyle dusting was chemical, the implication I presume being that he is failing to recognise it as chrysotyle because it has been chemically altered. You have recognised it by electron microscopy, I presume not by electron microprobe analysis; this is really a very elegant form of chemical analysis. The shape of the fibre is not readily apparent in sections. Presumably you used digests of the lung to distinguish between fibres. I have a second question - I wonder how you might explain the peripheral distribution of asbestos dust and therefore the peripheral distribution of asbestos related disease, and the development of mesothelioma.

BIGNON: To the first question concerning the data of Chris Wagner, I have only a hypothesis because they measured the silica content of lungs of animals. In the case of animals exposed to chrysotyle he found a low silica content and said that there was no chrysotyle left. It was an assumption point. To the second question I have no answer. Perhaps it is that the concentration of fibres in the periphery of the lung is due to the mechanical distribution of fibres during inspiration I do not know but it is clearly shown in the data of Morgan on animals. Perhaps it is related to the distribution of asbestosis, but in our study in cases with different diseases, we found that with severe asbestosis there was a greater concentration of fibres in the upper lobe even if the fibrosis was predominant in the lower lobe.

COHEN: My question is maybe in two or three parts. The first part was concerned with your mechanism of carcinogenesis, where you suggested that asbestos was not an initiator or it was not mutagenic in the Ames system. It is possible, based on what Allison proposed many years ago, that by causing cell lysis you release DNAase and DNAase could then affect DNA in target cells causing an alteration which you would not pick up by the method you used, and that would also be compatible with the fact that you did not find the fibres within the DNA, because it would not be needed by such a mechanism. The second question really is, has there been any work trying to show whether asbestos is a co-mutagen? In an analogous system to a co-carcinogen, there are also tests now to detect a co-mutagens.

BIGNON: I do not know a work showing that asbestos is a co-mutagen. To your first question, our result is only a suggestion, we do not know exactly the mechanism of carcinogenesis by fibres. In cats, after using radon, which is more an initiator than a promotor, it seems to indicate that it could work as a promotor - these findings are the first argument I gave.

COHEN: Yes, but it would be better to call it a co-carcinogen rather than a promotor, because a promotor by definition should not be carcinogenic itself.

BIGNON: Yes, it was an oversimplification for this lecture, I am sorry.

IMMUNOLOGY AND THE LUNG

Margaret Turner-Warwick

Cardiothoracic Institute

Fulham Road, London SW3 6HP

GENERAL OVERVIEW

It is reasonable to suppose that the survival of immunological responses in the phylogeny of man is due to their crucial role in protective immunity. Further, immunological responses, more usually associated with hypersensitivity tissue damaging disease states are fundamentally similar to protective ones. Indeed hypersensitivity can be viewed not as an independent and wholly undesirable aberrant reaction, but as an amplification of protective mechanisms.

If these statements are true then we should expect to observe immunological phenomena in the absence of disease, where antigen has primed the immunological defences but where further antigen has not yet triggered activation of various cells or systems to release tissue damaging mediators at an appropriate site. Thus in the study of disease, factors stimulating the release of inflammatory mediators becomes more important than the study of specifically primed systems per se. Further, observations on the immunological status in different compartments of the body at any one time may reveal very different information.

The consequence of the argument set out above is that we have to recognise and be much more critical about the compartment of the body that we have selected to study. Some compartments easily available are set out in Table 1.

It can be seen from this table that the cells and tissue responses observed in one compartment are not necessarily identical with those from another. In particular where local priming or

185

TABLE 1   THE 'COMPARTMENTS' OF THE BODY AVAILABLE FOR STUDY

'Compartment'	Effector cells or Systems studied
Blood	Serum or cells
Skin on antigen challenge	Vascular responses to extra vascular antigen, reacting with circulating antibody or sensitised lymphocytes or complement components, distant from the site of natural exposure.
Bronchial mucosa challenge	Vascular and smooth muscle responses to inhaled antigen reacting with intraluminal sensitised mast cells, local specific or circulating antibody, complement components and sensitised cells at site of 'natural' exposure.
Bronchial lavage	Local inflammatory cells and their mediators.
Lung biopsy extraction	Cell populations in tissues at the site of damage.

sequestration of cells has taken place in the lungs these cells
may not be represented in similar numbers in the peripheral blood.
Further, in view of the different nature of the target organ
(i.e. skin, airways or peripheral lung) the fact that certain mediators affect different tissues (e.g. vascular bed or smooth muscle),
means that information obtained from study of one area may have only
indirect bearing on events occurring at another. Studies of these
different compartments must be viewed critically, they are complem-
entary and are certainly not identical. For example recent studies
comparing the sub-classes of lymphocytes in broncho-alveolar lavage
and peripheral blood in sarcoidosis has demonstrated the far greater
numbers of T lymphocytes present in the lung (Weinberger et al,
1978) and their reduction compared to normal controls in the peripheral blood. Further, depression of lymphocyte function observed
from blood samples is not reflected in those obtained from the lungs
which appear to be 'activated'.

Studies of blood (cells or serum) can often be used to demonstrate the immunological characteristics of the primed individual
while those using material directly from the lungs may reflect
changes more immediately related to tissue injury and disease.
Alternatively immunological studies (from any source of material)
on a case control basis may distinguish features occurring as a
simple result of exposure from those more closely linked with
disease. For example, lymphocyte sensitisation to avian protein
antigen appears to be more closely associated with disease than
the presence of circulating precipitating antibody.

From these observations it can be seen that we are now moving
forward from an era where immunological phenomena are recorded
either in vitro or in vivo to that which studies the local changes
specifically related to local injury.

Table 2 lists some of the immunological abnormalities frequently
and easily identified in primed individuals but which do not
necessarily relate to clinical disease. In this respect they may
be regarded as epiphenomena. By contrast Table 3 summarises some
of the features relating to disease states.

In this talk I wish to consider some recent studies which I
believe have increased our understanding regarding pathogenesis of
the lung.

TABLE 2   IMMUNOLOGICAL PHENOMENA AND CLINICAL LUNG DISEASE

Those which may be unassociated with
clinical lung diseases

	Marker for:
Positive immediate skin tests	IgE
Circulating precipitating antibody	IgG
Circulating non-organic specific autoantibodies	IgG/M
Positive delayed skin reactions	Lymphocytes
Deposition of IgG in capillaries	IgG aggregates

TABLE 3   IMMUNOLOGICAL PHENOMENA AND CLINICAL LUNG DISEASE

Those which may be associated with
clinical lung diseases

Degranulation of mast cell by IgE or IgG and antigen

Activation of complement by the classical or
        alternative pathway

Activation of macrophages and neutrophils with
        release of lysosomal enzymes

Activation of lymphocytes with lymphokine release

IMMUNE COMPLEXES

    A number of techniques have been developed recently which are
purported to indicate the presence of immune complexes. Some depend
upon the property of some complexes to acquire additional amounts
of certain complement components (e.g. c1q binding), others identify
protein of a size that indicates their aggregated or complexed form
(e.g. ultracentrifugation or polyethylene glycol precipitation).
Others depend on demonstrating blockade of normal C3b or Fc receptor
sites on certain cell surfaces after prior incubation with serum
containing complexes (e.g. reduction of antibody dependent cyto-
toxicity). These different techniques demonstrate complexes of
different sizes and having different characteristics in terms of
complement content. Discordant results using different test systems
can often be explained on this basis and do not necessarily indicate
that one test is 'better' than another.

    Immune complexes have different potential depending on their size
and on whether they are formed in antigen or antibody excess.
Large complexes (greater than 19S) formed in antibody equivalence
rapidly precipitate and are phagocytosed by the monocyte/macrophage
system and are thence degraded. Small complexes of less than 19S
in antigen excess tend to remain soluble and commonly contain
relatively little complement and remain in the circulation with
little tissue damaging potential. Larger complexes in antigen excess
circulate and may be precipitated in the capillary endothelium
initiating damage which is then amplified by adherent platelets, thus
increasing permeability, with subsequent escape of complexes into the
tissue with consequential additional local damage. Under some
circumstances an additional independent event triggers capillary
permeability (e.g. histamine liberated from an activated mast cell)

and this allows tissue damaging complexes to permeate into the interstitial spaces.

Little is known about the movement of immune complexes formed in tissues (and particularly in the lung) back into the circulation. This might occur when endothelial junctions have widened by some independent mechanism such as Type I reaction. It is however interesting to observe that in those conditions where inhaled antigen meets local specific antibody in a primed individual as in extrinsic allergic alveolitis, circulating immune complexes are rarely detected.

Circulating immune complexes have been detected in serum using a variety of methods in a variety of connective tissue disorders of the lung (Fig. 1). These findings have now been confirmed by several groups of workers (Dreisin et al, 1978; Eisenberg et al, 1977). The question remains whether they are related to the clinical lung damage. The fact that the majority of cases do not have associated renal damage suggests that either they are not related immediately to pathogenesis or that some additional factor in the lung is needed to localise them preferentially to pulmonary tissues. Alternatively the complexes form in the lung, local inflammation alters pulmonary vascular permeability allowing escape of complexes into the circulation. This suggestion would not easily explain the fact that inflammatory changes at extra pulmonary sites (e.g. joints) frequently precede the lung changes (Courtenay Evans and Turner-Warwick, 1977).

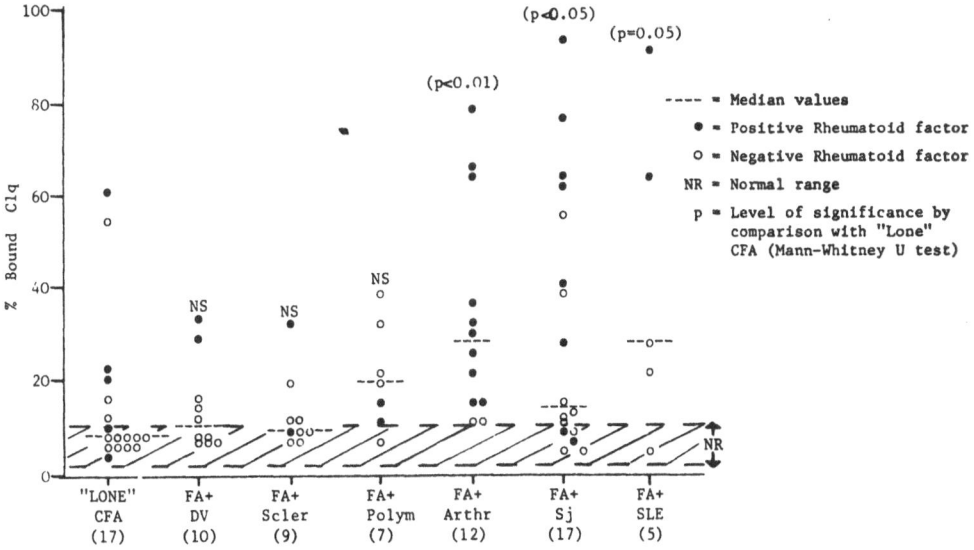

Figure I. Clq Binding Studies in 77 Patients with
Cryptogenic Fibrosing Alveolitis.

Whatever the sequence of generation of the complexes, evidence of local deposition is found from two types of studies. Immunofluorescence studies have demonstrated granular deposits of IgG and complement in some patients with a variety of connective tissue disorder including cryptogenic fibrosing alveolitis (CFA) (Dreisin et al, 1978; Turner-Warwick et al, 1971), Rheumatoid Arthritis (Turner-Warwick, 1967; de Horatius et al, 1972) and Systemic Lupus Erythematosus (SLE) (Turner-Warwick, 1974 and Eisenberg et al, 1973). Most studies however agree that electron dense deposits are rare. Studies of bronchoalveolar lavage have demonstrated immune complexes in supernatants and crossed electrophoresis of supernatants with added polyethylene glycol have demonstrated an early peak containing IgG and complement (Townsend et al, 1980).

These studies may be especially important in suggesting a possible mechanism for activation of alveolar macrophages. Earlier in vitro studies by Allison and coworkers (1974) have shown that immune complexes activate peritoneal guinea pig macrophages to liberate lysosomal enzymes. Other studies suggest that they may occur in vivo in man in CFA. Further support for this proposal is provided by studies reported from our laboratory by du Bois who has shown more rapid spreading of alveolar macrophages compared with controls (du Bois, 1980) and increased lysosomal enzyme content of the supernatant from these cases.

The complement component C5a has been identified in supernatants from bronchiolar lavage (Hensen et al, 1979). This has a number of important implications. Because there are at least two likely mechanisms for its development; either by activation through the classical or alternative pathway in relation to antigen or through splitting by collagenase, an enzyme known to be released from alveolar macrophages and which has now been identified in bronchoalveolar supernatants (Gadek et al, 1979), a finding now confirmed in our laboratory. Whatever the mechanism of its induction C5a has potent chemotatic properties for neutrophils and eosinophils as well as the capacity to activate mast cells to liberate their mediators. Several groups of workers have observed the preponderance of neutrophils and eosinophils in lavage samples from CFA and C5a may well be one explanation for their accumulation although it is likely this is not the only mechanism. For example a small molecular weight chemotactic factor has been demonstrated liberated from macrophages in these patients (Merrill et al, 1979)

The demonstration of C5a in supernatants has other importance. Following the suggestion that this complement component might act as an anaphalatoxin Haslam sought and found increased amounts of histamine in supernatants from patients with CFA. That this arises

from mast cells is supported by the easy demonstration of these
cells in 1μ sections from biopsy samples.

While the accumulation of granulocytes and mast cells in the
lungs of patients with chronic interstitial pneumonia is likely
to be important in explaining local tissue damage it must not
be forgotten that the predominant inflammatory cell on light
microscopy is the lymphocyte. Another well established mechanism
of macrophage activation is lymphokines derived from sensitised
and stimulated lymphocytes. While the presence of lymphokines has
not so far been detected directly in the supernatant this is a very
likely possibility particularly because under some circumstances,
morphological evidence of transformed lymphocytes have been found
in such conditions as extrinsic allergic alveolitis and sarcoidosis.

CONCLUSION

These studies on immunological events using samples obtained
directly from diseased human lung have opened the door to a far
greater understanding of the consequences of immunological events
directly relating to tissue damaging processes, and are indeed
beginning to highlight the differences observed between different
diseases. Definition of distinctive pathogenetic mechanisms in
pulmonary disease is important in view of the different responses
observed to anti-inflammatory agents and we are now beginning to
define patterns of inflammation which appear to be refractory to
corticosteroids. With this definition we have the opportunity of
developing and monitoring effective new and better drugs in the
treatment of these steroid resistant cases.

EFFECTS OF DRUGS

This summary would be incomplete if it failed to draw
attention to the close relationship between the presence of
lymphocytes and lung lavage and corticosteroid responsiveness
irrespective of disease and conversely in general the refractoriness
to this drug in cases with increases of neutrophils and/or
eosinophils. Studies should now be directed to the development of
anti-inflammatory agents which would control the presence of
granulocyte mediated tissue damaging reactions.

EXTRINSIC ALLERGIC ALVEOLITIS

With these newer concepts in mind it is proper to reconsider
the immunopathogenesis of hypersensitivity pneumonitis. In
contrast to CFA and connective tissue disorders the predominant
cell in lung lavage fluid is the lymphocyte. These may occur in
very large numbers and frequently show bilobed nuclei suggestive
of in vivo transformation. In this respect the findings are

similar but more marked than those in sarcoidosis. The fact that
both conditions are characterised by the development of granulomas
suggest that local accumulation of sensitised and activated
T lymphocytes may play an important role in their formation. The
fact that both diseases are often responsive to corticosteroids
and that this drug inhibits lymphokine mediated responses demonstrates further similarity. Oddly, study of the responses to
mitogens and recall antigens of circulating lymphocytes in EAA do
not seem to have been reported as extensively as in sarcoidosis.
Some however appear to show some diminution of delayed skin test
responses to recall antigens when careful tests using several doses
are used (Haslam, unpublished). There is also some indication of
a general reduction of PHA responses. Trends at least which are
similar to those found in sarcoidosis.

    Circulating precipitating antibody appears to be mainly a
marker of exposure. Many individuals have antibody without clinical
evidence of disease, and deposits of antibody or complement are
rarely observed in the lung although these have occasionally been
reported. Furthermore experimental models have usually also yielded
negative results. It is however of some interest that while total
IgG and IgG1 are raised in both healthy exposed and disease farmers,
IgG3 is significantly raised in those with disease compared to
matched exposed controls. This suggests that this sub class of
antibody may have a more important role in causing tissue damage
and that those with propensity to produce it (? possibly genetically
determined) may be more susceptible to disease. At this stage in
our knowledge this suggestion remains no more than speculative.

    Recent studies however have suggested that M.faeni may induce
tissue damage without interaction either with specific antibody or
sensitised lymphocytes. Edwards et al (1974), has shown that this
thermophylic M.faeni is capable of activating the alternative
pathway of complement and recent studies by Allison and Davies
(1974), have demonstrated that M.faeni together with macrophages
enclosed in a millipore chamber and placed intraperitoneally in
normal unsensitised mice will induce fibrosis. Since only soluble
non-cellular material can escape from these chambers it has been
concluded that M.faeni/macrophage interaction results in the
secretion of fibroblast stimulating factor. Thus the immunopathogenesis of Farmer's lung has to be reviewed in the light of
these newer studies.

REFERENCES

Weinberger, S.E., Kelman, J.A., Elson, N.A., Young, R.C.,
    Reynolds, H.Y., Fulmer, J.D. and Crystal, R.G., 1978,
    Bronchoalveolar lavage in interstitial lung disease.
    Annals of Internal Medicine 89, 459-466.

Dreisen, R.B., Schwartz, M.I., Theofilopoulos, A.N. and
    Stanford, R.E., 1978, Circulating immune complexes in idiopathic
    interstitial pneumonias. New England Journal of Medicine 298,
    353-357.

Eisenburg, H., Barnett, E. and Simmonds, H., 1977, Diffuse pulmonary
    interstitial disease: immune complex disease. Clinical Research
    25, 132 (abstract).

Courtney Evans, R. and Turner-Warwick, M., 1977, Pulmonary
    manifestations of rheumatoid disease. Extraarticular manifestat-
    ions of rheumatoid arthritis. Clinics in Rheumatic Diseases 3,
    549-564.

Turner-Warwick, M., Haslam, P. and Weeks, J., 1971, Antibodies in
    some chronic fibrosing lung diseases II. Immunofluorescent
    studies. Clinical Allergy 1, 209-219.

Turner-Warwick, M., 1967, Autoallergy and lung disease. Journal
    of the Royal College of Physicians 2, 57-66.

Eisenburg, H., Dubois, E., Sherwin, R.P. and Balchum, O.J., 1973,
    Diffuse interstitial lung disease and systemic lupus
    erythematosus. Annals of Internal Medicine 79, 37-45.

Turner-Warwick, M., 1974, Immunological aspects of systemic disease
    of the lungs. Proceedings of the Royal Society of Medicine 67,
    541-547.

Allison, A.C. and Davies, P., 1974, Increased biochemical and
    biological activities of mononuclear phagocytes exposed to
    various stimuli, with special reference to secretion of
    lysosomal enzymes in Van Furth R. (ed.)   The Mononuclear
    Phagocyte, pp. 487-506. Oxford, Blackwell Scientific
    Publications.

Edwards, J.H., Baker, J.T. and Davies, B.H., 1974, Precipitin test
    negative Farmers' Lung - activation of the alternative pathway
    of complement by mouldy hay dusts. Clinical Allergy 4, 379-385.

Merrill, W.W., Naegel, G.P., Matthay, R.A. and Reynolds, H.Y., 1979,
    Production of chemotactic factors by in vivo cultured human
    alveolar macrophages. Chest, 75, 224.

de Horatius, R.J., Abruzzo, J.L. and Williams, R.C., 1972,
    Immunofluorescent and immunologic studies of rheumatoid lung.
    Archives of Internal Medicine 129, 441.

Townsend, P. and Haslam, P.L., 1980, Detection of immune complexes
    using crossed immunoelectrophoresis and polyethylene glycol
    precipitation. Unpublished.

Henson, P.M., McCarthy, K., Larsen, G.L., Webster, R.O., Giclas, P.C.,
    Dreisen, R.B., King, T.E. and Shaw, J.O., 1979, Complement
    fragments, alveolar macrophages and alveolitis. American Journal
    of Pathology, 97, 93-105.

Gadek, J.E., Kelman, J.A., Weinberger, S.E. Horwitz, A.L.,
    Reynolds, H.Y., Fulmer, J.D. and Crystal, R.G., 1979,
    Collagenase in the lower respiratory tract of patients with
    idiopathic pulmonary fibrosis. New England Journal of Medicine,
    301, 737.

DISCUSSION

LECTURER: Turner Warwick   CHAIRMAN: Bonsignore

CANDURA: I would like to ask whether the speaker has any
                    experience of parenchymal sampling using the fibre
                    endoscope, and whether she thinks it is useful or
                    if there are limitations of an ethical nature.

TURNER-WARWICK: You mean a transbronchial biopsy? I think that if
                    one is trying to study living cells one has got to
                    have enough cells to separate and analyse. The
                    volume of material you get even from several
                    transbronchial biopsies is far too small in our
                    hands to form a useful body of material. For this
                    purpose therefore our tissue studies have been done
                    on open biopsies. I have clinical reasons for
                    preferring the open biopsy in this group of
                    conditions in any case but that is outside the
                    present context.

BONSIGNORE: The second lecture of the afternoon will be given
                    by Molina.

CLINICAL ASPECTS OF TYPE III IMMUNITY

C. Molina*

Clinique de Pneumologie
Hôpital Sabourin
63018 Clermont-Ferrand Cédex, France

Many pulmonary diseases are associated with demonstrable type III hypersensitivity reactions. But it seems an oversimplification to consider that a disease is relevant to only one immunologic mechanism.

The classic examples of type III hypersensitivity reactions in lung diseases are interstitial pulmonary fibrosis (IPF); collagen vascular diseases, such as Wegener granulomatosis, systemic lupus erythematosus, and rheumatoid arthritis; and, mainly since the contributions of Jack Pepys, allergic bronchopulmonary aspergillosis (ABPA) and extrinsic allergic alveolitis (EAA). The features of all these diseases are well known.

In IPF and collagen vascular diseases, the exact mechanisms are unknown and the pathogenesis only speculative (autoimmunity is highly suspected). But in ABPA and EAA, which are due to inhalant antigens (Aspergillus, actinomycetes, or other organic dusts), the immunologic disorders are gradually clarified.

We would like to emphasize the new insights provided in EAA by the technique of bronchoalveolar lavage (BAL) performed in some cases of farmer's lung, bird breeder's disease, and cheese-worker's disease, all conditions frequently observed in our region.

---

*With the help of A. Jeanneret and J.M. Aiache, Laboratoire d'Immunologie de la Clinique Pneumologique; P. Jouanel, B. Dastugue, and C. Motta, Laboratoire de Biochimie de la Faculté de Clermont-Ferrand; J. Brun, Clinique Médicale du C.H.U. de Caen (Pr. Lemenager); and D. Wahl, Clinique de Pneumologie, Hôpital Sabourin, Clermont-Ferrand.

I. CLINICAL ASPECTS

There were six major criteria for the diagnosis of EAA (20 subjects):

1. History of exposure to organic dusts.
2. Clinical symptoms such as dyspnea, fever, rales. (In some acute cases, the symptoms were severe and needed hospitalization with the diagnosis of Adult Respiratory Distress Syndrome in Intensive Care Unit.)
3. X-ray showing bilateral shadows, miliary or infiltrative type.
4. Restrictive ventilatory defect and impairment of CO transfer with drop of PaO_2. (In recent functional studies by flow-volume curves, we outlined the frequent involvement of small airways (fall of $\dot{V}_{max25}$ and $\dot{V}_{max50}$).)
5. Pathological features such as alveolar and interstitial granuloma.
6. Presence in the serum of specific precipitating antibodies, IgG type, to causal antigens (moldy hay dust, bird droppings, or cheese products).*

II. THE BALL TECHNIQUE

The BAL technique was used to study the cell populations and the biochemistry of this group of diseases and in a group of 13 controls. Bronchoalveolar lavage was performed with the use of a standard flexible fiber-optic bronchoscope and infusion of 250 cc of saline serum, heated to 37°C. The fibroscope was placed in "blocked catheterism" in a segmental or subsegmental bronchus of an upper lobe. The saline serum was infused in 30-ml aliquots, which were recovered immediately.

The lavage fluid was quickly sent to the laboratory, where, after centrifugation, the following tests were carried out:

- on the cell residue:
   • a study of cell viability (Blue Trypan test)
   • a morphological and immunological study of the cells, including:

   (a) Malassez cell count
   (b) Cell formula on colored smears according to May-Grunwald-Giemsa

---

*We reported cases of cheese-worker's disease elsewhere. During our study, we observed myriads of little creatures on the surface of the cheese. These were identified as mites (Acarus siro or Tyrophagus casei, and we consider them the main source of antigens.

(c) Study of lymphocyte subpopulations by E, EA, and EAC rosette techniques, and study of membrane immunoglobulins by immunofluorescence (a parallel study of lymphocyte population in peripheral blood was also carried out)

- on the supernatant:
  • protein dosage: albumin, immunoglobulins, C3, C4
  • a study of lipids and particularly phospholipids, by means of thin-layer chromatography, and evaluation of the degree of fluorescence polarization, a physical parameter quickly obtained, well correlated with the range of total phospholipids, with the microviscosity of alveolar lavage, and independent of the amount of fluid recovered.

III. RESULTS

1. Study of Cell Populations

The distribution of cell populations appears immediately to be very different in subjects suffering from EAA and in those of the control group (see Table 1).

The nonsmokers of the control group showed a high percentage of macrophages (80% on average) and a small number of lymphocytes (not exceeding 18%). The presence in small quantities of poly-nuclear cells and of bronchial mucous and ciliated cells was also noted.

In EAA, the data are completely different. There are strikingly large quantities of lymphocytes, which increase in the acute stage of the illness (44 ± 15.7%). Even the chronic forms show a far higher percentage of lymphocytes than the control group (33.8 ± 20) (Table 1).

Table 1. Cells in Lavage Fluid

	Macrophages (%)	Lymphocytes (%)
9 Controls	81.55 ± 4.75	13.10 ± 5.25
7 Acute EAA	50.30 ± 15.55	44. ± 15.75
9 Subacute EAA	57.50 ± 19.85	34.65 ± 21.10
5 Chronic EAA	55.40 ± 25.10	33.80 ± 20.05

Table 2. Pulmonary Lymphocytes and Tobacco Smoking

	Smokers	Nonsmokers
9 Controls	11.50 ± 9.50 (n = 6)	16.35 ± 3.10 (n = 3)
7 Acute EAA	52. ± 8.50 (n = 2)	40.80 ± 17. (n = 5)
9 Subacute EAA	25.25 ± 12.75 (n = 4)	44. ± 25.30 (n = 5)
5 Chronic EAA	18. (n = 1)	37.75 ± 20.75 (n = 4)

n = number of patients studied.

Tobacco smoking modifies the cell distribution (Table 2). The lymphocyte population decreases in the usual manner in smokers, whether from the control or pathological group. This is particularly so in subacute and chronic forms of the disease, but the difference does not appear in acute forms.

The study of lymphocyte subpopulations in lavage fluid (Table 3), e.g., B lymphocytes (identified by EAC rosettes, surface immunoglobulins) and T lymphocytes (identified by E rosettes), shows:

• a very small percentage of EAC rosettes (2.50 ± 4.55%)
• a high percentage of E rosettes (40.67 ± 19.53%)
• a large difference in the distribution of lymphocyte markers between the lavage fluid and blood (lower percentage of E and EA rosettes in the lavage fluid), which gives evidence of the immunological independence of the lung

Interestingly, thermophilic actinomycetes have been identified in macrophages by electron-microscopic study in the BAL of a case of farmer's lung (Romet-Lemonne).

Table 3. Distribution of Blood and Pulmonary Lymphocytes

	Blood (n = 12)	Lung (n = 9)
EAC Rosettes (%)	16.30 ± 6.15	2.50 ± 4.55
E Rosettes (%)	57. ± 8.55	40.70 ± 19.55

n = number of patients studied

All these results are similar to those found in the recent literature (Reynolds, Voisin).

2. Biochemical Studies

Biochemical studies were performed in the EAA group, in the control group (13 cases), and another group of sarcoidosis (25 cases).

A. Proteins. Among subjects suffering from EAA or sarcoidosis, the protein concentration was significantly higher than in the control group (see Table 4). In some cases in the EAA group, we found an increase of albumin, IgG, IgM, C3, and C4.

B. Lipids. The study of phospholipids was of greater interest. As shown in Table 4 and Figure 1, there is in the EAA group a very typical phospholipid profile, with:

- total absence of phosphatidyl choline (lecithin), which is considered as having the highest tensio-active property in the surfactant
- increase of the phosphatidyl-inositol fraction
- elevation the degree of fluorescence polarization
- elevation of the microviscosity of the fluid

All these parameters are significantly different from the controls and the sarcoidosis group. This biochemical characteristic phospholipid profile seems to provide a new diagnostic guide in EAA.

IV. PATHOGENESIS

EAA appears to be a combination of many immunologic mechanisms.

Evidence for type I hypersensitivity reactions is provided by the presence of positive skin test against certain antigens and history of atopy (in 30% of cases).

Evidence for type III reactions is given by

- Delayed symptoms after exposure
- Arthus-type skin reactivity
- Precipitating antibodies and immune complexes in the serum
- Positive challenge tests
- Pathological features
- Experimental studies showing the role of complement

Evidence for type IV is suggested by

- Negative delayed hypersensitivity responses to PPD
- Lymphocyte sensitization (lymphocyte transformation, MIF)

Table 4.   Biochemical Studies in Granulomatous Diseases

	Control (n = 13)	Sarcoidosis (n = 25)	Allergic alveolitis (n = 20)
Total proteins (mg/ml)	$\overline{m}$ = 0.18 (0.01 - 0.78	$\overline{m}$ = 0.45 (0.08 - 2.00) S	$\overline{m}$ = 0.66 (0.02 - 3.40) S
Phospholipids (nmol/ml)	$\overline{m}$ = 43 (10 - 110)	$\overline{m}$ = 30 (3 - 90) NS	$\overline{m}$ = 99 (12 - 270) S
Fluorescence polarization P	$\overline{m}$ = 0.16 (0.05 - 0.23)	$\overline{m}$ = 0.18 (0.10 - 0.27) NS	$\overline{m}$ = 0.25 (0.13 - 0.38) S
Microviscosity η (poises)	$\overline{m}$ = 1.15 (0.10 - 2)	$\overline{m}$ = 1.18 (0.55 - 2.8) NS	$\overline{m}$ = 2.45 (0.8 - 9.5) S

NS = Not significant; S = Significant; $p = 0.01$.

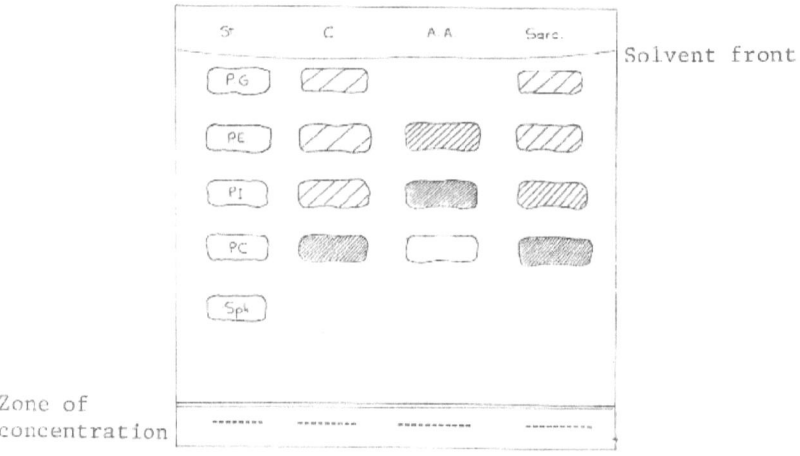

Figure 1.   Different fractions of phospholipids (thin-layer chromatography) in bronchoalveolar lavages.

- Bronchoalveolar lavages showing increased lymphocyte
  population and preponderance of T lymphocytes
- Granuloma formation

The characteristic changes observed in cell populations,
immunologic reactions, and biochemistry of the alveolar content in
EAA confirm the mediated hypersensitivity mechanisms and may result,
in the future, in new therapeutic approaches.

*(This work was supported by a grant of Fonds Spécial des Comités
Départementaux contre les Maladies Respiratoires.)*

REFERENCES

The 21st Aspen Lung Conference: Immunology of the Lung, 1979, Chest,
    75, 2, supplement.
Fujiwara, T., Maeta, H., Chida, S., Morita, T., Watabe, Y., and
    Abe, T., 1980, Artificial surfactant therapy in hyaline-membrane
    disease, Lancet, 1, 8159, 55.
Gadek, J.E., Kelman, J.A., Fells, G., Weinberger, S.E., Horwitz, A.L.,
    Reynolds, H.Y., Fulmer, J.D., and Crystal, R.G., 1979, Colla-
    genase in the lower respiratory tract of patients with
    idiopathic pulmonary fibrosis, New Engl. J. Med., 301, 14,
    737-742.
Hunninghake, G., Gadek, J., Weinberger, S., Kelman, J., Elson, N.,
    Young, R., Fulmer, J., and Crystal, R.G., 1979, Chest, 75, 2
    supplement, 266-267.
Jouanel, P., Motta, C., Brun, J., Roche, G., Dastugue, B., and
    Molina, C., 1979, Lipid analysis of alveolar lavage fluids from
    patients with extrinsic allergic alveolitis, in: "Les Colloques
    de l'INSERM, Le lavage broncho-alvéolaire," INSERM, 84, 73-84.
Lynn, W.S., Sahu, S., Giraldo, H., and Tanswell, A.K., 1979, Lipids
    and their associated enzymes found in secretions of diseased
    human airways and in cultured human type II cells, in: "Les
    Colloques de l'INSERM. Le lavage broncho-alvéolaire," INSERM,
    84, 65-72.
Martodam, R.R., Twumasi, D.Y., Liener, I.E., Powers, J.C., Nishino,
    N., and Krejcarek, G., 1979, Albumin microspheres as carrier
    of an inhibitor of leukocyte elastase: potential therapeutic
    agent for emphysema, Proc. Natl. Acad. Sci. USA, 76, 5,
    2128-2132.
Molina, C., Brun, J., Jeanneret, A., Betail, G., Chollet, P., and
    Roche, G., 1979, Etude des populations lymphocytaires et de la
    biochimie du liquide de lavage broncho-alvéolaire au cours des
    pneumopathies d'hypersensibilité, in: "Les Colloques de
    l'INSERM. Le lavage broncho-alvéolaire," INSERM, 84, 313-320.
Myrvik, Q.N., Leake, E.S., and Fariss, B., 1961, Study on pulmonary
    alveolar macrophages from the normal rabbit. A technique to
    procure them in a high state of purity, J. Immunol., 86, 128.

Perrin-Fayolle, M., Harf, R., Chevalier, J.P., Biot, N., Frobert, F.,
    and Kofman, J., 1979, Intérêt des L.B.A. itératifs dans la
    surveillance des sarcoïdoses, in: "Les Colloques des l'INSERM.
    Le lavage broncho-alvéolaire," INSERM, 84, 321-328.
Ramirez, R.U., Schultz, R.B., and Dutton, R.E., 1963, Pulmonary
    alveolar proteinosis: a new technique and rationale for treat-
    ment, Arch. Intern. Med. (Chicago), 112, 419.
Reynolds, H.Y., Fumer, J.D., Kazmierowski, J.A., Roberts, W.C.,
    Franck, M.M., and Crystal, R.G., 1977, Analysis of cellular
    and protein content of broncho-alveolar lavage fluid from
    patients with idiopathic pulmonary fibrosis and chronic
    hypersensitivity pneumonitis, J. Clin. Invest., 59, 165.
Reynolds, H.Y., and Newball, H.H., 1974, Analysis of proteins and
    respiratory cells obtained from human lungs by bronchial
    lavage, J. Lab. Clin. Med., 24, 559.
Romet-Lemonne, J.L., Lemarie, E., and Choutet, P., 1980, Ultra-
    structural study of bronchopulmonary lavage liquid in farmer's
    lung disease, Lancet, 5, 777.
Shinitsky, M., Dianoux, A.C., Betler, C., and Weber, G., 1971,
    Biochem. J., 10, 2106.
Turner-Warwick, M., 1979, in: "The 21st Aspen Lung Conference:
    Immunology of the Lung," Chest, 75, 2, supplement.
Voisin, C., 1979, in: "Les Colloques de l'INSERM. Le lavage broncho-
    alvéolaire," INSERM, 84.
Voisin, C., Tonnel, A.B., Aerts, C., Lafitte, J.J., and Ramon, P.,
    1977, Les populations cellulaires des espaces aériens broncho-
    alvéolaires dans la sarcoïdose, les alvéolites allergiques
    extrinsèques et les cancers bronchiques, Presse Med., 6, 2685.

D I S C U S S I O N

LECTURER: Molina   CHAIRMAN: Bonsignore

LENZINI: After looking at the slides about BAL I would say
                that we have more and more questions to answer. For
                instance what is the diagnostic precision of the
                methods and how does it help in the definition of the
                immunopathogenic mechanism. If we think of
                sarcoidosis as a granulomatous disease with delayed
                hypersensitivity, then in the bronchopulmonary lavage
                we see an increase in the number of T lymphocytes
                locally;   conversely when we look at a
                hypersensitivity disease, farmers lung for instance,
                there is also an increase in T lymphocytes. Finding
                increased T lymphocytes both in hypersensitivity
                disease and in sarcoidosis leads one to think that we
                still are very far from understanding what are the
                immunopathogenetic mechanisms of these two diseases.
                Another point concerns the diminution of A and C
                rosettes, which makes one think of a diminution in T
                lymphocytes. So, in farmer's lung T lymphocytes
                decrease if we accept the rosette evidence or they
                increase if we consider the lavage. Things being so
                we must admit that we have some problems of
                interpretation. Do monocytes not exist in these two
                diseases? Why we do not look at this? Is the sample
                we take with BAL a true image of what really exists
                at tissue level? All these data – no monocytes in
                sarcoidosis, no monocytes in hypersensitivity
                pneumonia, increase in T lymphocytes in sarcoidosis
                and in hypersensitivity, decrease in A–C rosettes in
                hypersensitivity – I would say that we still are in
                the realm of hypothesis.

MOLINA: Morphologically there is a great difficulty in
                differentiation of monocytes and lymphocytes. So we
                are obliged to use markers for lymphocytes, and among
                these cells which are called mononuclear cells maybe
                there are some monocytes. But all we can say is that
                B cells are in very small proportion, because we are
                aware of the fact and we are sure E and C cells are B
                cells and E rosettes are T cells. For the other
                cells we are not sure that they are not monocytes or
                different varieties of lymphocytes.

CORRIN: Among the various forms of evidence for type III hypersensitivity in extrinsic allergic pneumonitis you listed pathological features. Could you tell us what those pathological features are.

MOLINA: In experimental studies immune complexes may provoke granuloma, exactly the same as cell-mediated granuloma observed in type IV hypersensitivity reaction. Immune complexes in excess of antigens or equivalence between antigens and antibodies may result in pathological features exactly the same as granuloma formation due to type IV hypersensitivity reaction.

CORRIN: It seems a little unfair then to list the pathological features as evidence supporting type IV hypersensitivity as well as type III hypersensitivity. They really seem compatible with either.

MOLINA: Yes, this is the difficulty of interpretation that Turner Warwick said in her talk. We have two possibilities to diagnose these two types of hypersensitivity pneumonitis, besides immunological symptoms. Biopsy or lavage. And these two techniques provide somewhat different information, but in some cases the information is similar. And in lavage as in biopsy we see information about the presence of T cells and macrophages and giant cells and granuloma formation. There is a concordance between these manifestations but in some cases there is disparity in the symptoms observed in lavage and in biopsy, and this maybe explains your question.

COHEN: You have some very interesting differences in the phospholipids in the different disease states. Is your only evidence that the material in the allergic alveolitis is phosphatidyl inositol from thin layer chromatography, because it did not co-chromatograph with your authentic standard. And then secondly if you could care to comment on the possible sources of phosphatidyl inositol and its possible relationship to prostaglandins.

MOLINA: For the second question, phosphatidyl inositol, in our fluid there is a difference not only in the quantity of phospholipids, which are increased, but the quality of the alveolar lining of this fluid is different. Because phosphatidyl inositol is less viscous than phosphatidylcholine, lecith, you see. This fraction of phospholipids replaces phosphatidylcholine, which is absent. To increase the viscosity, I think this is the interpretation of this transformation. Phosphatidyl inositol replaces phosphatidylcholine, it is a change in quality of the phospholipids. Your first question, whilst phosphatidyl inositol did not co-chromatograph exactly the same, the difference is not significant.

COHEN: Do you have any other criteria to say it is phosphatidyl inositol? Other chromatographic systems or gas chromatography?

MOLINA: No, we have not. But all the studies show the constant increase of phosphatidyl inositol.

CUMMING: If you use only chromatography then you cannot say for certain that it is one or the other.

MOLINA: Only chromatography; but it seems a good method. You do not think so? A team of biochemists are studying the phospholipids in the amniotic fluid, and we think that the separation of different fractions is very sure.

COHEN: A single separation in one GLC system, where it even does not co-chromatograph is insufficient evidence to say it is phosphatidyl inositol, you require much more evidence than that.

MOLINA: Yes, but the problem is not phosphatidyl inositol, the problem is the absence of phosphatidylcholine. We think it is the mean feature.

COHEN: That is O.K., that is acceptable, but the phospholipids, perhaps phosphatidyl inositol, perhaps another phospholipid. The increase of total content of phospholipids, is correlated with the degree of fluorescence polarisation and the microviscosity of the fluid.

HEATH: I do not want Margaret Turner Warwick to go back to England feeling disappointed. So I am going to ask her the question that she suggested I would. I am still not certain how you can make certain that these cells which you see in BAL are macrophages. Everyone has admired the confidence of the speakers who talked about macrophages; I only wish I was confident as they. You see, Margaret says you can go to electron microscopy and it shows all the features of macrophages. When Corrin reported to the Journal of

Pathology, he was at great pains in those articles to point out the extreme difficulty you may have in distinguishing between macrophages and granular pneumocytes. I think it is true to say that he did the differentiation eventually on enzyme studies, in that paper. Now we know that in fibrosing alveolitis the granular pneumocytes are very active in the early states of the disease. Well, how does it come about, Margaret, that you are so confident that what you see in these fluids are macrophages?

TURNER-WARWICK: Of course no one is absolutely confident about what they choose to call any cell when it is perhaps in an atypical or intermediate form. And the criteria that we have applied, and then of course I am not an electron microscopist, so I lean heavily on Corrin and Drew to say that the majority of cells that they see have that sort of appearance. Some of those of course might be phagocytic cells, that you would choose to call by another name. But I am not sure that it is very fruitful to spend a lot of time arguing about the intermediate cells that you think you and I would agree, if we looked at the samples together, formed a minority. One can clearly get very typical neutrophils, that I think would convince you, as a mature cell on the one hand, and the mature macrophage with its normal specific stain on the other. What you would wish to argue about might amount to 5% of the total, but I doubt if that is profitable. It is well known, that for functional purposes, for much of the time they are operating in rather similar ways, so I doubt, even when one is talking in terms of pathogenesis, that it is going to be a very profitable argument.

HEATH: I am sorry to take up time, but I wonder if I could ask Corrin whether he regards the granular pneumocyte and the macrophage as interchangeable terms, or quite distinct.

CORRIN: No, I would not regard them as being interchangeable terms. I think the type II pneumocyte is an epithelial cell which is quite distinct from the macrophage, the macrophage is mesodermally derived and it is not derived from the epithelium. In the work that you quoted confusion arose because the macrophages were accumulating many lamellar bodies and in that special situation came to resemble type II pneumocytes. But in the alveolar washes such as we have considered today, from normals, from sarcoidosis patients, from idiopathic interstitial fibrosis and hypersensitivity pneumonitis, I think there are certainly many difficulties in distinguishing these mononuclear cells from epithelial cells, without sympathising with Lenzini, there is often some difficulty in distinguishing whether the cells are macrophages or lymphotyctes, not typical lymphocytes or typical macrophages, but cells which are intermediate in their ultrastructural appearances, they are possibly activated lymphocytes or they could conceivably be immature macrophages. That is where I think the difficulty lies with ultrastructural interpretation.

TURNER-WARWICK: Would you not agree that there, with that distinction, the non specific esterase stain is extremely valuable?

CORRIN: It should be.

HEATH: If I could without being tiresome return to my point, and what bothers me about this. We have seen various papers by Molina and hewhere there are a lot of phospholipid membranes lying about, where you have the trouble. In that last paper this is the very situation where there is difficulty in distinguishing them. I am sorry if I am being difficult, I do not mean to be.

NEWMAN TAYLOR: Molina has shown us that there is an excess of T lymphocytes in the wash of patients with extrinsic allergic alveolitis with a deficit of B lymphocytes. I wonder if he has been able to look at the responsiveness of these lymphocytes to the specific allergens which are thought to be responsible for the disease in farmer's lung or even serum proteins in the pigeon fanciers. Secondly, with the accumulation of information about the presence of T lymphocytes and the presence of granulomata, another feature of what is described as lymphocyte mediated hypersensitivity in the lungs in patients with allergic alveolitis, does he believe that there is convincing evidence for the role of immune complexes in producing tissue damage in allergic alveoltiis, or does he think that the presence of precipitins and the other features thought to be typical of an immune complex mediated disease are in fact epiphenomena and not related to the tissue damage.

MOLINA: The problem is to know if farmer's lung is a disease

caused by immune complexes or associated with immune
complexes. We have not studied the responses of T
cells to antigens in the lavage fluid, but we have
studied in blood by migration inhibition test. When
you compare this test with blood test,
immunoelectrophoresis, MIF and bidimensional
immuno-electrophoresis or other techniques, there is
not a striking difference between them.
Immunoelectrophoresis seems the easiest technique and
we think it gives the initial and easiest information
for diagnosis. The team of Jordan Fink in Wisconsin,
considered the presence of precipitins as not being
sufficient for the diagnosis of the disease and they
think that the presence of antibody fixing complement
protects against the disease. To answer your
queston, I have not yet studied the response of T
cells to the antigens in lavage fluid. In the same
manner we are beginning work to study the electron
microscopic aspects of lavage fluid and particularly
the presence of almellar bodies in the macrophages to
make a correlation between our findings of
phospholipids in the supernatant and the presence or
not of lamellar bodies in the cells.

NEWMAN TAYLOR: The work of John Salvaggio in patients with allergic
alveolitis due to serum protein hypersensitivity
showed that in the 24 to 48 hours after exposure to
pigeon dust, there were present in the alveolar fluid
cells, T lymphocytes which were responsive to the
serum protein at a time when you could not
demonstrate such responsiveness in the circulation,
which is why I was wondering whether or not you have
been able to look at the two together.

MOLINA: Yes, it is another proof of the immunological
indpendence of the lung.

DENISON: As an immunologically very ignorant man I very much
enjoyed your talk. I was particularly struck by the
biochemical findings in the lavage fluid in groups of
people distinguished between normals, patients with
sarcoid and patients with fibrosing alveolitis. I
noticed that there was a quite large spread in the
results and I wonder whether any single biochemical
test, or any combination of them, was sufficiently
sensitive to be useful in the diagnosis in individual
cases, rather than understanding mechanisms in groups
of people.

MOLINA: We use the classical method of diagnosis for

sarcoidosis or for allergic alveolitis, bot only one method. For allergic alveolitis our laboratory has been trained for many years in the diagnosis of this disease and for sarcoidosis the diagnosis was made by biopsy, bronchial biopsy or lung biopsy, open lung or transbronchial biopsy. Our pathologists are very skilled in the diagnosis of this disease.

DENISON: I think you misunderstand me, I did not in the least doubt your cinical ability. It was whether the biochemical tests were of diagnostic value in individual cases.

MOLINA: Yes. When you study with all our team, biochemist, pathologist and physicians, before the physicians give the diagnosis the biochemist is able to tell us; it is probably a case of allergic alveolitis, or it is probably a control, or a sarcoid group, only by biochemical study. We have very characteristic profiles, now and our biochemist can make the diagnosis only by the study of phospholipids.

MAPS OF PULMONARY RISK: POSSIBILITIES AND PROSPECTIVES

E. Zecca

Institute of Hygiene of Pavia University

Italy - PAVIA, Viale Forlanini,1

In the past few decades industrial development, increasing city population and dramatic increase in motorvehicle traffic have drawn more attention to the problem of air pollution and its effects on the environment and human health.

The employment of new and sophisticated technologies has made the problem itself more difficult because of the continuous evolution in the materials used in production and consequently the dynamic typology of airborne polluting agents. These technological and social changes have been accompanied by a qualitative and quantitative modification of the relationship between the environment and health, increasing sections of the population have found themselves exposed to higher and qualitatively different pollution levels because of the presence of substances until recently detectable in occupational environments only. The consequence has been a remarkable increase in the incidence of respiratory diseases, both acute and chronic. While assigning an unquestionable role of contributing causative agent to air pollution, this has made the adoption of measures capable of controlling its spread even more imperative. These control measures cannot be effectively taken in several places and over large areas simultaneously owing to the complexity of the economic and organizational problems involved, but they must be planned, sized, and applied according to a priority scale based on precise and full knowledge of the degree and type of pollution of any given area and of the risk level thus resulting for the population living there.

Therefore, planning these measures involves the preliminary

Table 1 - Substances that are able to cause damage to respiratory
         system.

. Acrolein   . Nitric Acid

. Allyl Chloride   . Nitrogen Oxides

. Ammonia   . Ozone

. Asbestos   . p-fenylenediamine

. Aromatic Polyciclic   . phosgene
   Hydrocarbons
                                     . Phatlic Anhydride
. Beryllium
                                     . Smog (photochemical reactions)
. Cadmium and compounds
                                     . Sulfur Dioxide
. Carbon monoxide
                                     . Sulfuric Acid
. Chlorine
                                     . Suspensed particulate matter:
. Chromium (non soluble salts)
                                       - amorphons and cristalline
. Cobalt and compounds   silica

. Dimethyl sulfate   - silicon compounds

. Fluorine   - mica

. Formaldeyde   - perlite

. Furfuryl Alchool   - Portland stone

. Hydrochloric Acid   - stone

. Hydrofluoric Acid   - tale

. Ketene   - natural graphite

. Inorganic fluorine   . Vinyl chloride

. Maleic Anhydride

. Nickel Carbonyl

identification of territorial areas differing from each other by the degree of exposure or, as is commonly said, by risk level. This means that one must plot on the chosen territorial area the so called 'risk-maps' (in our case 'pulmonary risk-maps') which can be obtained by singling out sub-areas characterized by a known spatial distribution of the quality and gradient of atmospheric chemical pollutants. In fact, as regards the relationship between the diseases of the respiratory tract and pollution, 'pulmonary risk' can be defined, albeit somewhat inaccurately, as the chance of occurrence of damage to the lung or the respiratory tract as a result of inhalation, over time, of chemical pollutants emitted into the atmosphere.

The severity of any risk is a function of two variables: toxicologic characteristics of pollutant substances (and therefore damage potential of single risk factors and length of exposure to a given concentration of them - inhaled dose -).

As to the former, Table 1 shows the list of the substances recognized as possible causes of damage to the respiratory tract, and includes gases and/or organic and inorganic vapours (acids and bases), dusts, polycyclic hydrocarbons and photochemical smog. Most are produced by specific industrial processes while only a few are more frequently found in residential areas (Table 2), being emitted by domestic heating plants and motorvehicle traffic as well as by reactions catalyzed by natural variables such as radiation, humidity and prolonged contact: this is typically the case for the polluting mixture referred to as 'photochemical smog'.

In addition, other factors such as climate, infectious and allergic factors and individual habits (cigarette smoking) may cause damage to the respiratory tract.

The pathogenetic model of bronchopulmonary disease is complex: there is a great variety of causes which interfere and interact with each other to produce these disease states slowly over time.

The evaluation of the risk involved allows us to plot 'maps', having them available allows us not only to carry out environmental control measures in the best possible way but also to achieve the following objectives:
-planning medical action, in its aspects of prevention diagnosis, treatment and rehabilitation, in terms of the population's real needs;
-rational territory management: choice of the areas where industrial and/or residential development should occur made of the basis of the distribution of pollution gradients;
-development of research on atmospheric forecast simulation

Table 2 - More frequently aerodisperse pollutants

SO_2

NO_2

O_3

CO

Aliphatic Aldehydes

Suspensed particulate matter

Asbestos

Aromatic Polycyclic Hydrocarbons

Smog (photochemical reactions)

Table 3 - National Ambient Air Quality Standards of Sulfur dioxide

Country	Short-term standard $(mg.m^{-3})$	Long-term standards $(mg.m^{-3})$
U.R.S.S. - Bulgaria	0.50	0.05
Czechoslovakia - D.D.R. Yugoslavia - Hungary	0.50	0.150
Canada	0.90	0.06
Argentina	-	0.07
U.S.A.	1.30	0.08
Sweden	-	0.125
Belgium	-	0.150
Spain	-	0.150
Italy	0.750	0.380
WestGermany	0.750	0.500
France	-	1.00
Ratio between values (the greater divided by the smaller)	2.6	15.00

through 'field' tests aiming at verifying the reliability of both
its design and application.

PRESENT POSSIBILITIES

    Whilst is may be advisable to have risk maps available,
their acquisition seems to be either impossible, or complex and
hazardous for the two following reasons:
a) lack, at least in Italy, of information about the rate and
territorial distribution of air pollution. As a matter of fact,
scanty scientific research combines with inadequate policies,
followed by public institutions in this field (Provincial Hygiene
and Prophylaxis Laboratories, Air Pollution Regional Committees
and Public Health High Commission). Indeed, the few
investigations carried out by them are limited in time and space
and therefore do not provide a sufficiently complete and organic
picture. Besides, they are often conducted merely for tax
purposes, and when this is the case, monitoring is limited only
to few specific pollutants associated with exceptional and
sporadic episodes. Thus, the lack of continuous and global
measurements of the air pollution level does not allow to know
the pollutant content of the air even in the areas with the
highest risk level.

b) Inadequate knowledge, in relation to most pollutants, of their
real potential for damage, at least with regard to their chronic
and long-term effects. That this is so becomes evident when one
considers the concentrations permitted in several countries of
the world as well as the threshold values reported by different
Authors. The data in question are shown in Tables 3,4,5,6,7. As
regard air quality standards, one cannot help observing: -
a great variation in the standards adopted, especially those
relating to prolonged or chronic exposures. For instance, for
sulfur dioxide (Table 3), standards vary from 0.05 mg/m^3 (Soviet
Union) to 0.75 mg/m^3 (France), a variation ratio of 1 to 15;
-for nitrogen oxides (expressed as NO_2) (Table 4), standards vary
from 0.04 mg/m^3 (German Democratic Republic) to 0.2 mg/m^3 (Italy)
with a variation ratio of 1 to 5;
-for suspended dusts (Table 5), standards vary from 0.06 mg/m^3
(USA) to 0.3 mg/m^3 (Italy) with a variation ratio of 1 to 5;
-for carbon monoxide (Table 6), standards vary from 1.0 mg/m^3
(Soviet Union) to 23.0 mg/m^3 (Italy) with a variation ratio of 1
to 23.

This heterogeneity seems to lack any technical or scientific
justification. Indeed, the only explanation that the examination
of the data reported above suggests seems to have to do with the
different conceptual frame works associated with countries with
different socio-economic systems. This accounts for the fact

Table 4 – National Ambient Air Quality Standards of Nitrogen Oxides
        (expressed as NO_2)

Country	Short-term standard (mg.m^{-3})	Long-term standard (mg.m^{-3})
D.D.R.	0.100	0.040
U.R.S.S. – Bulgaria Hungary – Yugoslavia	0.085	0.085
Czechoslovakia – Romania	0.300	0.100
Canada	0.400	0.100
U.S.A.	–	0.100
West Germany	0.300	0.100
Italy	0.600	0.200
Ratio between values (the greater divided by the smaller)	7.1	5.0

Table 5 – National Ambient Air Quality Standards of suspensed
        particulate matter

Country	Short-term standard (mg.m^{-3})	Long-term standard (mg.m^{-3})
U.R.S.S. – Bulgaria Czechoslovakia – D.D.R.	0.500	0.150
Canada	0.120	0.070
U.S.A.	0.150	0.060
West Germany	0.300	0.100
Italy	0.750	0.300
Ratio between values (the greater divided by the smaller)	6.25	5.0

that air threshold standards of countries with planned economies are generally lower than those of the other countries. But it does not provide a full justification of the heterogeneity of the data in question: in fact ratios are not constant (for example, the standards established in the Soviet Union for inert dusts are higher than the corresponding standards established in the USA), and then, and above all, even among countries with a similar economic system the policies adopted vary considerably from each other (see, for instance, the standards for SO_2, nitrogen oxides, dusts and carbon monoxide adopted in Belgium, Germany and Italy).

Finally, it must be added that, as Table 7 shows, for each pollutant no constant and precise relationship seems to exist between the air quality standards established with regard to occupational environments and the threshold values relating to experimental toxicology tests.

The only criterion that seems to have been followed in the definition of air quality standards is related to economic or productive considerations which have very little to do with health protection. It is clear that the chance of contracting lung diseases through inhalation of a given dose of pollutant is independent of both the geographical area and the socio -economic system to which one belongs.

c) If, the situation appears to be confused, for common pollutants it is even more difficult to assess the role of other substances which are not even mentioned in the legislative provisions which regulate pollution control. This is the case, for example, of asbestos fibres and certain organic compounds like monomeric vinyl chloride.

d) Finally, the rationale underlying 'air quality standards' does not allow then to provide sound and exhaustive criteria of risk evaluation. They refer to single pollutants for which they state the maximum permissible concentration for prolonged or short-term exposures and do not permit a global and integrated assessment of health risk resulting from the simultaneous presence of different substances as well as the modifications and interactions they may undergo in the air (synergistic, cumulative and integrated effects).

The conclusion is that at the present time it is impossible to obtain a reliable quantitative evaluation of the risk of damage to the respiratory tract. Therefore for the time being 'pulmonary risk maps' remain unfeasible, even though they are necessary and advisable.

Table 6 - National Ambient Air Quality Standards of Cabron Monoxide

Country	Short-term standards $(mg.m^{-3})$	Long-term standards $(mg.m^{-3})$
U.R.S.S. - Bulgaria D.D.R. - Yugoslavia	3.0	1.0
U.S.A.	40.0	10.0
Argentina	57.7	11.5
Canada	35.0	15.0
West Germany	40.0	10.0
Italy	57.7	23.0
Ratio between values (the greater divided by the smaller)	19.2	23.0

1) DIVISION OF THE URBAN AREA IN A SQUARE GRID PATTERN

2) GROUPING OF SMALL SOURCES CONTRIBUTIONS in X_a [S.A.C.]

3) X AS A FUNCTION OF

   (*) Quantity of emitted pollutant

   (*) Wind speed and direction

   (*) Stability class of low atmospheric layers

4)   BUILT VOLUMETRY

    METEOROLOGICAL DATA   S.A.S.   ($\mu g\, m^{-2}\, s^{-1}$)

    FUEL CONSUMPTION   for every grid square

5) σ_z : FUNCTION OF LOW ATMOSPHERIC LAYER STABILITY

6) GIFFORD AND HANNA'S FORMULA

$$X = \frac{1}{\gamma}\sqrt{\sqrt{\frac{2}{\pi}}\,\frac{(\Delta x/2)^{1-b}}{\bar u\, a\,(1-b)}}\left[Q_A(0,0)+\sum_{i=-4}^{4}\sum_{j=-4}^{4} Q_A(i,j)\,f(i,j)\left[(2r+1)^{1-b}-(2r-1)^{1-b}\right]\right]$$

7) SIMULATION OF SO_2 ISOCONCENTRATION

Fig. 2 Models of diffusion and dispersion from area sources (Mod.A.T.D.L. by Gyfford-Hanna) essential points.

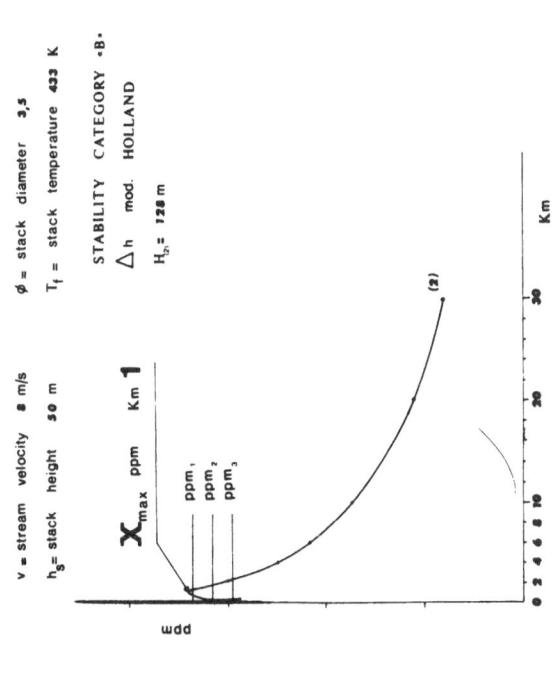

v = stream velocity **8** m/s ϕ = stack diameter **3,5**

h_s = stack height **50** m T_f = stack temperature **433** K

STABILITY CATEGORY «B»

Δh mod. HOLLAND

H_D = **128** m

X_{max} ppm Km

ppm₁
ppm₂
ppm₃

(2)

ppm

0 2 4 6 8 10   20   30   Km

STABILITY CATEGORY «B»

Δh mod. HOLLAND

H_D = **128** m

X_{max} ppm Km

ppm₃
ppm₁
ppm₂

x(Km)

y(Km)

Fig. 1 Forecast calculation of air pollutants fall-down from a standard industrial chimney (Holland-Sutton mathematical model).

Table 7 - Comparison between National Ambient Air Quality Standards of different pollutants

Substance	Short-term standards ($mg \cdot m^{-3}$)		Long-term standards ($mg \cdot m^{-3}$)		Threshold damage level ($mg \cdot m^{-3}$)			Work threshold limit ($mg \cdot m^{-3}$)
	Scatter of values	Ratio of scatter	Scatter of values	Ratio of scatter	Lauwerys	Sax	Lebowitz	
SO_2	0.50 1.30	2.60	0.05 0.40	8.0	0.075	0.1	0.10	13.0
NO_2	0.085 2.00	23.50	0.04 0.20	5.0	1.0	-	0.20	10.0
O_3	0.16 0.20	1.25	0.03 0.10	3.3	0.10	2.0	0.20	0.2
Particles	0.12 0.75	6.25	0.06 0.30	5.0	-	-	-	*10.0 (total) * 5.0 (respirable)
CO	3.0 57.7	19.2	1.0 23.0	23.0	55.0	110.0	20.0	55.0
Total Hydrocarbons	5.0 53.3	10.6	0.16 26.6	165.0	-	-	-	-

FUTURE PROSPECTIVES

These difficulties derive from lack of scientific
information and can be overcome only through an appropriate
research effort.

This calls for a series of longitudinal epidemiological
investigations of a prospective nature in areas with different
characteristics, but such investigations are by no means easy to
conduct owing to the multiplicity of contributing variables.
Besides air pollution, other factors are actually at work in the
genesis of chronic bronchopulmonary pathology: climate,
infections, allergic factors and individual habits (cigarette
smoking). The concomitant presence of two or more of these risk
factors suggests that the population under study be subdivided
into several sub-groups in order to isolate and assess the
specific influence of each factor.

Hence the need for involving in the research large samples
of population over long periods of time: because of this the
management of the investigation presents great difficulties which
can be reduced if suitable methodological criteria are adopted.
Because of the great number of the variables involved, the
research field can be compared to a closed system with several
components being present simultaneously. If one wants to
evaluate the specific influence of one of these on the whole, one
must arrange two homogeneous and comparable systems in each of
which the selected component (or variable) is kept constant. In
our case, therefore, one must select population groups which are
exposed to homogeneous pollution levels (the constant component
of the system) but present each different infectious climatic and
allergic factors (variable components). To do that it is
necessary to identify zones or areas of the territory under study
with known and uniform gradients of airborne pollutant content:
that is the areas which are called 'isoconcentration areas or
zones'. Within each of these areas one must then select
population groups to be studied longitudinally, with evaluation
of possible additional exposures to polluting agents (associated
with occupational and/or leisure activities) until the on set of
demonstrable respiratory diseases. To this end the investigation
methodology to be followed in order to arrive at a global or
integrated risk evaluation must aim at achieving the following
objectives:

A) constant and continuous monitoring of total inhalation risk
in ambient air (primary sources: motor vehicle traffic, domestic
heating, particular occupational activities and photochemical
mixtures);

B) evaluation of the degree and type of possible additional exposure associated with occupational activities;

C) quantitative evaluation of pollutant content in the air of indoor environments (at least winter and summer measurement of SO_2, nitrogen oxides and airborne dusts inside houses).

The assessment of these three components will make it possible to evaluate the global and integrated risk level over the whole day (the 24 hours period).

The identification of isoconcentration areas or zones which, as previously noted, seems to be a necessary prerequisite to the implementation of the operational methodology just illustrated. These can be achieved, at least as a first approximation, by means of theoretical-mathematical models whose elaboration provides the spatial and temporal distribution of isoconcentration gradient of airborne pollutants.

The models that can be adopted differ and can be distinguished from each other mainly in terms of two variables: characteristic of the source and chemical quality of the pollutant.

As far as the former variable is concerned, there are models of point sources (single chimneys), multiple sources (two or more chimneys) and area source (urban or industrial areas). The models can be applied to inert or non-reactive compounds and to reactive mixtures (photochemical smog). In the case of point sources, Sutton's equation (of the Gaussian type) is generally used. By means of it, it is possible to obtain the exact distribution under different meterological conditions, of isoconcentration values for increasing distances from the source and for time intervals varying from 10 minutes to 24 hours. This also allows identification of the point of maximum concentration, and ground level isoconcentration values. As an example, Figure 1 shows both the concentration source distance curves and the resulting isoconcen- tration curves in a hypothetical case.

The operational use of this model requires information about:
a) source characteristics: geometric or building; height from ground level, diameter of the chimney mouth, smoke emission rate, smoke temperature, ambient air temperature;
b) effluent characteristics: physical state concentration, emission frequency and duration;
c) meterological and climatic characteristics of the study area: wind velocity and direction, intensity and distribution of inversions, frequency of the categories of atmospheric stability (according to Pasquill).

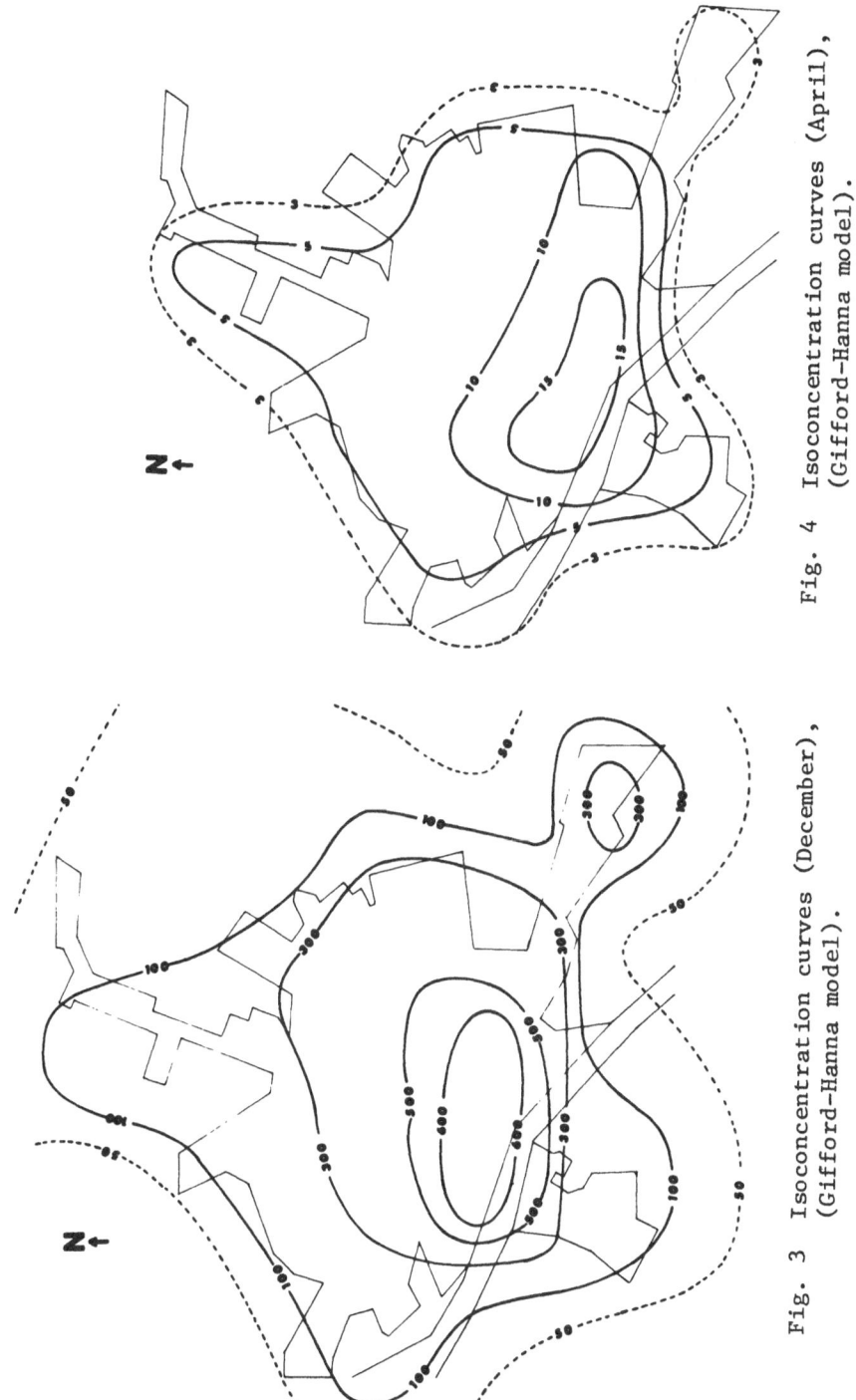

Fig. 4 Isoconcentration curves (April),
        (Gifford–Hanna model).

Fig. 3 Isoconcentration curves (December),
        (Gifford–Hanna model).

As far as area sources are concerned, the most used model is the Gyfford-Hanna model whose application in several towns, both foreign and Italian (Milan, Pavia) has shown that it can represent pollutant dispersion and deposition in urban areas with a satisfactory degree of reliability. The logical sequence of its application and its mathematical expressions are outlined in Figure 2. Finally, Figures 3 and 4 show SO_2 ground level isoconcentration curves relating to the months of December and April which are indicative of the different intensitites of domestic heating sources. The results concern the investigation carried out in Pavia urban area.

The parameters which are needed for the application of this model are:
-meteorological and climatic indexes: wind velocity and direction, distribution of monthly percent, distribution of the categories of atmospheric stability (according to Pasquill), monthly mean temperature of the air;
-technical and environmental indexes: housing density in each grid, mean built volumetry of buildings, quality of the fuel being used, potential of heating plants.

Because of its relatively simple elaboration, elastic application and satisfactory reliability the Gyfford-Hanna model can be used in several urban areas.

Finally, one has often to deal with 'mixed' urban areas in which industrial complexes occupy a large part of the territory. In such cases, isoconcentration areas can be obtained by using point source and area source models jointly.

As noted before, there are also models for chemically reactive substances (photochemical smog) as well as for working out case distributions over entire national territories. In both cases, investigations are under way which aim at verifying the reliability of their design and simulation.

To sum up, by using models for industrial chimneys and models for area or urban sources jointly, it is possible to forecast the description of isoconcentration areas with a good approximation to reality.

This in turn makes it possible to carry out epidemiological investigations following the methodology illustrated above.

CONCLUSIONS

Summarising, one notes that air pollution has now reached alarming levels with consequent marked effects on human health, which makes it necessary and sometimes very urgent to adopt control measures which, however, cannot be planned and carried out simultaneously over different and large areas for economic and organizational reasons.

Hence the need for identifying territorial zones with different pollution levels, that is for flotting 'pulmonary risk maps'.

This does not seem to be feasible for the time being not only because we still lack information about the degree and type of pollution but also because of the inadequacy of the evaluation criteria available. Air quality standards adopted in several countries appear to be suggested by economic or production reasons, and refer to the action of individual substances without taking their possible synergistic and integrated effects into account. This is why at present it is impossible to obtain an accurate and complete evaluation of damage potential and consequently plot 'risk maps'. In order to overcome these difficulties out it is necessary to carry out an extensive and spcecific research program based on epidemiological investigations of the prospective type. But research seems very complex because of the great number of the contributing parameters involved. The use of mathematical models can facilitate it as they allow to single out isoconcentration zones or areas. In this way it will be actually possible to select population samples exposed to the same pollution levels and therefore having as discriminating variables only the variables relating to climatic, infections and allergic factors.

By also evaluating possible additional exposures associated with occuptional and/or leisure activities one will then arrive at an estimate of the global and integrated risk for the whole day.

Only after obtaining these results will it be possible to plot risk maps reflecting the real situation and begin a serious prevention and control scheme on this basis.

REFERENCES

Checcacci, L. 1972, "Igiene e Medicina Preventiva", Casa Editrice
    Ambrosiana, Milano.
Goldsmith, J.R. and Frieberg, L.T. 1977, Effects of Air Pollution on
    Human Health, in "Air Pollution", vol. III, A.C. Stern, ed.,
    Academic Press, Inc., New York.
Hanna, S.R. 1971, A simple method of calculating dispersion from urban
    area sources, J.Air Pollution Control Association, 21, n. 12.
Izmerov, N.F. 1974, "Lutte contre la pollution de l'air en URSS",
    Cahiers de Sante Publique, O.M.S., Geneve.
Newill, V.A. 1977, Air Quality Standards, in "Air Pollution", vol. V,
    A.C. Stern, ed., Academic Press, Inc., New York.
Shy, C.M., Goldsmith, J.R. 1978, "Manual de gestion de la qualite de
    l'air de villes", O.M.S., Publication regionale, Serie
    europeenne n. 1, Copenhagen.
"The World's air quality management standards", 1974, vol. 1, E.P.A.
    650/9/75-001-a, W. Martin and A.C. Stern, eds., Washington.
Turner, D.B. 1967, "Workbook of Atmospheric Dispersion Estimates",
    U.S. Department of Health, Education and Welfare, Cincinnati,
    Ohio.
Zecca, E. et al. 1979, Epidemiologia delle broncopneumopatie:
    proposta metodologica per lo studio dell'inquinamento
    atmosferico nell'aria urbana di Pavia, Medicina Toracica,
    fasc. IV.
Zecca, E. et al., in corso di stampa, Calcolo teorico dell'emissione
    di SO_2 nell'aria urbana di Pavia e risultati dell'applicazione
    del modello di Gyfford-Hanna, Igiene Moderna.

DISCUSSION

LECTURER: Zecca   CHAIRMAN: Bonsignore

CANDURA: I would like to say a few things very briefly on this
                paper. Zecca is a well-known chemist and I myself
                have collaborated with him, so I think I may have
                understood something he did not say directly, but
                there may be people here who have not. The first
                thing is that when Zecca speaks of pulmonary risk
                maps, he does not mean maps of damage to the lung,
                but risk maps of harmful causes which penetrate the
                lung and can reach even a very distant target. The
                first slide showed carbon monoxide which is a poison
                which whilst absorbed through the lung has no direct
                action upon it. A second point is the reason why
                great differences exist between the maximum
                acceptable concentrations in socialist countries,
                those countries which look to the USSR as a model,
                and western countries. The difference is that while
                western people base themselves on the classical
                criteria of experimentsl toxicology and epidemiology,
                that is on animal experiments and the data collected
                from retrospective and prospective study of human
                populations, Soviet people prefer the model of
                behavioural toxicology, derived from Pavlov's work on
                conditioned reflexes. This is why gases such as CO
                and ethylene oxide are considered by the Soviets as
                having a very low acceptable concentration, whereas
                dusts have a higher acceptable concentration. If you
                expose an animal to CO you see that the variation of
                the conditioned reflexes is very marked. Conversely,
                if we expose the same animal to a dust things are
                completely different. And this explains the great
                difference.

ZECCA: I would like to stress that in the slides the
                difference between "socialist countries" was also
                shown, so this type of explanation is not absolutely
                applicable. There must be something else which we
                have not grasped.

PAOLETTI: Apart from the pessimism expressed by Zecca about
                pollution control, in Italy the CNR has promoted an
                epidemiological study to evaluate the incidence and,
                if possible, also the natural history of chronic
                bronchopulmonary disease, whereby in five towns we

explanation is not clear enough to me. I did not understand why SO_2 or CO would have a certain effect on behaviour and dusts would not.

CANDURA: I did not say it was obvious, I said that this is what happened historically. We considered this carefully and I was a member of the committee set up by the Department of Labour in Italy to fix the concentrations, and we found these differences between the values fixed by the Americans and those fixed by the Russian Academy of Sciences. The Russians do in fact use models of behavioural toxicology and they published their data based on it. If I may, I would like to add that when we speak of risk we must bear in mind that risk means probability. It is a relationship between two types of variables; the number of cases favouring a certain event - positive or negative - and the number of possible cases. This means that when we consider certain concentrations as acceptable we must implicitly consider that this acceptability is only valid for a certain number of the population exposed, although a large number. Indeed, I would say that by the very definition of risk as probability we can admit that it is valid for 95% of the exposed population, but not for all the population exposed. This is an important consideration which sets a further limit to the possibilities of interpreting the situation, because even if we draw these maps we then have to bear in mind that prevention can only be effective for a certain amount of the population, not all the population.

CUMMING: You will all remember the great London fog of 1952 and the excess deaths from bronchopulmonary disease of some 4000 which resulted from it. The probability here was quite clear, and the political will to do something about it rapidly appeared in the famous Beaver report of 1955, subsequent to which the prevalence and incidence of bronchopulmonary disease in Great Britain fell progressively and continued so to fall. I was therefore surprised to hear that in Italy the prevalence of bronchopulmonary disease is still increasing, and I would like Zecca to comment on that first.

ZECCA: I was not referring specifically to the case of Great Britain. I was stating the fact that the level of pollution is in any case increasing, at least in industrialised countries, including Italy. As a are trying to measure pollution in the same way. You spoke of epidemiological studies on a population sample in which certain risk factors are absent, in order to evaluate the effects of atmospheric pollution. This is a very straightforward approach to assess the effects of pollution on the lung. The best sample is in children. As we all know, chronic obstructive lung disease, the natural history of which is not known clearly, is presumably due to a series of risk factors - socioeconomic, familial, smoking, occupational exposure and pollution - therefore to evaluate all these factors in the adult population leads to what you called integration. Perhaps in this way it will be possible to obtain results with regard to the adult population. One question I wanted to ask is what do you think of the methods for measuring pollution in relation to respiratory epidemiology. You know that there is the problem of where the measurement is made and where the subjects go to in the towns where these studies are made. They are probably not always in the same place.

ZECCA: I would like to comment on what the CNR is going to do in various Italian towns to study the relationship between pollution and COLD. The reference I made to national bodies, which are rather ignorant of the way in which these investigations should be carried out, only included those bodies which are responsible for this under the law. This is not so in the case of the CNR. The initiative is certainly admirable, but it is promoted by a research institute. As to the criteria for monitoring, it seems to me that there is some confusion internationally, especially if we compare the different approaches to the problem, and in addition to that it is greatly difficult to measure the different exposures throughout the day - different exposures in terms of both quality and quantity. What I wanted to underline is the need to look at total exposure throughout the day in the population sample, considering the working environment, the home and so on. Of course this raises enormous difficulties, and this is why I suggested the use of selected groups, to reduce these difficulties and make the study feasible.

SANT'AMBROGIO: I would like to ask Candura to tell us something about the criteria chosen by the Soviet authorities and by the western authorities to fix tolerable concentrations for SO_2, CO or dusts. The behavioural consequence, then, the incidence of bronchopulmonary disease is also increasing. The specific, acute case of England that you were referring to was not in my mind when I said that.

CUMMING: My second point concerns the fact that the pollutants you are mainly concerned with were oxides of nitrogen, SO_2, CO and so forth, that is the products of fossil fuel combustion. Are you therefore in favour of the policy that replaces fossil fuel combusion with nuclear power?

ZECCA: No, this is not automatically said.

BONSIGNORE: Thank you then to all the speakers.

DISCUSSION

LECTURER: Fiocco   CHAIRMAN: Cumming

CUMMING: May I begin with a question myself. You referred to the rising level of CO_2 which is present in the atmosphere, can you tell us something about the balance between the output of CO_2 from humans, which I compute to be around 1000 million tons each year and the consumption of CO_2 by plants and its conversion into carbohydrates? How is that balance going?

FIOCCO: I do not have numbers on this, but I know for sure that numbers do not come back that well. If you compute what the sources are you find that in fact the atmosphere increases its CO_2 content at much less a rate you would expect. So our estimates of these things are not terribly correct; the way CO_2 interacts with the environment is certainly with the photosynthetic plant growth mechanism, but also with the surface of both the solid earth, because of the action with the rocks, and also because the CO_2 diffuses in water and then by sedimentation produces the bed of the oceans. And this is not very well known, so the estimates are probably off by 50%, and this is why we lack the precise content of CO_2 in the air.

JEFFERY: Can I just ask our Chairman, is that computation made with regard the results of cellular respiration, or the amount of CO_2 exhaled with every breath?

CUMMING: The latter.

JEFFERY: The amount produced by cellular respiration might be the more appropriate thing.

CUMMING: But much more difficult to calculate.

JEFFERY: Quite certain. Can I ask our speaker a question that is in many people's mind. What affects do the propellants in aerosol canisters have on the upper atmosphere?

FIOCCO: This is a question of current interest and has been
                     debated in the best scientific circles. It has also
                     been a way to attract funds in atmospheric research
                     from industry and governments, so that all of us have
                     been enjoying this subject of research. One has to
                     realise that for physicians disease is a subject of
                     interest, for the geophysicist, earthquakes and other
                     disasters are the reason of existance of the
                     profession. So, perturbation is essentially needed
                     in order to see how the environment performs. Going
                     now to the question of aerosols, intended here as
                     aerosol propellants which belong to the family of
                     freons. These are essentially methanes where the
                     hydrogen has been replaced by either chlorine or
                     fluorine or a mixture of the two, and the aerosols
                     used in the spray cans are either F11 or F12, which
                     means Cl_3F or Cl_2F_2. These are not the only freons
                     which we have to bother with, there are those used in
                     refrigeration systems which are of different
                     composition vaporisation temperature. To return to
                     the issue, the production is immense and the reason
                     why they are used so much is that they are so stable.
                     But stability means that once you have injected them
                     into the atmosphere they will stay there practically
                     for ever, so long as they stay in the lower
                     atmosphere. When they diffuse up, and it will take
                     time for them to reach stratosphere, there is an
                     upper limit of altitude beyond which even the
                     structure of the stable freon is broken down by solar
                     radiation, so that in the end out of freon molecules
                     we begin to obtain Cl molecules and whatever. Cl.
                     reacts very promptly with O, produces ClO, ClO reacts
                     with ozone and produces ClO_2, ClO_2 again reacts with
                     ozone and gets back to ClO once again and the final
                     result of this is that ClO remains always at the same
                     level, but ozone is removed continuously by this
                     catalytic process. The prediction of the effects of
                     this on the stratosphere are rather terrifying if one
                     were to assume that the production of these gases
                     were to increase with the increasing size of
                     industry. If we assume no increase in production,
                     then an estimate made by ten different laboratories
                     using the same algorithms gave similar results that
                     by 2020 or 2030 we would have a depletion of O_3
                     content of 18.5%. Wheich begins to have a
                     substantial influence on the level of radiation which
                     reaches the lower atmosphere as if we were all going
                     to shift from where we live now to a somewhat more
                     mountainous location. One wonders if this is a
                     correct estimate. There has been an estimate on the

accuracy of this estimate, and this would be probably something like another 7% uncertainty, which if added would be 25%. But the main problem insofar as the accuracy of the model is concerned relates to the general circulation of the atmosphere which was not included in a very sophisticated way in these models because of the difficulty of including it. So these are one-dimensional models in which the vertical motion of the atmosphere is simulated by means of a vertical diffusion coefficient based on the best estimates available so far.

COHEN: You showed a considerable production of free radicals which of course are extremely reactive and have very short half lives and as such perhaps would not create a hazard for human health. But are there atmospheric conditions which prolong the half-lives of those free radicals and then cause a much greater danger?

FIOCCO: The radicals enter into our technique of computation, because we would not be able to encompass the wide range of equations which we use in finding the evolution of the atmospheric chemistry. As I said, they are very difficult to measure. If we have a petrochemical plant downwind you find all sorts of doses, in large cities you find large levels of pollutants. To go back to your specific question, they would react with each other, and the production of O_3 is in essence the removal of a radical, I do not see what you really mean by that.

COHEN: If you could stabilise the radical by interaction, say by absorption onto material such as stabilises natural radicals.

FIOCCO: This would be essentially a process which would take place when there is heterogeneous nucleation, the formation of aerosols is probably one of the mechanisms by which these radicals would find a way of getting into a more stable form. It is not that the creation of an aerosol is less dangerous than the radical itself, it is probably more dangerous, but from the point of view of chemistry, it is essentially stabilised.

CUMMING: I think then we should end the session at that point. Thank you very much for your very interesting presentation.

THE PARTICULATES IN THE ATMOSPHERE AND THEIR

INTRAPULMONARY DEPOSITION

Vittorio Prodi, Carlo Melandri, Giuseppe Tarroni

Health Physics Laboratory of CNEN

Via Mazzini 2, 40138 BOLOGNA (Italy)

INTRODUCTION

The respiratory tract is, of the body structures, one of the most exposed to environmental agents, in view of the extremely high surface area and of the air volume which is exchanged daily.

The substances contained in the air can come into contact with the walls of the respiratory tract and can enter in that way the human body. Only substances in the form of particles will be considered here since their behaviour while airborne and in the interaction with the airways is quite peculiar and has to be treated separately.

The particles in the atmospheric environment comprise an extremely wide size range: if we limit our consideration to the aerosols that have a minimum of stability, the sizes range from a few nanometers to several tens of micrometers, that is, through four orders of magnitude. Since the size is by far the ruling parameter in the aerosol behaviour it is understandable that aerosols have widely different properties.

Aerosols have to be treated as distributions; if a single source is acting, generally the aerosol produced follows a log-normal distribution [1], characterized by a median or geometric mean size and a width, the geometric standard deviation σ_g.

In the free atmosphere, many processes are at work and we must expect different distributions or "modes".

If we count the number of particles in each logarithmic [2] size

235

interval, a maximum is found around 10 nanometers, and the concentration in other size ranges is very low as compared to the 3 to 100 nm range (Fig. 1).

The surface carried by the particles per logarithmic size has quite a different aspect: a prominant peak appears between 0.1 and 1 µm (100 and 1000 nm).

If the volume (mass) of the particles is measured as the function of size a peak appears around 10 µm, but the peak between 0.1 and 1 still retains a considerable fraction of the total volume carried by the particles.

These three peaks reflect different production mechanisms in the atmosphere. If we follow the nomenclature proposed by Whitby[2] these result in modes: the smallest particles called the "nuclei" are produced during combustion or photochemical conversion; because of their high diffusivity they coagulate very rapidly to form larger particles. These, together with condensational growth and chemical reactions, form the "accumulation" made that gives the highest contribution to the surfaces carried by the atmospheric aerosols.

The mode carrying the larger part of the aerosol mass is called the "coarse" mode and is generally originated by mechanical processes both natural and anthropogenic. There is very little exchange between the "accumulation" and the "coarse" mode so their chemical composition is markedly different.

The relative magnitude of these modes, depends on the particular conditions: proximity to the source, age, dilution processes, scavenging by precipitations. The median size and the geometric standard deviation may also vary with the specific conditions.

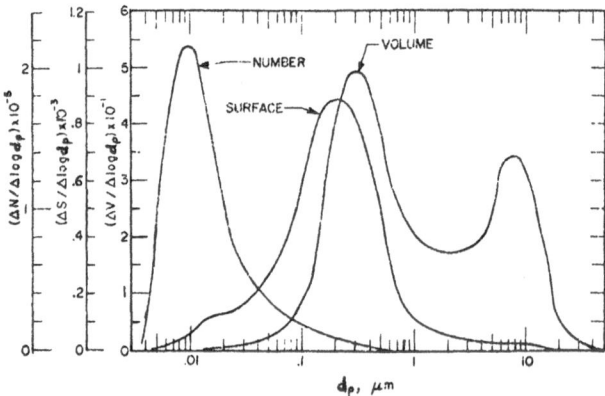

Fig. 1.   Normalized number, surface, volume distribution of an urban aerosol. (From Ref.[2]).

INCORPORATION

Incorporation of airborne particles can result from their interaction with the airways: it is a step process that comprehends inhalation of the aerosol, deposition of the inhaled aerosol, clearance or translocation of the deposited aerosol and the final result is the retention of the contamination with the resulting toxic action.

Each one of these steps has measured or measurable efficiencies that depend on aerosol properties, on airways configuration and on respiration patterns. Very detailed reviews have been published by the ICRP Task Group [3] on Lung Dynamics and by Lippmann[4]. Therefore here only the most recent results will be reported, with emphasis on the parameters relevant to the risk of incorporation.

Definitions

Inhalability: is the probability for an airborne particle to enter the airways[5].

Deposition: is the probability for an inhaled particle to touch a surface of the respiratory tract and to adhere to it[5].

Total deposition: is such a probability referred to the entire respiratory tract.
If a more detailed picture of the contamination is needed, the respiratory tract is divided into regions, characterized by predominant deposition and, for slowly soluble particles, clearance mechanism. It is generally accepted the definition of region [3,5]:

Extrathoracic airways, in which the deposition is mainly due to inertia and particles are cleared within minutes either by mechanical transport of particles or secretions.

Tracheobronchial airways, in which particles are deposited by inertia and settling and from which particles are removed within hours by mechanical transport of secretions.

Alveolar air spaces, characterized by small particle-to-wall distances, in which particles are deposited mainly by gravitational settling and brownian diffusion. The removal takes months or even years, mainly by phagocytosis, and subsequent cell transport to the mucociliary excalator or to the lymphatic system, or solubility.

Regional deposition is the probability for an inhaled particle to reach a surface of the given region and adhere to it. Total deposition is the sum of regional depositions. Deposited particles are cleared from the respiratory tract or translocated to other body regions.

Retention: total retention is the probability for a deposited parti-

cle to be retained in the body, while reçional retention is such a probability referred to a given region.
Deposition mechanism: particles are transported accross streamlines
by a number of mechanisms; if they touch a surface,
they deposit there. The main mechanisms are gravitational settling, brownian diffusion, inertial impaction
and electrostatic attraction due to image forces. All
of these mechanisms arre strongly size dependent. Where
aerodynamic affects prevail (settling and impaction),
the
aerodynamic diameter, d_a of a particle, defined as the diameter of a
1 g cm^{-3} density sphere having the same settling velocity as the particle, is a satisfactory and unique
parameter accounting for deposition[3]; aerodynamic techniques should be preferred for characterisation in this
range. Where diffusion prevail, since the diffusion
coefficient is independent of density, the aerodynamic
diameter is not representative and a
diffusive diameter should be introduced. This could be defined as
the diameter of the sphere that has the same diffusive
properties of the unknown particle. It is therefore
important that characterisation of an aerosol in this
range should be based on diffusion or on techniques
that can be uniquely related to diffusion. In case of
very large densities the range of overlap of diffusion
and aerodynamic effects varies and one can have still
aerodynamic deposition with appreciable diffusive deposition.

AIR FLOW IN THE RESPIRATORY TRACT

Deposition depends strongly on the flow which is effective in
the airways, because it can affect the time-dependent mechanisms
through the mean residence time, MRT, and the "prompt" mechanisms
(impaction) through the Volumetric Flow Rate [5],VFR

Respiratory frequency and tidal volume rule the MRT and the VFR.
Mixing between tidal and residual air has a very important role in
transferring inhaled particles into the residual volume and thereby
strongly increasing their residence time.

The respiratory tract is generally modeled after Weibel [7] as a
series of tubes branching out in regular dicotomy in each generation
from trachea (generation 0) down to the 23rd generation.

The air flow in the respiratory tract is very complex[4,8] characterized by a wide range of Reynolds numbers (Re). At a VFR of 1000
cm^3 sec^{-1}, Re is larger than 2000 down to lobar bronchi included,
which means turbulent flow and good mixing in this section of the

conductive structures. In addition the flow can be turbulent even
at lower VFR because of the corrugated walls and of the short cylindrical sections (a few diameters only) which prevent a fully developed laminar flow to be reached: at a VFR of 1000 cm^3 sec^{-1} there
can be turbulence down to segmental bronchi.

A laminar regime in the whole respiratory tract takes place
only at low VFR, corresponding to a minute volume of 3500 cm^3.

Schroter and Sudlow[46] have also shown that even in laminar conditions, at the bifurcations double vortices are stablished and
propagate into the daughter tubes during inhalation, and conversely
two double vortices develop during exhalation which propagate into
the parent tube. This can happen at Re down to several times unity,
corresponding, at 1000 cm^3 sec^{-1} VFR, to the 15th generation (bronchioles without cartilage).

The general effect is that a well developed laminar flow pattern
is never reached and mixing is practically extended to the whole anatomical dead space.

Even below the 15th generation the velocity profile is fairly
flat because the flow is not confined by smooth walls, but rather by
the alveolar openings: therefore at the opening the velocity of the
air is finite and it persists during the expansion of the alveolus;
tidal air penetrates into the alveolus like a tongue tangential to
the wall. This pattern is not reversed during exhalation.

An important mixing mechanism, active in the whole respiration
tract, is due to the non uniformity, both geometric and dynamic, of
the airways. Their diameter in each generation is considerably
scattered; branching angles, also, are asymmetric; this produces
a non uniform flow in the following bifurcations which is not reversed in the exhalation.

This non uniformity can be enhanced by the relatively rigid
structures (blood vessels and bronchi) of the airways: during expansion and contraction, geometric similarity is not preserved.
The mixing effect is enhanced at lower values of the functional
residual capacity and this can account for the increased deposition
with decreasing expiratory reserve volume,[9,10,11] ERV

The mixing effect has been shown for an array of branching
tubes also by Ultman and Blattman [6] by injecting a pulse of tracing
gas and following it down the generations.

If the pulse is at the boundary between tidal and residual air,
part of it may be left behind: its residence time is therefore increased by at least one respiratory period.

This problem has been treated also theoretically by Yu and Taulbee [10,13,14] through an effective diffusion coefficient and the overall deposition behaviour is accounted for.

For a detailed description of calculation steps the reader is sent to in the original works.

As it will be shown later, a considerable scatter of deposition data [11,15] is found among subjects inhaling in the same controlled conditions.

Unfortunately accurate morphometric data are not available for populations; it is not possible therefore to correlate the scatter of deposition values to any distribution of morphometric parameters, but there is little doubt that this scatter should be due to anatomical as well physiological or pathological differences[4].

AEROSOL BEHAVIOUR

Aerosol particles are affected mainly by inertial, gravitational and diffusive mechanisms. If the particles carry electric charges, their deposition is enhanced by image charges induced on the wall.

Inertial effects play a role whenever the streamline carrying the particle is forced to velocity changes. The particles tend to persist in their velocity because of their inertia. For spherical particles it is expressed by the relaxation time τ (the time required for a particle to reach 1/e of its terminal velocity under the action of a constant force)[1].

$$\tau = d_p^2 \, \rho_p K_s / 18\eta$$

where d_p is the particle diameter, ρ_p its density, K_s is the slip correction, important below some micron, and η is the gas viscosity.

The upper limit of τ in the range of sizes we are interested into is about 100 μsec: this is therefore a prompt mechanism since it is much shorter than the transit time through bifurcations. The deposition by inertia is proportional to the stokes number

$$St = \tau u / d$$

where u is the flow velocity and d is a characteristic dimension of the obstacle; it is therefore proportional to the particle size squared, to instantaneous VFR and inversely proportional to the radius of curvature of the streamlines.

Gravitational settling and diffusion have much longer characteristic times and even longer than a respiration period. Settling

velocity is u_g = τg, where g is the acceleration due to gravity.
Therefore the time[13] required for a particle to cover a characteristic airway size λ (of the order of 0.3 mm) is T_s = λ/ u_g. Settling velocity increases with the square of size: for 1 and 10 μm
particles it is $3.37 \cdot 10^{-3}$ and 0.302 cm sec^{-1} respectively.

Brownian diffusion can bring the particles in contact with the
walls: in that case they adhere to them. Their mean square displacement in the time t is

$$\overline{x^2} = 2\ D_p\ t$$

where D_p is the particle diffusion coefficient, D_p = kTB, where k
is the Boltzmann's constant, T is the absolute temperature and B
is the particle mobility, the asymptotic velocity per unit force.
Since B = K_s/3$\pi\eta d_p$ the diffusion coefficient is inversely proportional to particle size.

The characteristic time in the air ways is therefore λ^2/D_p:
1., 0.1 and 0.01μm particles have times of 3000, 100 and 2 sec respectively.

Settling and diffusion are therefore time-dependent mechanisms.

The walls of air spaces are practically conducting and no charge
can be accumulated: only charges carried by the particles induce
image charges on the walls[16]. If q is the particle charge, the
force is proportional to q^2, inversely proportional to the square of
the particle-to-wall distance and to particle size. The velocity
toward the wall is

$$v = \frac{B\ q^2}{16\pi\varepsilon x^2}$$

The electrostatic effect acts then in the vicinity of the wall,
when a new boundary surface is produced at a bifurcation or when
the effects previously mentioned bring the particles close enough
to the wall for the charge to be effective. At least in the range
0.3 to 0.1 μm also electrostatic effects are time-dependent.

Non spherical particles can be treated by the same considerations, through the introduction of a shape factor that relates the
property of interest of a particle (e.g. number, surface, mass) to
a diameter distribution (geometric or aerodynamic)[1].

A particular case is determined by fibrous particles, which are
important from the point of view of industrial hygiene and environmental health.

Fibers can be defined as particles having one dimension exceed-

ing the other two. They can be modeled either by cylinders or pro-
late spheroids. If their length exceeds the diameter by more than
a factor of 10, their aerodynamic behaviour is ruled by the fiber
diameter and the ratio of aerodynamic to fiber diameter is in the
range of 2.5 to 6 for asbestos or glass fibers[17,18].

Diffusion nevertheless is affected also by the length and even
with very small diameters diffusion displacement is much lower than
for compact particles of the same diameter.

Interception, though, becomes very important in connection with
the shear flow in high velocity gradients in the airways[8,19]. Fibers
undergo a periodic motion in a shear flow; this motion is slow when
fibers are aligned with the flow velocity and tend to tumble to re-
align with a 180° rotation. If, while tumbling, one end of the fiber touches a wall of the airways, it adheres to it and is intercepted.

Brownian rotation may have some part in disaligning the fibers
from the velocity and then starting a rotation.

At equal diameter, fibers can carry a much larger number of
charges than compact particles. This may be important for electrostatic deposition since the fibers are then aligned with the field[20]
and electric mobility may be very high.

This has been also utilized[21] for the separation of fibers from
a population of compact particles.

Another complication may arise if the particles are water soluble or even wettable. In that case they grow by water vapour condensation in the high humidity environment of the respiratory tract.
Dautrebande and Walkenhorst[22] showed, for sodium chloride in the
airways, a size increase of a factor of seven. However direct size
measurement[23] as a function of relative humidity did not show such
an increase.

In any case more data is needed on particle dynamics in humid
environments for the hygienic and therapeutic implications.

INHALABILITY

Very little is known on the efficiency of intake of the airways
for aerosol particles as a function of size, respiratory conditions,
wind speed, orientation of the head. The only extended work has been
done by Ogden and Birkett[24,25]. Experiments were run on a taylor
dummy in a wind tunnel and with wind speeds of 0.75 and 2.75 m sec^{-1}
and orientation from 0° to 180°.

With 2.75 m sec^{-1} wind speed, already at 5 μm diameter the aerosol is strongly oversampled for nose and mouth breathing at 84 cm^3 sec^{-1} VFR, is sampled faithfully at 335 cm^3 sec^{-1} and undersampled for higher VFR. At lower wind speed the portal of entry is not as important and the deviation of the sampled concentration from the true concentration is at most of the order of 25% above 15÷20 μm.

Orientation between 0° and 90° is very important with a 84 cm^3 sec^{-1} VFR at 90° and with a wind speed of 0.75 m sec^{-1} the nose can sample 5 μm particles with an efficiency almost halved with respect to 0°.

At a higher wind speed and at 45° the nose can strongly oversample, while the mouth samples always with a reduced efficiency.

Ogden and Birkett conclude that if the exposure is averaged on all wind directions, to simulate a worker uniformly exposed, the differences largely disappear and for winds between 0.75 and 2.75 m sec^{-1}, minute volumes between 20 and 40 l for nose and mouth breathing, the efficiency ranges as a functions of the aerodynamic diameter are shown in Table I.

Table I.   Range of Entry Efficiencies for a Head Randomly Oriented to a Wind Between 0.75 and 2.75 m sec^{-1} Breathing with Minute Volume Between 20 and 40 liters

Aerodynamic Diameter	0	5	10	15	20	25	30
Efficiency Range	100	68-83	46-72	39-69	33-60	31-55	30-52

(From Ogden and Birkett[24])

TOTAL DEPOSITION

Total deposition has received greater attention than regional deposition, since it can be studied in vivo with simple approaches, using monodisperse aerosols. Two techniques will be considered here: one[26,27,5] is based on the measurement of inhalation and exhalation flow rates and particles concentration just at the entrance of the airways. The number of inhaled and exhaled particles and the volumes are then computed by moltiplication and integration (Fig.2).

The other[11] is based on the measurement of the particle concentrantion averaged over several cycles with the volunteer (exhalation) and without the volunter (inhalation) and on a separate measurement of respiratory volumes (Fig. 3).

The two techniques have been compared and gave coincident results[28].

Heyder and coworkers[5,29] have throughly investigated total deposition, DE, as a function of route of inhalation, particle size,
respiratory parameters.

Size

The behaviour of DE as a function of size is shown in Fig. 4
(from Heyder et al[5]) for mouth breathing. The general trend shows
a minimum around 0.5 μm and increases both for decreasing and for
increasing size. Below 0.5 μm the increase is due to increasing
diffusion coefficient. Unfortunately in the ultrafine range experimental data is scarse[30]. Above 0.5 μm deposition is due to gravi-
tation and impaction.

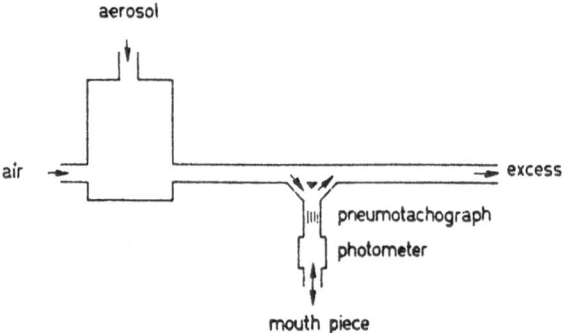

Fig. 2. A light scattering photometer is used for monitoring con-
         centration and a pneumotachograph is used for monitoring
         flow rate (Ref.[28])

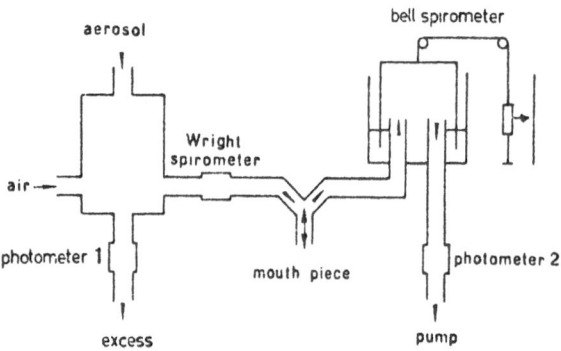

Fig. 3. Aerosol concentration and respiratory volumes are measured
         separately.(Ref.[28])

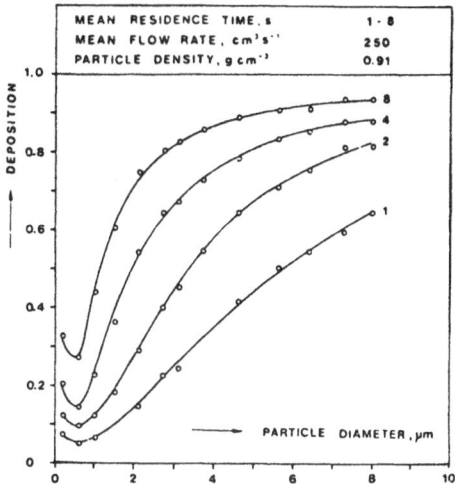

Fig. 4. Effect of particle size and MRT on total deposition.
(From Heyder et al.[5])

MRT

Fig. 4 shows also the effect of MRT on total deposition[5]: DE
increase with increasing MRT because both diffusion and settling
are time dependent. Impaction becomes important at higher sizes and
this is shown by the smaller effect of MRT at high VFR.

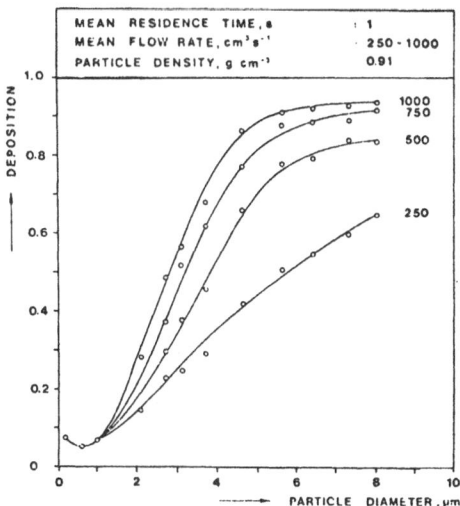

Fig. 5. Effect of particle size and VFR on total deposition.
(From Heyder et al.[5])

<u>VFR</u>

Total deposition is independent of VFR up to 1 ÷ 1.5 μm since impaction is not effective for the range of VFR encountered. At higher sizes VFR begins playing a more and more important role, as shown in Fig. 5 (from Heyder et al.[5]). The importance of impaction is depicted in Fig. 6 where MRT and VFR are varied while keeping the Tidal Volume constant. There is a definite cross-over of the curves, which is even more dramatic for nose breathing, as shown in Fig. 7 (also from Heyder et al.[5]), where it takes place around 1 μm, showing the contribution of impaction to nose deposition.

<u>Biological variability</u>

It is now generally accepted[4] that even under strictly controlled breathing conditions and residual volumes there is a definite intersubject variability of total deposition. An example of this is given[11] in Fig. 8 where DE is plotted as a function of particles size for six volunteers between 0.3 and 1.5 μm unit density spheres, breathing at 1000 cm^3 TV and 15 resp/min., each at his own expiratory reserve volume, ERV. This is interesting since in this range total deposition is also alveolar deposition.

The scatter of data reaches a factor of 2 and cannot be explained on the basis of respiratory parameters.

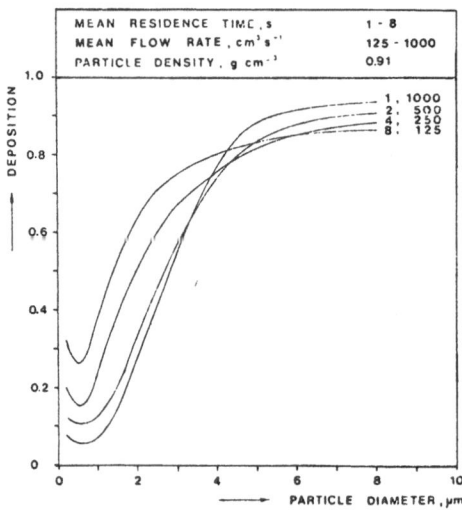

Fig. 6.   Effect of particle size on total deposition for mouth breathing at 1000 cm^3TV.(From Heyder et al.[5])

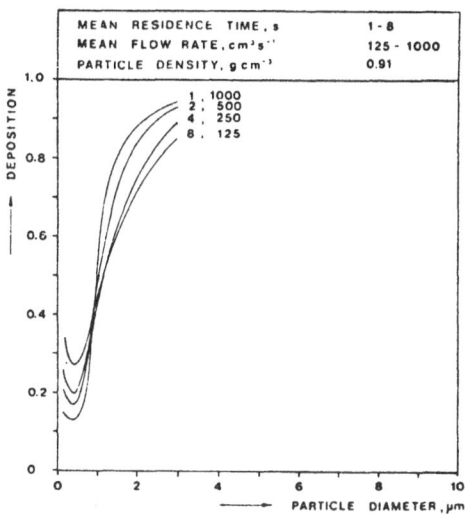

Fig. 7. Effect of particle size on total deposition for nose
         breathing at 1000 cm³TV.(From Heyder et al.[5])

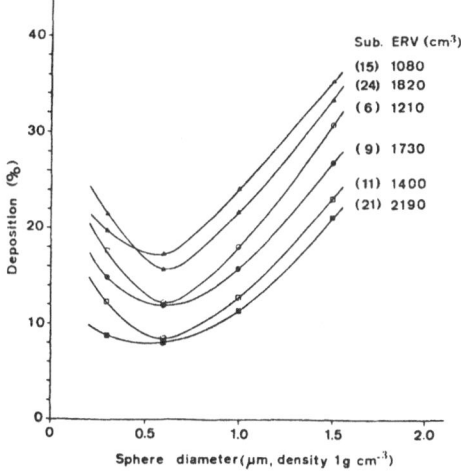

Fig. 8. Total deposition for a group of 6 volunteers, at 1000 cm³
         TV and 15 resp. min[-1].(From Tarroni et al.[11])

     For each volunteer instead, with 0.6 μm aerosols, a marked de-
pendence on ERV is found. The relative DE can be expressed as a
- 1/3 power of the ERV relative to normal, probably due to a stronger
mixing with smaller volumes[14]: to a 30% variation of ERV around the

normal value, a 10% variation of DE corresponds[11].

Electric charges

Electric charges carried by aerosol particles have a definite effect on deposition, leading to an increased efficiency. This has been found in deposition measurements of aerosols produced by atomization; when the aerosol was not neutralized, deposition was higher and poorly reproducible.

Quantitative measurements[16,11] have been performed only with monodisperse particles charged with positive elementary charges in a narrow number distribution and in controlled breathing conditions. The electrostatic deposition, shown in Fig. 9 increases monotonically with increasing charge number for 0.3 and 0.6 μm unit density spheres and is due to image forces[16] and therefore connected with the charge individually carried.

This has been confirmed theoretically[31] in the concentration range considered.

Quantitative information on charge distribution of actual aerosols is lacking: for aerosols freshly generated by disruption (gringind and atomization) the absolute charge can be very high and its contribution to total and regional deposition should be evaluated.

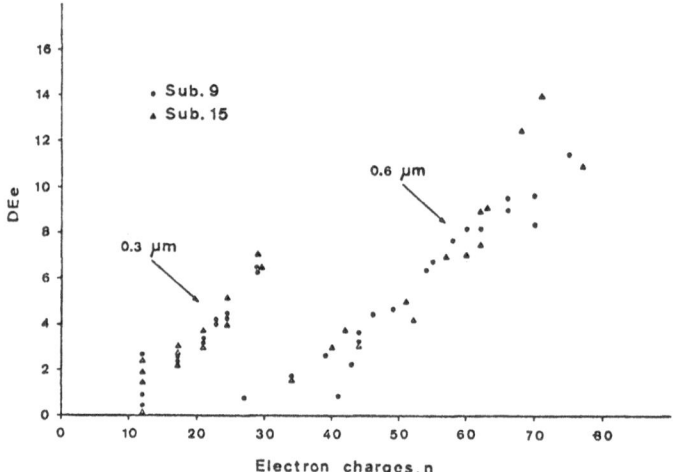

Fig. 9.   Contribution to total deposition from electric charges
           carried by 0.3 and 0.6 μm particles. (From Tarroni et
           al[16])

DEPOSITION PARAMETER

Recently, Heyder and coworkers[32] have proposed a deposition parameter X_M that summarizes the dependence of deposition DE on the various breathing and aerosol parameters. This parameter is limited to the size range where aerodynamic effects prevail and takes into account inertial deposition through the term

$$lg \left[\frac{\rho_p}{\rho_o} \left(\frac{d_p}{d_o} \right)^2 \right]$$

where ρ_o is 1 g cm^{-1} and d_o = 1 μm, the effect of time, through the term t/t_o where t_o =1 sec since the distance traveled by a particle under gravity is $\rho_p d_p^2 t$. The curves are empirically fitted with an exponent k = k(F) function of the volumetric flow rate F. The effect of flow rate is empirically fitted by introducing the parameter F/F_o where F_o is 1 cm^3 sec^{-1}. The complete form of the deposition parameter is

$$X_M = (lg \frac{F}{F_o} - 1.43)1 \left[\frac{\rho_p}{\rho_o} \left(\frac{d_p}{d_o} \right)^2 \left(\frac{t}{t_o} \right)^{24} \sqrt{F/F_o} \right]$$

Deposition date for three subjects in the range 0.5 X_M 2.5, that is for sizes larger than the minimum in deposition, can be described satisfactorily by the expression

$$DE = 1 - exp \left(- \frac{X_M^3 + 1}{6.5} \right)$$

presented in Fig. 10.

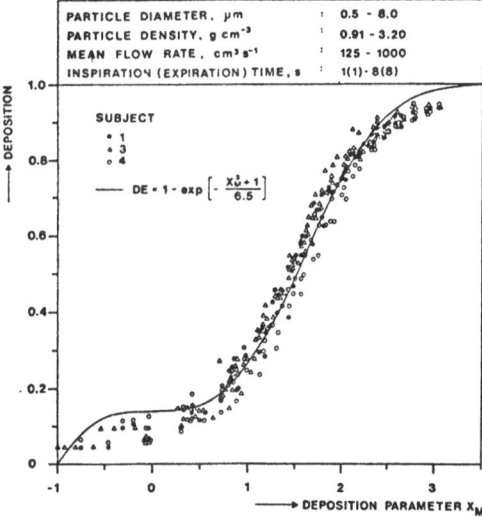

Fig. 10. Deposition as a function of the deposition parameter[32]

    This does not include, though, intersubject variability that,
in the group of three subjects that volunteered in the measurements,
was much lower than what can be expected from any group of the general population.

REGIONAL DEPOSITION

<u>Nose Deposition</u>

    Nose deposition must be treated separately since the breathing
route can be to some extent a matter of choice for the subject, and
has peculiar clearance pathways; this is important both for total
and regional deposition.

    Nose deposition data[33,34,35,36,37] have been recently summa-
rized by Lippmann[4] and shown in Fig.11.

    The deposition is a simple function of the logarithm of the
inertia parameter $\rho_p d_p^2 F$, where F is the VFR, and is practically quantitative for 9 μm particles at a flow rate of 30 liters per minute.

    The results are in fairly good agreement with the ICRP Task
Group deposition curve, based on Pattle's[38] data.

    Heyder and Rudolf[37] have also successively studied in detail
the deposition in the nose during exhalation, which is fairly close
to deposition during inhalation, although this efficiency applies
only to the transmitted fraction.

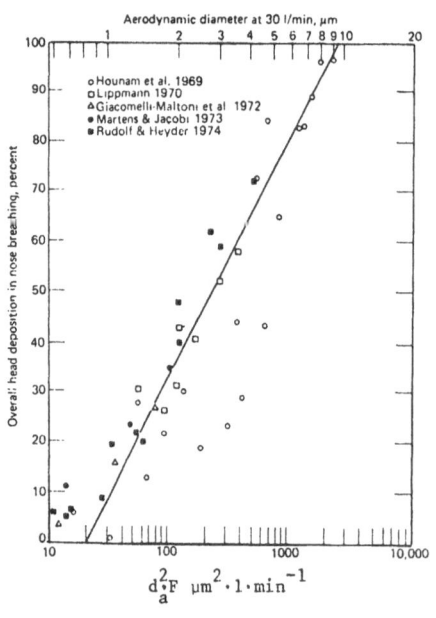

Fig. 11. Head deposition during
          inhalation via the nose
          vs the impaction para-
          meter $d_a^2 F$. (From
          Lippmann[4]). The solid
          line is the ICRP[3] curve
          based on Pattle's[38]
          data.

Mouth Breathing

Extrathoracic deposition

Regional deposition in mouth breathing has been studied by Lippmann and Albert[39], Chan and Lippmann[41], and by Stahlhofen et al.[40] by external counting of labeled monodisperse particles deposited in the airways.

Head deposition too can be linearly fitted with the lg $\rho_p d_p^2 F$ parameter. Lippmann's[4] data have been extrapolated to obtain the size for quantitative head deposition, that is around 17 μm for a 500 cm³ sec⁻¹ VFR.

Stahlhofen et al's[40] data show a slightly higher efficiency pointing to 100 percent deposition around 11 μm at the same flow rate.

The inertia parameter is not fully representative of deposition since the geometry of the airways may be dependent on the flow.

In Fig. 12 the average extrathoracic deposition of three subjects[40] is reported for 2 flow rates together with Lippmann's curve[4]

In Fig. 13 the effective total and regional depositions are shown as the average of three subjects[40], for two breathing patterns:

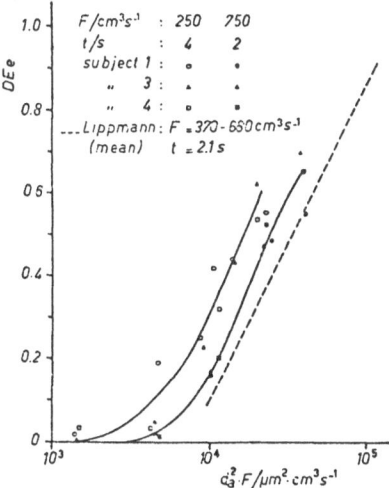

Fig. 12.   Head deposition during inhalation via the mouth vs impaction parameter. The solid lines and points are for three subjects at two VFR.(From Stahlhofen et al.[40]) while the broken line is Lippmann's[4] fitted curve.

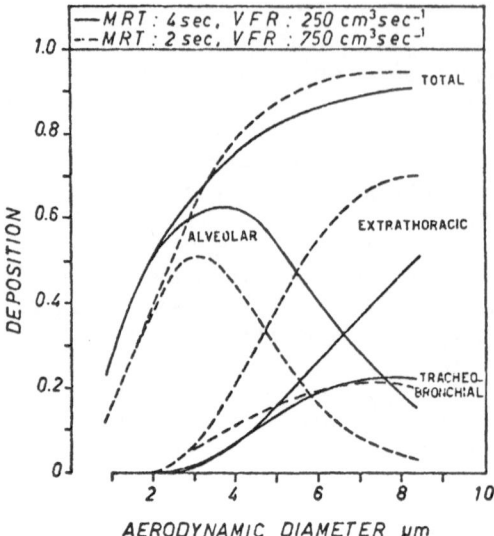

Fig. 13.   Average total and regional deposition for three subjects
           at two different respiratory patterns and for mouth breath-
           ing (based on data of Stahlhofen et al.[40])

TV = 1500 cm^3, 15 resp. min^{-1} (MTR = 2 sec, VFR = 750 cm^3 sec^{-1})and
TV = 1000 cm^3,7.5 resp. min^{-1} (MTR = 4 sec, VFR = 250 cm^3 sec^{-1}).
For the extrathoracic deposition the curves are derived from the same
data points of Fig.12.

Trachebronchial deposition

        Trachebronchial deposition has been studied in vivo by Lippmann,
Albert and Peterson[42], on a large number of volunteers. Lately it has
been studied by Chan and Lippmann[41] both in hollow casts and in vivo
and by Stahlhofen, Gebhart and Heyder[40] on three healthy subjects.
In addition, detailed studies have been performed on hollow casts of
human bronchial tree by Chan, Schreck and Lippmann[43], that have pointed out the flow pattern in the trachea, and preferential deposition
sites in connection with air flow and turbulence. In addition the
effect of electric charges on hollow cast deposition has been examined
[44] and the dependence on image forces has been confirmed.

        The studies of Chan and Lippmann[41], have shown a remarkable bio-
logical variability of deposition data even in the trachebronchial
tree.

        Fig. 14 shows the TB deposition expressed as a function of
the aerosol entering the trachea. The straight lines represent the
average and the scatter of the values found by Lippmann[4], while the

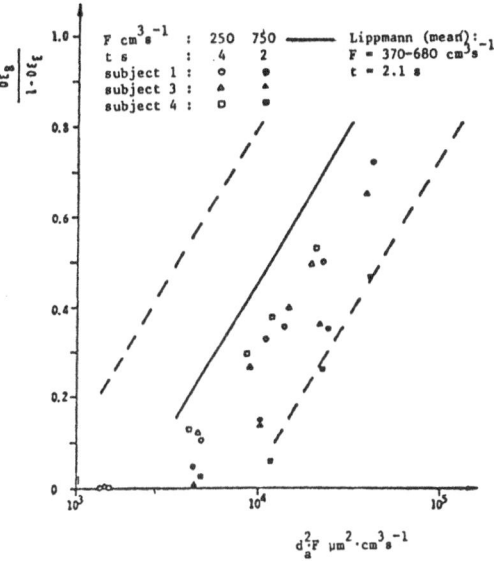

Fig. 14. Deposition in the ciliated trachebronchial region during mouth breathing, in percent of the aerosol entering the trachea. The straight lines represent the average and the scatter of Lippmann's[4] data while the points are the values obtained by Stahlhofen et al.[40]. (From Stahlhofen et al.[40])

data of Stahlhofen et al.[40] are shown with the points. These show a smaller scatter of data and two distint behaviours at two rates as well as values of deposition slightly lower than Limmpann's average.

The actual trachebronchial deposition is a bell-shaped curve that departs from zero around or slightly above 2 μm aerodynamic size and reaches a maximum, according to the flow conditions, between 6 and 10 μm.

In Fig. 13 the average actual TB deposition for three subjects[40] is plotted as a function of particle size for two respiratory pattern.

It has been pointed out that deposition depends strongly on the health conditions and increases for smokers and again for bronchitic patients. Chan and Lippmann[41] have proposed a parameter, called Bronchial Deposition Size, BDS, derived by expressing trachebronchial deposition as a function of the Stockes number. This was found 1.20 cm for healtly non smokers, 1.02 for smokers, 0.9 for patients under treatment for obstructive lung desease and 0.6 for severely disabled patients.

This effect may be important when establishing sampling guide-
lines and in evaluating health hazards of airborne particles since it
points out pupolation groups particularly at risk in some circumstances.

Alveolar Deposition

The gas exchange region of the airways is characterized by a
very large surface area and therefore by a small average particle-to-
wall distance and a large cumulative cross-section. Therefore deposition in the aerodynamic size range is practically due to gravitational settling.

The alveolar deposition therefore increases with increasing
MRT at constant VFR. Because of the behaviour of extrathoracic and
tracheobronchial deposition, also alveolar deposition follows a
bell-shaped curve: the relative maximum is aroung 3 μm and can be
shifted to smaller sizes both for increasing MRT at constant VFR
and for increasing VFR at constant MRT. In the first case the
alveolar deposition values increase since the extrathoracic and
trachebronchial deposition do not vary appreciably and gravitational deposition is more effective. Fig. 15 from Heyder et al.[5],
shows this effect in detail.

Instead, the increase in VFR causes a higher deposition by
impaction in the higher regions and therefore transmits a lower
fraction of large particles to the alveolar region and the effect
of increased VFR takes over the effect of decreased MRT.

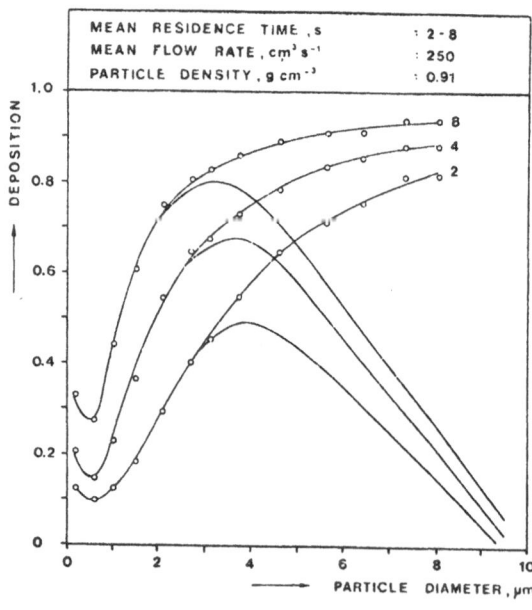

Fig. 15. Effect of particle
size and MRT on
total and alveolar
deposition.
(From Heyder et
al.[5]).

The data of Stahlhofen et al.[40] for alveolar deposition are
also summarized in Fig. 13 as the average of their three subjects.
These are in good agreement with Lippmann and Albert's[39], Chan and
Lippmann's[41], data for comparable respiration patterns. These data
are plotted together for comparison in Fig. 16 (from Chan and Lippmann[41]). The scatter of these are larger probably because of the
larger number of subjects. Consequently intersubject variability
is more considerable and the respiration is not as strictly controlled. This, though, makes them more representative of the range
of values that one can find in a population.

Deposition of Fibers

No data exists on fiber deposition in the human airways. The
only extended work on this subject is a theoretical calculation
performed by Harris[19], Harris and Fraser[8]. They combined the
mathematical expressions of particle behaviour with mathematical
expressions describing the air flow in the airways, considering
both the diameter and the lenght of the fibers.

In fact, as already pointed out, the aerodynamic behaviour is
ruled by the diameter while in interception and diffusion the length
plays a major role, as well as in the mechanisms of toxicity.

For the detailed calculation, the interested reader is sent
to the original work: here only the regional deposition curves as a
function of the equivalent aerodynamic diameter are shown in Fig.17
for a tidal volume of 750 cm^3.

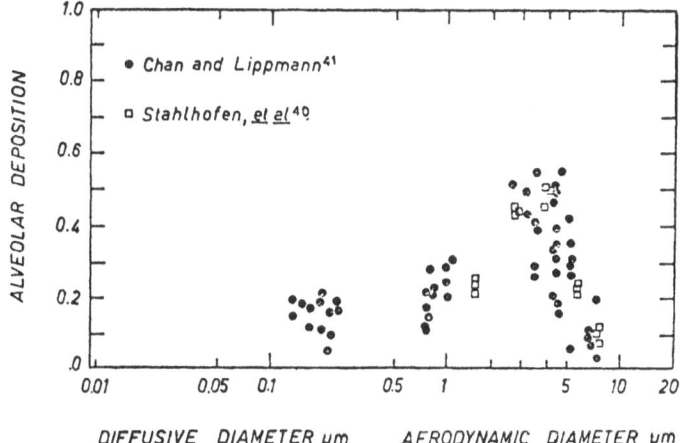

Fig. 16.   Alveolar deposition data of Chan and Lippmann[41] (dots)
            and of Stahlhofen et al.[40] (open squares).

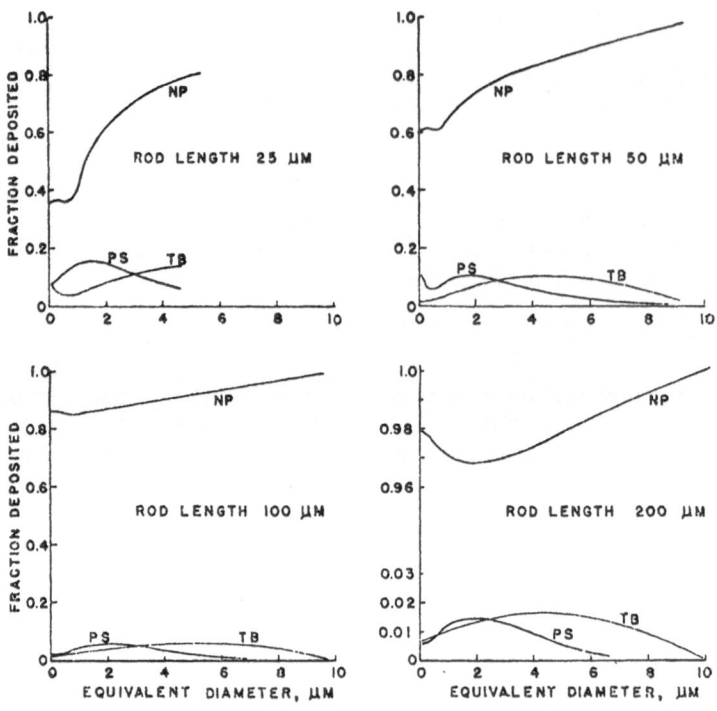

Fig. 17. Fractions of rods of selected lenghts which are predicted
          to deposit in each respiratory compartment. (tidal volume
          750 cm^3) NP = nasopharynx, TB = tracheobronchial, PS =
          pulmonary spaces. (From Harris and Fraser[8])

     The general trend   characteristic of compact particles is
followed, with the exception of Nasopharyngeal deposition that maintains a relevant efficiency even at very low equivalent sizes and
pulmonary deposition that practically has no pronounced minimum
around 0.5 μm.

     Even very long fibers (100 μm and over) have a finite proba-
bility of depositing in the pulmonary spaces.

CONCLUSIONS

     Particle deposition in the airways depends, as shown, on many
parameters. Most of the measurements refer to laboratory conditions
and their bearing to industrial or environmental hygiene is not
straightforward.   It is therefore interesting to study the sensitivity of total and regional deposition to aerosol parameters, respi-

ration pattern, inhalation portal, individual variability.

The geometric standard deviations of aerosols in the free atmo-sphere for each of the modes described above[2], ranges from 1.7 to 2.3. Therefore calculation[45] have been performed for a polydisperse aerosol of σ_g= 2, on the basis of deposition data for monodisperse aerosols.

The influence of individual variability on total deposition has been computed on the basis of the data of Fig. 8 and is shown in Fig. 18 as the 80% confidence interval.

Another example is given in Fig. 19 where the alveolar depo-sition is computed on the basis of Heyder et al's[5], Lippmann's[4] data for an aerosol of σ_g = 2 and for different respiration patterns and inhalation routes.

The results of the calculations are summarized in Table II for total deposition and in Table III for alveolar deposition.

In each case the mean deviation is shown, i.e. the average of the errors one can expect if a reference value is taken in place of actual value,in the size range 0.6 to 10 μm.

For particle size the reference value is 1 μm: a 46% mean devi-ation for total deposition implies that by taking 1 μm as the median diameter of an otherwise completely unknown aerosol, the average er-ror in the indicated ranges is 46%.

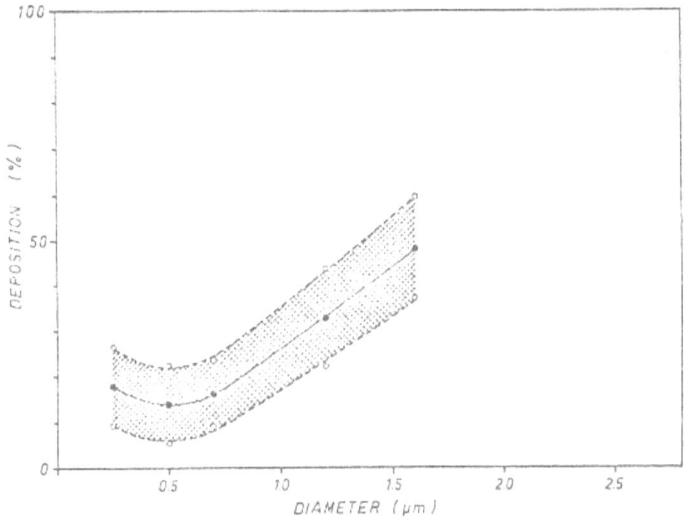

Fig. 18. Intersubject variability of total deposition[45]

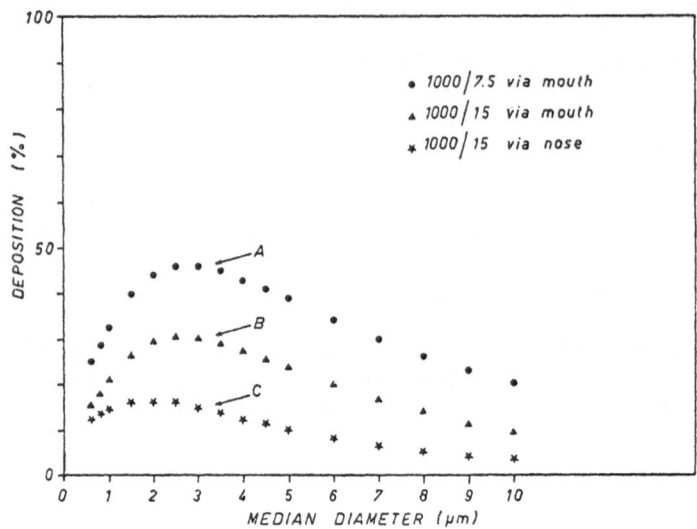

Fig. 19. Computed[45] alveolar deposition for polydisperse aerosols
of $\sigma_g = 2$

TABLE II.   Influence of various parameters on Total
            Deposition

	Mean deviation
Particle size	46%
Geometric standard deviation	17%
Electrostatic charge {0.3 μm; 29 e.c.	15%
Electrostatic charge {0.6 μm; 70 e.c.	26%
Respiration pattern	10%
Inhalation route	22%
Intersubject variability	19%

TABLE III. Influence of various parameters on alveolar
           deposition

	Mean deviation
Particle size	22%
Geometric standard deviation	30%
Respiration pattern	24%
Inhalation route	31%

Correspondingly, the reference values for the other parameters are: 2 for the σ_g, charge zero for electric effects, 1000 cm^3 TV and 15 min^{-1} for the respiration pattern. The average of mouth and nose for the inhalation portal, and the data of Fig. 19 for the inter-subject variability.

More data is needed, expecially on sizes below 0.6 μm, on charge effects on regional deposition, on particle growth in the high humidity environment of the airways. Nevertheless it can be stated that the uncertainties of the inhaling subject (respiration pattern, route of entry, biological variability) are comparable to the uncertainties of the aerosol parameters expecially for alveolar deposition.

The complete picture of incorporation implies a knowledge of translocation and clearance efficiencies of the deposited materials. Unfortunately these efficiencies are not yet known with comparable accuracy and more efforts should be devoted to a more detailed understanding of these mechanisms.

REFERENCES

1. T. T. Mercer, Aerosol Technology in Azard Evaluation, Academic Press, N.York, 1973.
2. K. T. Whitby and B. Cantrell, "Atmospheric Aerosols - Characteristics and Measurements",in International Conference on Environmental Sensing and Assessment. Session 29, Fine Particles, Las Wegas NV, Spet. 14-17, 1975, IEEE, New York, 1976.
3. P. E. Morrov, Chairman, ICRP Task Group Report on Lung Dynamics: Deposition and Retention Models for Internal Dosimetry of the Human Respiratory Tract, Health Physics 12: 173-208, 1966.
4. M. Lippmann, Regional Deposition of Particles in the Human Respiratory Tract, in Handbook of Physiology Sect 9: Reaction to Environmental Agents D.H.K. Lee, H.L. Falk, S.D. Murphy and S.R. Geiger Eds. Bethesda, MD : Am. Physiological Society, 1977, 213-232.
5. J. Heyder, J. Gebhart and W. Stahlhofen, Inhalation of Aerosols Particle Deposition and Retention,Aerosol Generation and Exposure Facilities, K.W. Willeke, Ed.: Ann. Arbor Science, Ann Arbor, 1980.
6. C. Melandri and V. Prodi, Simulation of the Regional Deposition of Aerosols in the Respiratory Tract, Am. Ind. Hyg. Ass. J. 32: 52-57, 1971.
7. E. R. Weibel, Morphometry of the Human Lung,Academic Press, New York 1963, pp. 136-140.
8. R. L. Harris Jr., D. A. Fraser, A Model for Deposition of Fibers in the Human Respiratory System, Am. Ind. Hyg. Ass. Journal, 37: 73-89, 1976.

9. E. D. Palmes, R. M. Goldring, Chiu-Sen Wand and B. Altshuler,
    Effect of Chronic Obstructive Pulmonary Disease on Rate of
    Deposition of Aerosols in the Lung During Breath Holding,
    in Inhaled Particles III, W. H. Walton, Ed. pp. 123-129.
    Unwim Brothers Limited, The Gresham Press. Old Woking,Surrey
    1971.

10. D. B. Taulbee and C. P. Yu, Theory of particle Deposition in the
    Human Lung, Ann. Meeting of the Gesellschaft für Aerosol-
    forschung, Bad Soden/Ts, 1976.

11. G. Tarroni, C. Melandri, V. Prodi, T. De Zaiacomo, M. Formignani
    and P. Bassi, An Indication on the Biological Variability of
    Aerosol Total Deposition in Humans, Am. Ind. Hyg. Ass. Con-
    ference, Chicago, Ill., May 27-June 1, 1979 - Submitted to
    Am. Ind. Hyg. Ass. Journal.

12. J. S. Ultman and H. S. Blatman, Longitudinal Mixing in Pulmonary
    Airways, Analisys of Inhert Gas Dispersion in Symmetric Tube
    Network Models,Respiration Physiol. 30: 349-367, 1977.

13. D. B. Taulbee and C.P. Yu, A Theory of Aerosol Deposition in the
    Human Respiratory Tract, J. Appl. Physiol. 30: 77-85, 1975.

14. C. P. Yu and D. B. Taulbee, A Theory for Predicting Respiratory
    Tract Deposition of Inhaled Particles in Man, Inhaled Particles
    IV, W. H. Walton, Ed. Pergamon Press, Oxford, 35-46, 1977.

15. G. Giacomelli Maltoni, C. Melandri, V. Prodi and G. Tarroni,
    Deposition Efficiency of Monodisperse Particles in the Human
    Respiratory Tract, Am. Ind. Hyg. Ass. Journal, 33: 603-610,
    1972.

16. C. Melandri, V. Prodi, G.Tarroni, M. Formignani, T. De Zaiacomo,
    G. F. Bompane, and G. Maestri, On the Deposition of Unipolarly
    Charged Particles in the Human Respiratory Tract, Inhaled
    Particles IV, W. H. Walton, Ed. 193-203, Pergamon Press,
    Oxford, 1977.

17. V. Timbrell, The Inhalation of Fibrous Dusts, Ann. N. Y. Acad.
    Sci., 132: 255-273, 1965.

18. W. Stöber, H. Flachsbart and D. Hochrainer, Der Aerodynamische
    Durchmesser von Latexaggregaten und Asbestfasern, Staub, 30:
    277-285, 1970.

19. R. L. Harris Jr., A Model for Deposition on Microscopic Fibers
    in the Human Respiratory System, Ph. d. Thesis, Sch. of Public
    Health, Univ. of North Carolina (1972), Chapel Hill, N.C.27514.

20. W. Stöber, Chr. Boose, V. Prodi, Uber die Orientierung und den
    Dynamischen Formfactoren von Kettenförmigen Aerosolteilchen in
    Ladung Spectro metern, Water, Air Soil Pollution, 3: 496-506,
    1974.

21. G. Zebel, D. Hochrainer, Chr. Boose, A Sampling Method with
    Separated Deposition of Airborne Fibers and Other Particles,
    J.Aerosol Sci., 8: 205-214, 1977.

22. L. Dautrebande, W. Walkenhorst, Inhaled Particles and Vapours,
    pp. 110-120, C.N.Davies Editor, Pergamon Press, Oxford 1961.

23. J. Porstendörder, Üntersuchungen zur Frage des Wachstums von Inhalierten Aerosolteilchen im Atemtract, J. Aerosol Sci., 2: 73-79, 1971.

24. T. L. Ogden and J. L. Birkett, The Human Head as a Dust Sampler, Inhaled Particles IV, W. H. Walton, Ed. 93-105, Pergamon Press. Oxford, 1977.

25. T. L. Ogden and J. L. Birkett, An Inhalable-Dust Sampler for Measuring the Hazard from Total Airborne Particulate, Ann. Occup. Hyg. 21: 41-50, 1978.

26. D. C. F. Muir, C.N. Davies, The Deposition of 0.5 µm Diameter Aerosols in the Lungs of Man, Ann. Occup. Hyg. 10: 161-174, 1967.

27. J. Heyder, J. Gebhart, C. Heigwer, C. Roth and W. Stahlhofen, Experimental Studies of the Total Deposition of Aerosol Particles in the Human Respiratory Tract, J. Aerosol Sci. 4:191-208, 1973.

28. J. Heyder, J. Gebhart, C. Roth, W. Stahlhofen, G. Tarroni, T. De Aiacomo, M. Formignani, C. Melandri and V. Prodi, Intercomparison on Lung Deposition Data for Aerosol Particles J. Aerosol Sci. 9: 147-155, 1978.

29. J. Heyder, L. Armbrüster, J. Gebhart, E. Grein and W. Stahlhofen, Total Deposition of Aerosol Particles in the Human Respiratory Tract for Nose and Mouth Breathing, J. Aerosol Sci. 6: 311-328, 1975.

30. D. L. Swift, F. Shanty and J. T. O'Neil, Human Respiratory Deposition Pattern of Fume-Like Particles, Am. Ind. Hyg. Conference, New Orleans, La., May 26, 1977.

31. C. P. Yu and K. Chandra, Precipitation of Submicron Charged Particles in Human Lungs Airways, Bull. Math. Biol. 39, 471, 1977.

32. J. Heyder, J. Gebhart, G. Rudolf, W. Stahlhofen, Physical Factors Determining Particle Deposition in the Human Respiratory Tract, 3rd Congress of the International Society for Aerosols in Medicine, Salsomaggiore Terme, Italy, May 7-10, 1980.

33. R. F. Hounam, A. Black and M. Walsh, Deposition of Aerosol Particles in the Nasopharyngeal Region of the Human Respiratory Tract, Nature 221: 1254-1255, 1969.

34. M. Lippmann and R.E. Albert, Deposition and Clearance of Inhaled Particles in the Human Nose, Ann. Otol. Rhinol. Laryngol. 79: 519-528, 1970.

35. A. Maertens and W. Jacobi, Die in vivo Bestimmung der Aerosolteilchen Deposition in Atemtract bei Mund-bzw-Nasen Atmung, Annual Meeting of the Gesellschaft für Aerosolforschung, Bad Soden/Ts, Oct. 17-18, 1973.

36. G. Rudolf and J. Heyder, Deposition of Aerosol Particles in the Human Nose, Annual Meeting of the Gesellschaft für Aerosolforschung, Bad Soden/Ts, Oct. 16-18, 1974.

37. J. Heyder and G. Rudolf, Deposition of Aerosol Particles in the Human Nose, Inhaled Particles IV, W. H. Walton, Ed. Pergamon Press, Oxford, 107-125, 1977.

38. R. E. Pattle, The Retention of Gases and Particles in the
     Human Nose, in Inhaled Particles and Vapors, C.N.Davies,
     Ed., Pergamon Press, Oxford, 302-309, 1961.
39. M. Lippmann, R. E. Albert, The Effect of Particle Size on the
     Regional Deposition of Inhaled Aerosols in the Human respi-
     ratory Tract, Am. Ind. Hyg. Ass. J. 30: 257-275, 1969.
40. W. Stahlhofen, J. Gebhart and J. Heyder, Experimental Determi-
     nation of the Regional Deposition of Aerosol Particles in
     the Human Respiratory Tract, Am. Ind. Hyg. Ass. Conference,
     Chicago, Ill. May 28-June 1, 1979, Submitted to Am. Ind.
     Hyg. Ass. Journal.
41. T. L. Chan and M. Lipmmann, Experimental Measurements and Em-
     pirical Modelling of the Regional Deposition of Inhaled
     Particles in Humans, Submitted to the Am. Ind. Hyg. Ass.
     Journal.
42. M. Lippmann, R.E. Albert and H. T. Peterson, The Regional Depo-
     sition of Inhaled Aerosols in Man, in Inhaled Particles III,
     W. H. Walton, Ed., Unwin, London, 105-120, 1971
43. T. L. Chan, R. M. Schreck and M. Lippmann, Effect of Turbulence
     on Particle Deposition in the Human Trachea and Bronchial
     Airways, 71st Ann.Meeting Am. Inst. Chem. Eng., Miami, Fl.
     Nov. 12-16, 1978.
44. T. L. Chan, M. Lippmann, V.R. Cohen and R. B. Schlesinger, Ef-
     fect of Electrostatic Charges on Particle Deposition in a
     Hollow Cast of the Human Larynx-Tracheobronchial Tree. J.
     Aerosol Sci. 9: 463-468, 1978.
45. G. Tarroni, C. Melandri, V. Prodi, T. De Zaiacomo, M. Formignani,
     Peso dei parametri ventilatori e degli aerosol sulla deposizio-
     ne nell'apparato respiratorio umano, AIRP Annual Conference,
     Palermo, Oct. 1979.
46. R. C. Schroter and M. F. Sudlow, Flow Patterns in Models of the
     Human Airways, Respiratory Physiol 7: 431-455, 1969.

DISCUSSION

LECTURER: **Prodi**   CHAIRMAN: Cumming

CUMMING: Thank you very much. When we prescribe our aerosols
                we perhaps do not think of the great complications
                that ensue.

FAGAN: Could I start off asking about your use of terms. In
                my youth I was brought up with the idea that a
                suspension of solid particles in a gas was a smoke,
                and that a dispersion of large droplets in a gas was
                a mist, and that an aerosol was a fine dispersion of
                droplets in a gas. Have the definitions all changed
                now?

PRODI: The term aerosol is a general term to mean a
                suspension of any particulate in a gaseous phase.
                So, aerosol would be the general term that
                comprehends smoke, mist, dust, fog and everything
                else you like.

FAGAN: Fine. Because where I got a little confused was in
                your equations about the sedimentation of the
                particles. You were using viscosity. Was this the
                viscosity of the aerosol, the viscosity of the
                suspended particle or the viscosity of the suspending
                gas?

PRODI: Sorry, I was not precise on that. It is the
                viscosity of the gas.

FAGAN: Fine, because when you went on to consider your
                sedimentation results you showed the percentage
                deposition plotted against the particle size, for
                different flow rates and for the minute volume
                variation, and then you showed first breathing
                through the mouth and then through the nose.
                Presumbly you were simply measuring the difference
                between inhaled concentration and exhaled
                concentration.

PRODI: Yes, deposition would be defined as the inhaled
                concentration minus the exhaled concentration divided
                by the inhaled concentration.

FAGAN: So you did it with different known particle sizes.

PRODI: Of course, yes, monodisperse aerosols.

FAGAN: That was a direct experimental data, but when you
                moved to the difference between tracheal and alveolar
                deposition, how did you do that?

PRODI: We did not do that experiment ourselves. These were
                done with data from Frankfurt and partially of New
                York University. This was done with radioactively
                tagged particles, measuring the alveolar deposition
                by extrapolation of the long term retention to zero
                time and then measuring the amount of radioactivity
                in the body by external gamma counting.

FAGAN: So it would be correct to say that your curve should
                be labelled large air passage deposition and small
                air passage deposition. You do not know that it was
                alveolar deposition.

PRODI: The longer term deposition is taken as an operative
                definition of alveolar deposition.

CUMMING: Perhaps even better would be central and non-central.

FAGAN: The reason I wanted to get at this is because I
                wanted to know how much agreement there is between
                the theoretical equations and the experimental work.
                It would seem to me that the precise localisation of
                deposition is of such critical importance that there
                must be very careful control of whether the equations
                are correct by practical work, because it would seem
                to me that the equations are very heavily dependent
                on the model of gas flow throughout the lung tree.

PRODI: Yes, definitely. A very extended theoretical work
                has been done at Buffalo. They have come up with
                data quite close to these, although they have based
                their calculations on Weibel's model which is the
                approximation we know and you know.

FAGAN: That is what I wanted to hear.

PRODI: So, there is no other way than making experiments,
                and these experiments were done on volunteers,
                because the load on subjects is very low. With
                fibres, of course it would be a completely different
                story.

FAGAN: I think there is a very interesting study to be done here with some precise control, counting the peripheral particles compared with the prediction equation, but thank you very much indeed.

PRODI: Well, there is a lot of work to do.

SANT'AMBROGIO: In one of your slides there was deposition with two different modes of respiration, and it varied much more in the extrathoracic airways compared with the alveolar region or the tracheobronchial tree. Is that because in different patterns of respiration the turbulence at the level of the mouth or of the naso-pharynx also varies? In other words, what is it which changes deposition with different patterns of respiration?

PRODI: You are right. The turbulence in the oropharynx is of great importance in determining deposition. This difference in deposition can be explained by a greater deposition efficiency in the tracheobronchial tree, which is due to a much higher flow capacity.

SANT'AMBROGIO: But is it deposition at the level of the upper airways?

PRODI: Yes, I would say so. Turbulence can, especially in subjects with bronchial pathology, be very important. And in fact Litman has measured the size of bronchi by measuring the efficiency of deposition of labelled aerosols - he calls it bronchial deposition size - and he found 1.2 cm in healthy people, non smokers, 1.02 in smokers, 0.9 in patients with obstructive lung disease and 0.6 in severely impaired patients. So, certainly turbulence affects deposition.

SANT'AMBROGIO: In subjects with bronchitic lesions is deposition in the tracheobronchial tree also affected by the mode of respiration.

PRODI: Yes, this is seen clearly using the radiolabelled albumin technique and the gamma camera, when there is very much increased deposition.

PAOLETTI: Central deposition, where the diameter of the large airways is less implies an increase in total resistance. This is perhaps more typical of asthmatics. In chronic emphysematous bronchitis deposition is probably more accidental, what is called spotty deposition. And it is much more difficult to say what, in these subjects, is the mechanism of deposition. Do you agree?

PRODI: But it might well be linked to turbulence. We must
              know clearly the pattern of flow. I am not so
              familiar with this subject.

CUMMING: Ladies and gentlemen, we have consumed all the time
              and must draw the discussion to a close.

BRONCHIAL HYPERREACTIVITY

G. Bonsignore, A. Rizzo, and V. Bellia

Institute of Respiratory Physiopathology
University of Palermo - 90146 via Trabucco 180
Italy

Bronchial hyperreactivity is generally considered as an abnormally high sensitivity of the airways to physical, chemical, pharmacological and biological stimuli whose main epiphenomenon is represented by an abnormal response expressed by a contraction of airway smooth muscle. The occurrence of this response may also be accompanied by reactions involving mucus secretion, vasomotor tone and breathing pattern, although broncho-constriction is always the prevalent feature.

Factors involved in eliciting this hyper-responsiveness include receptors, afferent and efferent pathways, mediators and bronchial muscle.

Evidence has long been produced about the identification within bronchial mucosa, of neurofibrillar endings which have been interpreted as representing the sensitive terminals of receptors that respond to inhaled irritants and allergens (Widdicombe, 1977).

As far as nervous pathways and related mediators are concerned, most investigations deal with the role of parasympathetic system, namely its increased activity: in addition many researches have highlighted the role of increased x-adrenergic activity (Prime et al., 1972; Bianco et al., 1974; Henderson et al., 1979) that of b-blockade (Szentivanyi, 1968), and, more recently, the role of non-adrenergic system which acts as stimulant of smooth muscle relaxation (Richardson and Bouchard, 1975; Richardson, 1977).

The resting tone of airway muscle and baseline airway calibre are also involved as determining factors of airways responsiveness, demonstrated on one hand by the decreased reactivity when vagi are cut and, on the other, by the increase in response when muscle hypertrophy and hyperplasia are present (Parker et al., 1965· Makino, 1966; Hossain and Heard, 1970; Benson, 1978).

In addition an important role is played by epithelial damage which determines increased permeability to irritants and an exaggerated response to them (Empey et al., 1976; Golden et al., 1978; Holtzman et al., 1979).

Whatever the contribution of each factor may be, the detection of bronchial hyper-reactivity in clinical physiology is usually accomplished by broncho-provocation tests. They consist in the administration of proper stimuli in order to detect a response of airway smooth muscle: the abnormality of such a response depends upon individual susceptibility and its degree is influenced by the technique of functional investigation used for its assessment.

Let us now consider each of these determinants of bronchial response to provocation.

Stimuli they may be either specific or non specific, the former being represented by allergens, the latter by pharmacological agents (histamine, cholinergic drugs etc.), irritants (e.g. noxious gases) and factors other than irritants (cold, fog, exercise, psychological factors, etc.).

The use of specific inhalation challenge must be submitted to a careful scrutiny because of their possible risk. In general, when history and/or results of skin test are unquestionable, the use of specific tests is not required, whereas they are indicated as means of comparison with other tests when definition of the etiologic agent is otherwise impossible. Further indications may be represented by the evaluation of therapeutic effect of specific hyposensitization and by the identification of the role of new allergens particularly in occupational diseases.

In any case, eliciting bronchospasm by the inhalation of specific allergen allows an insight into the mechanisms and time course of immunological responses and therefore distinction between immediate, late and dual reactions (Pepys and Hutchcroft, 1975).

In non specific provocation tests, routes of administration of pharmacological stimuli may influence the response: parenteral route in fact may result in inactivaction of the agent, as in the case of acetylcholine or of prostaglandin F2 x, or may cause unwanted systemic effects, or may influence the site of bronchial response.

The aerosol route may evoke different responses depending on

the different sites of deposition of particles: the latter depends
on the size of particles, on the timing of delivery and on the
presence of airway narrowing before inhalation.

Susceptibility, expressed in terms of reduced threshold to
challenge, is connected with many factors, some of which are
genetically determined (as atopic status) whereas some, not yet
well characterized, are of acquired nature.

The measurement of bronchial response is of critical
importance in the evaluation of these phenomena because results
depend on the reliability, sensitiveness and specificity of the
chosen technique. Investigation on this topic have mostly been
carried out either by means of tests using forced expiratory
manoeuvres (e.g. FEV_1) or of the measurement of specific airway
conductance (SGaw). It has been claimed that tests of forced
expiration are more specific than SGaw in distinguishing subjects
without hyperreactivity from the hyperreactive ones, whereas SGaw
would represent a more sensitive test in the detection of airway
response in spite of a wider scatter of its distribution (Fish and
Kelly, 1979). Nevertheless this topic is still a matter of
controversy because it has been demonstrated that, because of
previous deep inspiration, forced expiratory manoeuvres may
determine either a bronchodilation (through a stretch receptor-
induced reflex) or a bronchoconstriction (mediated by stimulation
of irritant receptors), predominating in subject with irritable
airways (Widdicombe and Nadel, 1963; Fish et al., 1977; Gayrard et
al., 1975). Obviously the latter phenomenon by enhancing the
magnitude of bronchial response to stimuli, allows an easier
identification of hyper-responsive subjects, whereas the former may
take the challenge-induced bronchoconstriction not detectable. In
order to prevent such a misleading effect, it has been suggested
that it is better to avoid maximum inspirations to TLC and to
perform forced expiratory manoeuvres starting from volumes near FRC
(Fish et al., 1977).

Whichever method is employed, two distinctive features of the
phenomenon are usually investigated: 1) how the response begins and
develops, once the mechanism is triggered; 2) the location of
smooth muscle constriction along the bronchial tree.

As far as the first problem is concerned, there is a general
agreement about the need for assessing the threshold dose of the
challenge, i.e. the smallest dose which is effective in determining
the minimum significant variation of the test chosen for detection
of airway response. By increasing the dose of the stimulus, either
cumulatively or not, beyond the threshold, a dose-response curve
may be constructed.

According to Orehek et al (1977), the minimum provocative dose has been interpreted as accounting for the "sensitivity" of the subjects, whereas the slope of the dose-response curve has been advocated as representative of their "reactivity".

When assessing these peculiar aspects of the hyperreactivity, a careful standardization of the procedure is necessary in order to allow a reliable evaluation of both dose and response. One of the most critical and difficult points is concerned with establishing the exact magnitude of the stimulus. Different approaches to this problem have included: a) keeping volume and concentration of each inhalation constant and changing the number of inhalations by using either a reservoir or a metered aerosol; b) changing concentrations in a constant volume and c) keeping a constant concentration and changing the time of delivery.

The peculiar aspects of dose-response curves of normal and hyperreactive subjects are shown in fig. 1 taken from our experimental observations. Taking into consideration as limits on one hand the threshold, i.e. the dose which decreases SGaw by 25% of baseline, and on the other the dose which causes a 60% decrease of the same value, it can be seen that in hyperreactive patients, when compared with reference subjects, both doses are lower and the slope of the relevant curve is steeper.

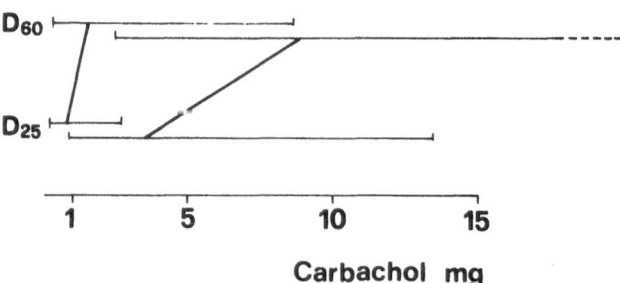

Fig. 1 - Carbachol dose-response curves detected in a group of 8
         reference subjects (on the right) and in one of 14
         latent asthmatics (on the left). Logarithmic mean + 2SD
         of doses effective in reducing SGaw by 25% (D25) and 60%
         (D60) are represented.

A more careful observation of the scatter of data (expressed
by the standard deviation from the mean) introduces a limitation in
the possibility of a clearcut distinction of the two samples on the
basis of the above mentioned approach: in fact a considerable
overlap between the extremes of the two groups can be seen. This
data, which can be interpreted by the well known variability of
factors interfering with airways patency even under normal
conditions, does not prevent extensive application of this test for
clinical use: in fact the applicability of the method is confirmed
by the evidence of the satisfactory reproducibility of its results,
as demonstrated in fig. 2. The meaning of this reproducibility is

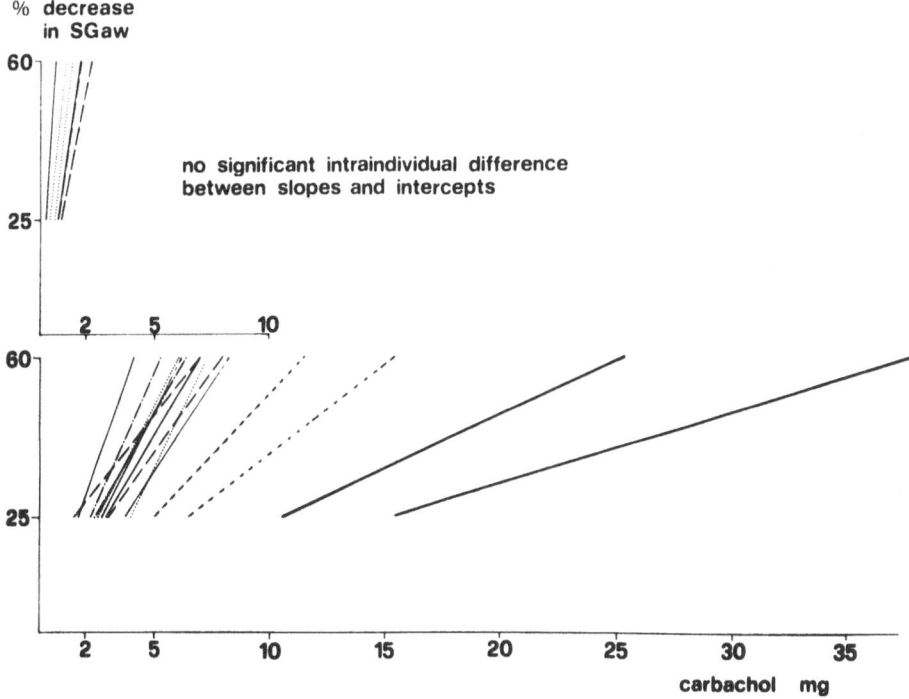

Fig. 2 - As in fig. 1: data relevant to the reproducibility of
         dose-response curves in reference subjects (lower
         diagram) and in latent asthmatics (upper diagram) are
         represented.

that the problems connected with the variability of different factors interfering with the response are mostly overcome.

Nevertheless it has been claimed that inhalation challenge tests suffer from some drawbacks, as they are time-consuming and, as previously shown, do not allow categorization of the response in all cases.

Special attention has been paid to muscular exercise as a factor inducing bronchoconstriction, as this method avoids the intervention of most of the mentioned factors of variability connected with the delivery and inhalation of aerosols. Recently, evidence has been produced indicating that this test is a very sensitive and highly specific means of detecting bronchial hyper-reactivity (Chandler Deal jr. et al., 1980). Factors involved in determining bronchial smooth muscle constriction in response to exercise have long been discussed and more recently identified as connected with heat loss from respiratory mucosa caused by hyperventilation: however, there is not yet agreement about the mechanism by which hyperventilation and cooling trigger this phenomenon. The intervention either of vagi or of mediators released from mast cells has been suggested: in fact is is not yet known what is the functional basis of the different behaviour of normals and asthmatics toward the same physical stimulus, although it has been recently suggested that the difference is inherent to the degreee of hyperventilation achieved by the two groups.

Let us now discuss the problem of the site of bronchial muscle constriction as a consequence of airway irritability.

Experimental investigations about the influence of autonomic nervous system were carried out by Woolcock et al. (1969): partitioning pulmonary resistance into central and peripheral components by a retrograde catheter, they found an influence of vagal tone at both levels, whereas the effect on peripheral airways was masked by adrenergic activity.

In clinical respiratory physiology the attempt to identify the location of airway obstruction in response to stimuli and then of establishing the mechanisms of broncho-constriction has been developed by the method of density-dependence of maximum flows at isovolume and by the distribution of inhaled particles labelled by radio-isotopes. Referring to the first method some investigations demonstrasted that in general in asthmatics the induction of broncho-constriction does not change the type of response, i.e. a predominant central or peripheral site of airflow limitation before challenge remains unchanged after bronchoprovocation test. By the same approach, McFadden et al. (1977) located the broncho-constriction in postexertional asthma either in large or in small airways: in the first case the broncho-constriction was abolished by pre-treatment with an anticholinergic agent, whereas in the second case the response was significantly diminished by disodium chromoglycate. This evidence suggests the intervention of multiple mechanisms and in some cases the participation of mediators.

However, the question about the detection of the topography of challenge-induced broncho-constriction and thus of deriving pathophysiological implications remains unanswered at least as far as clinical investigations are concerned. In particular, referring to the density-dependence method, we found (Bonsignore et al., 1980) that its reproducibility was very poor, and its application unreliable.

A contribution to the identification of topography of airway obstruction has been given by the analysis of distribution of radioactive aerosols. It has been demonstrated that either in the case of increased peripheral airway resistance (Dolovich et al., 1976), or in the case of a reduction in the patency of larger airways (Santolicandro and Giuntini, 1979), central deposition of particles occurs. In any case, no correlation is detectable between the altered pattern of deposition and the changes of parameters relevant to airway caliber: in fact, in subjects submitted to inhalation challenge, the former has been demonstrated to be an earlier indicator of airway response (Dolovich et al., 1976).

In conclusion, we have seen that clinical implications of bronchial hyperreactivity are concerned with diagnostic, prognostic and therapeutic aspects.

As far as the first aspect is concerned, the identification of aetiological factors is aimed at the early detection of changes induced by allergenic and/or irritating factors on the bronchial tree.

In general, non specific-hyperreactivity is always detectable in asthmatic subjects and a relationship between its degree and duration and severity of the syndrome has been demonstrated. It is less frequent and of lower degree in atopic conditions other than asthma: in any case, no correlation can be found with the degree of immunological changes (in that, differing from specific bronchial hyperreactivity) (Townley et al., 1979). The phenomenon can also be detected in chronic bronchitis and emphysema where to some extent it correlates with the severity of obstruction (Barter and Campbell, 1976). In addition to these conditions, the presence and magnitude of the bronchial hyperreactivity can be elicited in the case of upper respiratory infections and of exposure to irritating materials (Empey et al., 1976; Golden et al., 1978; Holtzman et al., 1979).

The specificity and sensitivity of the non-specific provocation tests are still aspect of interest in the perspective of identifying, among population exposed to risk factors, individuals susceptible of developing bronchial obstructive diseases.

The important implications of the latter indicate a need for using bronchial provocation for the following purposes: a) a careful screening on subjects before the employment in occupation exposed to environmental risk factors; b) a close surveillance of people living at risk; c) the follow-up of patients whose bronchial hyperreactivity is supposed to influence the course of chronic airflow limitation.

Finally, performing these tests in clinical medicine enables to investigate about the protective and/or therapeutic effects either of drugs or of allergen extracts, thus giving information about the kynetics of phenomena.

Summarizing, from the foregoing it is clear that the performance and interpretation of bronchial hyperreactivity tests call for an extensive background of knowledge on bronchial neuro-anatomy, on the physiology of smooth muscle, on the regulation of secretion. In addition, the application of bronchial provocation tests in respiratory medicine is made difficult by the technical requirements, so that the bulk of the problem is presently an object of research in progess.

REFERENCES

Barter, C.E. and Cambell, E.H. 1976. Relationship of constitutional factors and cigarette smoking to decrease in 1-second forced expiratory volume. Am. Rev. Respir. Dis., 113:305.

Benson, M.K. 1978. Bronchial responsiveness to inhaled histamine and isoprenaline in patients with airway obstruction. Thorax, 33:211.

Bianco, S., Griffin, J.P., Kamburoff, P.L. and Prime, F.J. 1974. Prevention of exercise-induced asthma by indoramin. Br. Med. J., 4:18.

Bonsignore, G., Bellia, V., Ferrara, G., Mirabella, A., Rizzo, A. and Sciarabba, G. 1980. Reproducibility of maximum flows in air and in $He-O_2$ and of $Vmax_{50}$ in the assessment of the site of airflow limitation. Eur. J. Resp. Dis. 61 suppl., 106:29.

Boushey, H.A., Holtzman, M.J., Sheller, J.R. and Nadel, J.A., 1980. Bronchial hyperreactivity: State of the Art, Am. Rev. Resp. Dis., 121:389.

Chandler Deal, E. jr., McFadden, E.R. jr., Ingram, R.H. jr., Breslin, F.J. and Jaeger, J.J. 1980. Airway responsiveness to cold air and hyperpnea in normal subjects and in those with hay fever and asthma. Am. Rev. Respir. Dis., 121:621.

Dolovich, M.B., Sanchis, J., Rossman, C. and Newhouse, M.T. 1976. Aerosol penetrance: a sensitive index of peripheral airways obstruction. J. Appl. Physiol., 40:468.

Empey, D.W., Laitinen, L.A., Jacobs, L., Gold, W.M. and Nadel, J.A. 1976. Mechanisms of bronchial hyperreactivity in normal subjects after upper respiratory tract infection. Am. Rev. Respir. Dis., 113:131.

Fish, J.E. and Kelly, S.F. 1979. Measurement of responsiveness in bronchoprovocation testing. J. Allergy Clin. Immunol., 64:592.

Fish, J.E., Peterman, V.I. and Cugell, D.W. 1977. Effect of deep inspiration on airway conductance in subjects with allergic rhinites and allergic asthma. J. Allergy Clin. Immunol., 60:41.

Gayrard, P., Orehek, J., Grimaud, C. and Charpin, J. 1975. Bronchoconstrictor effects of a deep inspiration in patients with asthma. Am. Rev. Respir. Dis., 111:433.

Golden, J.A., Nadel, J.A. and Boushey, H.A. 1978. Bronchial hyperirritability in healthy subjects after exposure to ozone. Am. Rev. Respir. Dis., 118-287.

Henderson, W.R., Shelhamer, J.H., Reingold, D.B., Smith L.F., Evans, R.III and Kaliner, M. 1979. Alpha-adrenergic hyperresponsiveness in asthma. N. Engl. J. Med., 300:642.

Holtzman, M.J., Cunningham, J.H., Sheller, J.R., Irsigler, G.B., Nadel, J.A. and Boushey, H.A. 1979. Effect of ozone on bronchial reactivity in atopic and non-atopic subjects. Am. Rev. Respir. Dis. 120:1059.

Hossain, S. and Heard, B.E. 1970. Hyperplasia of bronchial muscle in chronic bronchitis. J. Pathol., 101:171.

Makino, S. 1966. Clinical significance of bronchial sensitivity to acetylcholine and histamine in bronchial asthma. J. Allergy, 38:127.

McFadden, E.R. Jr., Ingram, R.H. Jr., Haynes, R.L. and Wellman, J.J. 1977. Predominant site of flow limitation and mechanisms of postexertional asthma. J. Appl. Physiol: Respirat. Environ. Exercise Physiol., 42:746.

Orehek, J., Gayrard, P., Smith, A.P., Grimaud, C. and Charpin, J. 1977. Airway response to Carbachol in normal and asthmatic subjects. Am. Rev. Respir. Dis., 115:937.

Parker, C.D., Bilbo, R.E. and Reed, C.E. 1965. Methacoline aerosol as test for bronchial asthma. Arch. Intern. Med. 115:452.

Pepys, J. and Hutchcroft, B.J. 1975. Bronchial provocation tests in etiologic diagnosis and analysis of asthma: State of the Art, Am. Rev. Resp. Dis., 112-829.

Prime, F.J., Bianco, S., Griffin, J.P. and Kamburoff, P.L. 1972. The effects on airways conductance of alpha-adrenergic stimulation and blocking. Bull. Physiopathol. Resp., 8:99.

Richardson, J.B. 1977. The neural control of human tracheo-bronchial smooth muscle, in: "Asthma: physiology, immunopharmacology and treatment", M.A. Lichtenstein, K.F. Austen, eds, Academic Press, New York.

Richardson, J. abd Bouchard, T. 1975. Demonstration of a nonadrenergic inhibitory nervous system in the trachea of the guinea pig. J. Allergy Clin. Immunol., 56:473.

Santolicandro, A.M. and Guintini, C. 1979. Patterns of deposition of labelled monodispersed aerosol in obstructive lung diseases. J. Nucl. Med. All. Sci., 23:115.

Szentivanyi, A. 1968. The beta-adrenergic theory of the atopic abnormality in bronchial asthma. J. Allergy, 42:203.

Townley, R.G., Bewtra, A.K., Nair, N.M., Brodkey, F.D., Watt, G.D. and Burke, K.M. 1979. Methacholine inhalation challenge studies. J. Allergy Clin. Immunol., 64:569.

Widdicombe, J.G. 1977. Some experimental models of acute asthma. J. Roy. Coll. Phycns., 11:141.

Widdicombe, J.G. and Nadel, J.A. 1963. Reflex effects of lung inflation on tracheal volume. J. Appl. Physiol., 18:681.

Woolcock. A.J., Macklem, P.T., Hogg, J.C. and Wilson, N.J. 1969. Influence of autonomic nervous syusem on airway resistance and elastic recoil. J. Appl. Physiol., 26:814.

D I S C U S S I O N

LECTURER: Bonsignore   CHAIRMAN: Cumming

WIDDICOMBE   I think Bonsignore is to be congratulated on an
                excellent review of a very complicated subject - I
                found it most informative. I think there is a slight
                semantic problem, but not a very important one,
                because Bonsignore did what we should all do, and
                that is define his terms very carefully. Having
                defined them, I would disagree with one of his words,
                namely sensitivity, because to the physiologist
                sensitivity means the slope of a stimulus-response
                curve, in other words what you describe as
                reactivity. However, that does introduce an
                important consideration about mechanisms, because you
                went on to stress and to illustrate the importance of
                plotting the dose-response curve, and not simply
                taking a single observation, which would be of very
                limited value in studying hyper-reactivity. If you
                plot a dose-response or a stimulus-response curve
                then you can draw some tentative conclusions about
                mechanism, just as the pharmacologist does with his
                in vitro preparations, in which he studies
                dose-response curves and comes to conclusions about
                competitive and non-competitive interactions. I
                would not develop that point because it is a
                complicated one; however, in the results that you
                showed, all the asthmatic patients have both a
                decrease in threshold or what you would call an
                increase in sensitivity, and an increase in
                reactivity. In other words the curve was shifted to
                the left and was steeper, and one could draw some
                conclusions from that. But my question is: did you
                study any other asthmatics, which for example might
                have had the same threshold, or sensitivity, and an
                increased reactivity, or which might have a different
                saturation of level of response if you increase the
                dose to a sufficient value? Because, again, one
                might draw conclusions about mechanisms from those
                kinds of hyper-reactive responses.

BONSIGNORE: We have previously discussed the semantic point
                concerning sensitivity and reactivity. Probably the
                terms are not well used, but they indicate the
                phenomenon from the point of view of physiology.

What is very important is that subjects with
different sensitivities, if I may use this term
meaning the lowest reaction to the provocative dose,
may have a different reactivity, and this is why the
study of hyper-reactivity should not be limited to
the detection of the first sign of
bronchoconstriction. We should also plot a
dose-response curve, because it has been observed
that a subject with a normal sensitivity may have a
high reactivity, and this is also shown by the large
standard deviation of the results. Therefore it is
important not to stop at the first point but to study
the dose response curve in order to evaluate
reactivity, because studying only sensitivity would
lead to wrong conclusions. The study of reactivity
and the dose response curve give a more precise
definition of the state of hyper-reactivity of the
subject. What Widdicombe has stressed is very
important, we also have observed subjects with normal
sensitivity and high reactivity.

NEWMAN TAYLOR: I wonder if I could ask you two questions in relation
to some of the points which you raised in your paper.
First of all the question of variation of bronchial
reactivity with time, which you alluded to in the
discussion. The reason that I ask this question is
that studies of those with occupational asthma due to
Western Red Cedar have shown that bronchial
reactivity could be measured by the provocation
concentration required to produce a 20% fall in FEV_1
with metacholine. The study showed that this could
be increased by as much as a hundredfold in those who
had elicited a late asthmatic reaction by exposure to
Western Red Cedar. Secondly, it has been shown that
in those with occupational asthma who are removed
from exposure to Western Red Cedar have their
metacholine reactivity decreased to normal, whereas
those who continue to be exposed maintain bronchial
hyper-reactivity as measured by metacholine challenge
tests. I think this is important not only in terms
of reproducibility, which is relevant when one is
looking at bronchial hyper-reactivity, particularly
in relation to comparison of the effects of
therapeutic agents, but it is also useful in relation
to the possible predictive aspect of bronchial
hyper-reactivity, in which the assumption would be
that bronchial hyper-reactivity precedes the
development of asthma related to occupational agents.
The evidence to date suggests that it is caused by it
rather than being the cause of it.

BONSIGNORE: I agree. These studies come mostly from the
                 experience of occupational physicians and I have not
                 sufficient experience to speak about the relationship
                 between non-specific hyper-reactivity and the
                 development of asthma. In general, occupational
                 doctors stress the link between these two
                 hyper-reactivities and the importance of studying
                 non-specific hyper-reactivity to evaluate
                 occupational asthma especially.

NEWMN TAYLOR: The second point was the relationship between the
                 effect of a drug on a challenge test reaction and the
                 relationship to its effectiveness in clinical
                 practice. The only study that I know of in the
                 English literature is that of Simon Godfrey in which
                 he looked at the effect of disodium chromoglycate on
                 exercise induced asthma in children and showed that
                 it bore very little relation to its clinical
                 effectiveness in treatment of those children.

BONSIGNORE: Yes, in general hyper-reactivity tests are very
                 useful for the evaluation of therapeutic effect and I
                 hinted at the possibility of studying the
                 pharmacodynamics of bronchodilators. There are
                 studies which show that prevention of
                 broncho-constriction by a beta stimulator has very
                 little effect if it is given one hour before, whereas
                 it becomes effective when it is given a few minutes
                 before or together with the challenge. This
                 experimental model shows that it is possible to
                 increase the amount of antigen and therefore shift
                 the dose-response curve upwards and to the left.
                 There are also studies showing that certain
                 challenges have a vagal effect which is well
                 controlled by vagolytic agents, and is not controlled
                 by other drugs. On the other hand, conditions of
                 allergic bronchostimulation due to allergic
                 hypersensitivity are prevented by beta stimulators.
                 So, there is an experimental model allowing
                 evaluation of the tests, especially in relation to
                 the therapeutical effectiveness of bronchodilators or
                 protective drugs.

CUMMING: Thank you. At that point I think I will end this
                 particular discussion and spend the last few minutes
                 discussing the general problems that you have
                 identified earlier. May I briefly set the scene.
                 When a symposium has the exact sciences meeting with
                 the life sciences there are many problems at the

interface. Some of the problems we began to see this morning and I think this is very much an interface problem, and having said this the topic is open for discussion.

BELLIA: My question is for Prodi. From what we heard from Bonsignore we can imagine a relationship between the type of challenge and the size of the particles. In other words in terms of site the deposition of a particulate is certainly influenced by the diameter. We normally use equipment which is available on the market and sometimes wonder whether these aerosols may undergo some changes. Changes which are due to the way we deliver the aerosol, the use of valves of different type, and so on. In his talk Prodi showed two different models and I would like to ask whether, from the point of view of physics, there is the possibility that the way in which we handle the aerosol may affect the size of particles. The second question is: is the magnitude we find in all formulas of sedimentation a fixed magnitude, or does it change? Lastly in terms of prognosis and treatment, the question of fresh aerosol with different charges, can this be important? Because variability and deposition coefficients were important.

PRODI: To your first question on the physical configuration of the system of nebulization on size, certainly one important point is pressure in the nebulizer; then surface tension is also important of the fluid used, apart from the constant size of the nozzles. Then of course the lines for the aerosol should be as short and as straight as possible, as any curvature and longer distance tend to move away the bigger particles. Then, as these are aerosols generated by rupture of a continuous body, the electrical charge is always rather high and can cause deposition in the stream lines and change the effectiveness of deposition in the lung. So it would be advisable, from time to time, to verify if the aerosol gets out in the right way. We should also stress that there is a certain level of humidity in the airways and therefore soluble or even wettable particles have their own dynamics which is not yet very well known and should be studied. We need to acquire a better knowledge of the aersol itself and of its dynamics in the airways. Uniform, non-wettable particles as we have seen this morning are something very particular and obviously cannot be used in practice. So, in particular, the electrical charge should be studied and we intend to study the influence of the electrical charge not only on total deposition, but on the picture of distribution.

CORRIN: Ventilation in the alveolar region is by diffusion, so we might expect deposition of dust in the alveoli to be limited to diffusion, rather than sedimentation. Are we only contaminating our alveoli with dust particles in the 1 to 5 micron range when we exercise, or perhaps at - rest is the advancing gas front not flat-faced but parabolic?

PRODI: In the alveoli, in that size range, you have only sedimentation, no doubt about that. You start having diffusion below that lower limit you mentioned. More or less, diffusion and sedimentation are of equal importance around 0.5 microns, 0.4 to 0.6, the point of minimum efficiency means an equal importance of these two facts. So, around 1 micron diffusion is much lower in importance than sedimentation and of course sedimentation increases in importance as sizes increase. In the alveoli you have just sedimentation.

CORRIN: But at rest our tidal volume is a little greater than our dead space.

CUMMING: May I just make a comment on that? Certainly, ventilation by fresh gas, both at rest and on exercise, is brought about by diffusion between the two concentrations of gas. However, bulk flow occurs into all areas at all times. The number of molecules entering an alveolus are determined by bulk flow, their type is determined by diffusion. Since the diffusion coefficient of particles is many orders of magnitudes greater than the diffusion coefficient of gases, diffusion plays a very small role and Prodi is quite right, it is mainly deposition. So, the bulk flow carries along the particles the whole way to where both flows stop. The alveolar wall.

PRODI: Well, actually, below 0.5 everything is deposited in the alveoli by diffusion. So, that depends on the mixing mechanism. You have this mixing between tidal and residual air, and to that extent you can have an increase of the residence time of the particles within the respiratory tract and you can have deposition of these particles in the alveoli. I have to be more precise about the size. While the aerodynamic size is a good parameter about 0.5

microns, below 0.5, since the deposition is by
diffusion, and the diffusion coefficient is
independent of density, that means that only the
geometric size matters in diffusion, at equal size
you can have different densities behaving in the same
way.

CUMMING: I think this is an extremely complex topic: what is
                bulk flow, what is diffusive mixing, what is
                turbulent mixing and so on. And I think that at this
                hour, before lunch, it would be wise not to embark
                upon it. So I will close the meeting.

FOLLOW-UP IN OCCUPATIONAL RESPIRATORY DISEASES

Emanuele Capadaglio

Istituto di Medicina del Lavoro
Universita di Pavia
Italy

The term "follow-up" can be used in different ways, either a
restrictive one, meaning a particular type of epidemiological
study ("prospective cohort study"), or an extended one; the
latter being more suitable when applied to surveys in the field
of occupational medicine and industrial hygiene. It is therefore
justified to include in the word "follow-up" the whole of
longitudinal procedures employed to take under control some
environmental and/or biological situations ("environmental
monitoring", "biological monitoring", "aimed surveys", "repeat
screening"). All these procedures can be used either separately
for descriptive purposes or appropriately correlated among them
for analytical ones.

The right use of these techniques gathered under the term
"follow-up" can give valuable help in solving many of the
problems concerning either occupational disorders known from the
aetiologic point of view ("occupational diseases") or unknown
risks (effects of work places pollutants coming from recent
technological progress, effects of new occupational risk factors
replacing others recognized noxious, results of surveys carried
out to verify the "ineffective doses"). Also the requirement of
planned longitudinal studies in the field of industrial hygiene
is often emphasized in provision of the law and contractual ties.

Accepting as valid the notion that occupational diseases
evolve chronically over time (stage of susceptibility, stage of
presymptomatic disease, stage of clinical disease, stage of
disability) and the hypothesis that their natural history is
dependent on exposition to risk factors, the contribution that
the above-mentioned procedures gives to primary and secondary

283

prevention can be summarized in the following way:

1) Prospective epidemiological surveys (prospective cohort studies or follow-up studies) are a step preceding the real primary prevention: in fact they give a valuable aid in recognizing new risk factors (if such a recognition is a compulsory step on the way to primary prevention).

2) Environmental monitoring can give a direct contribution to the primary prevention whenever essential work places modifications are rapidly carried out as necessary; the aim being to restore environmental concentration of noxious agents at safe levels.

3) Biological monitoring and aimed surveys are placed halfway between primary and secondary prevention. In fact they are mostly employed in preventing those risks for which it is possible to prove a biological exposition before the appearance of functional or morphologic lesions (privileged risks).

4) Repeat screening contributes, by definition, to secondary prevention.

−   If we consider respiratory risks and risk factors specifically it is worth going a bit more into practical details. First of all, from the epidemiological point of view, it is generally advisable to use retrospective surveys (case-control studies, retrospective cohort studies) or to employ this kind of surveys (simpler, less expensive and more rapid) to formulate causal association hypotheses that could be proved afterwards with more reliable prospective surveys (follow-up). In fact, retrospective surveys do not provide a direct measure of the risk associated with the postulated cause is often less accurate than in prospective surveys. (1)

Factors making practically unfeasible prospective surveys are long induction times of occupational respiratory diseases (cancer, COLD, chronic effects on airways of presumed irritant agents etc.) and the rarity of some of them (cancer). (2)

−   "Biological monitoring" and "aimed surveys" are certainly very useful investigating privaleged risks but this is not the case with respiratory risks; nevertheless the important question exists concerning the search of "ineffective doses" (.) (the ones
----
(.) Dose is the result of exposition during daily working time for a given extended period to an environmental pollutant; it might be quantified either as weighted average concentration over time, or as suitable parameters from a log-normal distribution (median value and geometric standard deviation) or as inhaled, absorbed, retained quantity.

not affecting the natural age-declining trend)· since it is
presently utopain to speak in terms of "null risk". At any rate
it is not without significance to emphasize that the
ineffectiveness of a dose depends also on the sensitivity of
available diagnostic tests (chest xrays, respiratory function
tests, sputum cytology, MRC or CECA questionnaires etc.). There
is a trend towards studying, suggesting and using more and more
sensitive diagnostic tests; this is almost always obtained at the
expense of specificity (a quite negative factor). The natural
history of airway disorders is affected by individual
susceptibility to some intrinsic factors (FEV_1 losers) and by the
effects of extrinsic (occupational) risk factors. These effects
may be different from case to case an during the follow-up their
distribution might take a pattern departing more and more from a
normal distribution.

Some cases can be exemplified:
a) effects of an effective dose in the past followed by
ineffective doses or by doses towards which a good tolerance is
expected (O_3, NO_2). In this case a curve representing a given
parameter for exposed subjects plotted verus age after an initial
divergence would show a trend quite similar to that of a normal
population
b) effects of an effective dose in the past followed by doses
working only on the more sensitive subjects (being on the
contrary ineffective on the rest of exposed subjects). In this
case after the initial divergence a slowly increasing spread
between the curves for exposed subjects and normal population
should appear.
c) Effects of an effective chronic dose working on the whole
group or on a part of it. In this case there should be a more or
less rapid spread between the curves from the very beginning.

At any rate the exposed population will show, in the long run a
distribution of effects widely departing from a normal one
(asymmetrical distribution) and a statistical procedure valid for
normally distributed populations will no longer be applicable.
(3)

-    The purpose of "repeat screening" is to sort out as early as
possible people suffering from respiratory disorders at the very
early symptomatic stage or at an asymptomatic one (apparently
healthy individuals); the assumption is that, if a disease has
not yet reached the threshold of clinical expression its
prognosis should be better. (4)

    In the case of occupational COLD, for which there is not a
true therapy, the early detection (at the stage of asymptomatic
OB or of the earliest signs of the disease) is, at least
theoretically the most advisable approach presently.

Not many years ago the "OB tests" were introduced some of which could be employed as screening tests thanks to their speed of application: others eventually could be employed as diagnostic tests (to sort out false positives).

The discussion about the validity of these tests and about availability of reference values is beyond the scope of this report. On the contrary it is worth while spending some words about their predictive value, either in a statistical sense (.) or in a prognostic one (likelihood that the group sorted out develops in the future a gravely disabling COLD).

In the first case it is known that relationship between predictive value and prevalence of disease exists; then if the test has a good yield and if the incidence of disease is not high, surveys following the first (reflecting prevalence) are, under the same conditions of validity of the test, less predictive. (4)

In the second case it must be emphasized that so far, results about prognostic value of OB tests are lacking.

This is also the case with chronic bronchitis questionnaires, mostly after recent criticism addressed to the unified hypothesis of chronic bronchitis. (5)

- Lastly the matter of environmental monitoring is worth mentioning.

It is absolutely inadequate to take into account as usually happens, only the quality of air pollution without reference to dose; on the other hand, speaking in terms of follow-up, the fact that, in the long run, not negligeable qualitative environmental changes, dose being equal, could arise, must not be underestimated. (6)

Also, letting aside the analytical problems encountered when dealing with noxious agents that are airway risk factors, it is important to emphasize that only with accurate simplification (without significantly affecting the validity of collected data) longitudinal prospective surveys are practically feasible.

The environmental pollution of work places can be regarded

(.) Predictive value = true positive/total positive = likelihood
                       that an individual with a positive test
                       has the disease.

as a function of two variables, time and space / f (t,s,) /. The two variables act simultaneously when it is a matter of personal samplers (moving points). These are more and more often employed when group-surveys are made. In case of fixed measuring points only the time variable is involved / f (t,k) /. With reference to this last case the use of the long-term detectors (giving weighted average concentration over a period of several hours) allows to get results comparable with weighted thresold limits (TWA). However the possibility to detect istantaneous concentration changes (peaks), as well as the range of fluctuation of the concentration and the occurrence of off-range values for short times (STEL) is excluded. These informations are obtained by means of short-term detectors giving a number n_1 of measures for every fixed point taken at fixed time intervals, or in a simplified way, a number n_2 of random samplings (random grab samples) with n_2 n_1. (7)

The randomized selection of the measuring time intervals is a big advantage in follow-up surveys without negatively affecting the inference.

Other simplifications are the hypothesis of a log-normal frequency distribution and the use of graphic evaluation of the results of measurement (entering in a probability network with the frequency sum and concentration as coordinates). (8)

By virtue of such simplifications it becomes easy to check an average weighted concentration and to relate it with exposed groups in comparison with standard limits; the obtaining range of fluctuation of the concentration, and identification of risk areas (homogenous areas) included between a maximum and a minimum of pollution are also made easier.

Results obtained with fixed measuring points are adequate when related to workers moving around, during working time, inside a homogenous area or from an area to another. For these workers respiratory area corresponds to the environmental area (or areas) of stay; for those working in fixed position (fettlers, welders etc.) and exposed to sources directly affecting the respiratory area, reference to environmental risk area is less accurate; it would be much better to refer to data obtained with personal samplers, more pertaining to their respiratory area.

We are persuaded that a normalisation of environmental monitoring techniques (in terms of environmental follow-up) should be made such as it has been done for risks.

REFERENCES

1) ABRAMSON J.H., 1974, "Survey methods in community medicine".
       Churchill Livingstone, Edinburgh, London: 5.
2) MAC MAHON B., PUGH T.F., 1970, "Epidemiology (principles and
       methods)".
       Little Brown and Co Boston: 209.
3) ESMEN N.A., HAMMAD Y.Y., 1977, "Log-normality of
       Environmental sampling data".
       J. Environ. Sci. Health, 12: 29.
4) MAUSNER J.S., BAHN A.K., 1974, "Screening in the detection of
       disease and maintenance of health" (from:
       "Epidemiology": an introductory text) : 237.
       W.B. Saunders Co. - London.
5) FLETCHER C., PETO R., TINKER C., SPEIZER F.E., 1976, "The
       natural history of chronic bronchitis and emphysema".
       Oxford Univ. Press, New York.
6) KENNETH A., BUSCH K.A., LEIDEL N.A., 1979, "Statistical
       design and data analysis requirements" - (from: Patty's
       Ind. Hyg. and Toxicol., vol. IIIth - J. Wiley and sons,
       New York).
7) CORN M., ESMEN N.A., 1979, "Work place exposure zones for
       classification of employer exposure to physical and
       chemical agents".
       Am. Ind. Hyg. Ass. J., 40:47.
8) LEICHNITZ K., 1979, "Air analyses at work places by means of
       short-term and long-term detector tubes".
       Drager Revue, 42:3.

DISCUSSION

LECTURER: Pezzano   CHAIRMAN: Cumming

COHEN: You pointed out the difficulty of relating lung
                    cancer to occupational exposure, because the disease
                    would be of very late onset, would have a very long
                    latent period, and the incidence would be fairly low.
                     Yet, there are many occupational lung cancers; how
                    would you go about assessing possible risks and know
                    the doses of exposure for these individuals?

PEZZANO: As there are people exposed to risk factors for
                    cancer and therefore, unfortunately, there are case
                    series available to the researchers, it is possible
                    to think of cause-effect relationships
                    retrospectively, with case control studies, between
                    pulmonary neoplasias and occupational risk factors.
                    For example methylether dichloride has proved to be a
                    powerful carcinogenic for the lung mainly through
                    animal experiments, but also by retrospective
                    studies. I think of all departments where ion
                    exchange resins are handled there is the possibility
                    of pollution of this kind in closed-circuit systems.

CUMMING: You mentioned that in the 300 normal people whom you
                    are following up, the signs of obstruction in the
                    large airways were preceded by bronchiolar
                    obstruction. What test did you use to determine the
                    presence of bronchiolar obstruction?

PEZZANO: Initially, this was an ambitious research, intending
                    to use not only one test for bronchiolar obstruction,
                    but several tests: flow-volume curves, and terminal
                    flow parameter, and iso-flow volume and closing
                    volume. Then, due to technical difficulties, this
                    was no longer possible and at the moment we get our
                    results every six months using flow-volume curves and
                    iso-flow volume.

NEWMAN TAYLOR: I wonder if I could come back to the interesting
                    problem raised by Cohen about identifying risks of
                    cancer in relation to occupational agents, and low
                    exposures. The problem is the difficulty of looking
                    at a disease that is very prevalent in a community
                    when the differences in relation to low
                    concentrations of occupational agents are relatively

small. I wonder if it would be of interest in this context to look at the relationship between those who are smoking and those who are not because certainly in the case of those exposed to asbestos and those exposed to radiation there appears to be a multiplicative effect in relation to smoking, and I wonder if this isn't the way into this very difficult problem, to compare the rates in smokers with non-smokers, in those who are occupationally exposed.

PEZZANO: That cigarette smoking is a risk factor for pulmonary neoplasia is well-known. That smoke associated with exposure to other risk factors, such as asbestos dust, has a multiplicative effect is also known. What I do not agree with is the statement that as smokers occupationally exposed to asbestos are at a high risk for pulmonary neoplasia they should not smoke, without considering that in fact asbestos is itself a powerful carcinogen. We are in a situation in which occupational agents are not considered harmful before their actual effect is proved, these substances being considered as harmless as long as the danger is not proved. This has very little to do with prevention because the environmental limit is fixed at the level of occurrence of. the risk, whereas this occurrence should be the extreme limit of exposure to an agent. If it is impossible to eliminate exposure to risk factors, then we are in a stalemate situation. If, on the contrary, some action is possible it is worthwhile to take it.

CUMMING: You mentioned the difficulties in the statistical handling of the data using parametric statistics. Do epidemiologists currently use non-parametric statistics?

PEZZANO: Yes

CUMMING: I think, ladies and gentlemen, we have come to the end of our allotted time. Can I thank Pezzano and hand the floor to Brian Corrin, who will talk about the air-blood barrier.

THE AIR-BLOOD BARRIER

Bryan Corrin

Cardiothoracic Institute
Brompton Hospital
London, SW3 6HP

By light microscopy it is not generally apparent whether cells
in the interalveolar septa are endothelial, epithelial or
interstitial. In pathological states which widen the septal
interstitium, a distinct epithelium can sometimes be recognised,
but its existence in normal lung remained a matter of some
controversy until the advent of the electron microscope. Fine
structural studies established that there was a very thin but
complete epithelium which extended throughout the alveoli, alveolar
sacs and ducts and into the respiratory and terminal bronchioles. At
a variable point within the bronchioles there is an abrupt
transition from flattened to columnar epithelium. The epithelium of
the respiratory portion of the lung is therefore continuous with
that in the airways, this whole lining being derived from the
original endodermal bud. The epithelium is separated from
underlying mesenchymal derivatives by a supportive basement membrane.

The capillary wall similarly consists of a thin endothelium
resting on a basement membrane. On one side the capillary is very
closely applied to the alveolar epithelium and the endothelial and
epithelial basement membranes fuse to form a single lamina. Here
the air/blood barrier is at its thinnest, in places measuring as
little as 0.15 μm with a mean value of 1.25 μm. The alveolar
epithelium, interstitium and capillary endothelium constitute about
30, 40 and 30% respectively of the barrier (Weibel and Knight, 1964).
On the opposite side of the capillary,interstitial tissue separates
endothelium from epithelium and the air/blood barrier is much
thicker.

The alveolar capillaries do not run independently of one another but form an intercommunicating network of short cylindrical segments 10 - 15 μm in length, a little longer than their 5 - 10 μm diameter, with an arteriolo-venular distance of up to 500 μm (Weibel, 1963). The capillaries therefore form a mesh work with quite small intercapillary spaces filled with interstitial tissue. The alveolar capillary endothelium is of the non-fenestrated type. In contrast to the alveolar epithelium, the endothelial cell junctions readily permit the passage of small proteins such as horse-radish peroxidase (M.W., 40,000 daltons)(Schneeberger-Keeley and Karnovsky, 1968). Larger molecules such as albumin (M.W., 70,000 daltons) are retained by the cell junctions, but small amounts of albumin cross the endothelium to reach the interstitium by pinocytotic transport (Feldmann, Chahinian, Leturcq and Bignon, 1973). Pinocytotic vesicles are numerous in the endothelial cells, and some which open onto the luminal aspect are claimed to contain the enzymes concerned in the metabolism of angiotensin I and bradykinin (Smith and Ryan, 1973).

The alveolar epithelium consists of two main cells, known as type I and II pneumocytes. The type II cell is slightly the more numerous but, because of marked differences in cell shape, the type I cell covers about 95% of the interalveolar septum. The type I - cell has few organelles and a very attenuated cytoplasm which extends long distances from the nuclear region. The cytoplasmic processes of this cell may even spread through the interalveolar septum, in collar-stud fashion, to line the opposite side (Weibel, 1971). On one side of the septum alone, a type I cell may cover over 2000 μm^2 yet measure as little as 0.2 μm in thickness. Its function is to provide as thin a barrier to gas exchange as is commensurate with a complete epithelial lining. Type I cells are connected to each other and to type II cells by tight junctions (zonulae occludentes) and this epithelium forms the major barrier to fluid movement into and out of the alveolus. Nevertheless, pinocytotic fluid transport is represented by numerous small vesicles within the cytoplasm. Macromolecules such as horse-radish peroxidase and probably albumin and globulin are absorbed from the alveolar lumen by pinocytosis (Gonzalez-Crussi and Boston, 1972). The phagocytic potential of type I cells is very limited, but extremely small dust particles gain entry to the alveolar interstitium by a similar vesicular transport mechanism (Corrin, 1969). The long lateral cytoplasmic extensions are very susceptible to damage by quite diverse agents and together with the capillary endothelium the type I cells represent the point most vulnerable to injury within the lung.

Type II alveolar epithelial cells, although slightly the more numerous, cover little of the interalveolar septum. This is fortunate from the point of view of gas exchange as the type II

cells are cuboidal or low columnar in shape. They are generally
found in the corners of the alveoli and often all but the apical pole
is covered by neighbouring type I cells. The free surface bears
blunt microvilli and the cytoplasm contains distinctive osmiophilic
lamellar vacuoles which are occasionally observed opening on to the
alveolar lumen (Bensch, Schaefer and Avery, 1964). The vacuoles
first appear in foetal life at a time when surfactant can first be
identified and they are regarded as the secretory vacuoles of
pulmonary surfactant. This has been established by electron
microscopic autoradiography tracing the incorporation of dipalmitoyl
phosphatidyl choline precursors in a sequential fashion through the
usual secretory organelles (Chevalier and Collet, 1972), and by cell
separation techniques which enable pure type II cells suspensions to
be studied in vitro (Kikkawa and Yoneda, 1974; Kikkawa et al., 1975).
A variety of lysosomal enzymes in the lamellar vacuoles derive from
fusion of multivesicular bodies with the vacuoles (Corrin and Clark,
1968; Corrin, Clark and Spencer, 1969; Meban, 1972). One of these
acid hydrolases, phosphatidic acid phosphatase, controls an essential
step in surfactant synthesis and the presence of others is not
necessarily indicative of the usually degradative processes associated
with lysosomal activity. The type II cells are not phagocytic to any
significant degree.

Fixation by vascular perfusion preserves a continuous extracellular biphasic alveolar lining layer which smooths off epithelial
surface irregularities (Gil and Weibel, 1969). The lining is
relatively thick in the corners of the alveoli and thin over the
lateral extensions of the type I cells. It consists of an electron-
lucent aqueous hypophase in contact with the epithelial cells and a
thin surface layer of osmiophilic lattice-work material which
probably represents the surface active lipids. The means of disposal
of this lining is uncertain, but the alveolar lining is continuous
with that of the airways and may thereby contribute to the formation
of sputum. Alternatively some spent surfactant may be removed by
alveolar macrophages (Naimark, 1973).

Alveolar injury often presents a non-specific pattern of
"diffuse alveolar damage" (Katzenstein, Bloor and Liebow, 1976),
consisting of collapsed alveoli lined by hyaline membranes. The
latter represent a mixture of epithelial debris and fibrin. Diffuse
alveolar damage is seen following injuries as diverse as shock, viral
infections, irradiation and many chemicals, both inhaled and
ingested. The most susceptible cells in the lung to such non-
specific injury are the thin delicate capillary endothelial and type
I alveolar epithelial cells. The type II epithelial cells appear to
be more resistant to all injurious agents and play an important role
in epithelial regeneration; indeed they represent the progenitor
cells of the alveolus (Evans et al, 1973; Adamson & Bowden, 1975).
After alveolar damage there are found occasional cells which possess
features of both main epithelial cell types. These are generally

The Air-Blood Barrier

flattened squames with the long thin cytoplasmic processes of type
I cells, but the surface microvilli and osmiophilic lamellar
inclusions of type II cells. Such intermediate cell forms are
explained by thymidine labelling experiments which indicate that
type II cells differentiate into type I (Evans, Cabral, Stephens and
Freeman, 1973). The type II cell is therefore to be considered not
only as the source of surfactant but also the progenitor cell from
which the alveolar wall is relined after injury has damaged the more
delicate type I cell. With chronic injury type II cells proliferate
but do not differentiate into flat type I cells, and the alveolar
wall becomes lined by a cuboidal epithelium recognisable with the
light microscope, a process of type II cell hyperplasia rather than
cuboidal metaplasia as it is usually termed. In the normal lung the
turnover time of type II cells is 25 days and transformation of type
II to type I cells takes two days (Evans, Cabral, Stephens and
Freeman, 1975).

Alveolar cell necrosis is frequently accompanied by interstitial
cell proliferation and organisation of the hyaline membranes. This
may result in either an interstitial fibrosis indistinguishable from
the chronic idiopathic variety or in a totally obliterative pattern
of alveolar fibrosis. The latter differs from the usual post-
pneumonic fibrosis in which granulation tissue polyps protrude into
the air spaces. Instead the alveoli are entirely obliterated by new
fibrous tissue with the position of the totally enveloped interalveolar septa marked only by persistant capillaries and preformed
reticulin and elastin fibres.

REFERENCES

Adamson, I.Y.R., and Bowden, D.H., 1975, Derivation of type I
    epithelium from type 2 cells in the developing rat lung, Lab.
    Invest.,   32:736.
Bensch, K., Schaefer, K., and Avery, M.E., 1964, Granular
    pneumocytes: electron microscopic evidence of their exocrinic
    function, Science, 145:1318.
Chevalier, G., and Collet, A.J., 1972, In vivo incorporation of
    choline-^{3}H, leucine-^{3}H and galactose-^{3}H in alveolar type II
    pneumocytes in relation to surfactant synthesis. A
    quantitative radioautographic study in mouse by electron
    microscopy, Anat. Rec., 174:289.
Corrin, B., 1969, Phagocytic potential of pulmonary alveolar
    epithelium with particular reference to surfactant metabolism,
    Thorax, 24:110.
Corrin, B., and Clark, A.E., 1968, Lysosomal aryl sulphatase in
    pulmonary alveolar cells, Histochemie, 15:95.

The Air-Blood Barrier

Corrin, B., Clark, A.E., and Spencer, H., 1969, Ultrastructural
    localisation of acid phosphatase in the rat lung, J.Anat.,
    104:65.
Evans, M.J., Cabral, L.J., Stephens, R.J., and Freeman, G., 1973,
    Renewal of alveolar epithelium in the rat following exposure
    to NO_2. Amer.J.Path., 70:175.
Evans, M.J., Cabral, L.J., Stephens, R.J., and Freeman, G., 1975,
    Transformation of alveolar type 2 cells to type I cells
    following exposure to NO_2. Exp.Mol.Path., 22:142.
Feldmann, G., Chahinian, P., Leturcq, E., and Bignon, J., 1973,
    Localisation ultrastructurale par anticorps couples a la
    peroxydase de l'albumine extra-vasculaire dans le poumon de
    rat, Comptes Rendues Academie de Science de Paris, 277:251.
Gil, J., and Weibel, E.R., 1969, Improvements in demonstration of
    lining layer of lung alveoli by electron microscopy, Resp.
    Physiol., 8: 13.
Gonzalez-Crussi, F., and Boston, R.W., 1972, The absorptive
    function of the neonatal lung: ultrastructural study of
    horseradish peroxidase uptake at the onset of ventilation,
    Lab.Invest., 26:114
Katzenstein, A-L.A., Bloor, C.M., and Liebow, A.A., 1976, Diffuse
    alveolar damage - the role of oxygen, shock, and related
    factors, Amer.J.Path., 85:210.
Kikkawa, Y., and Yoneda, K., 1974, The type II epithelial cell of
    the lung. I. Method of isolation, Lab.Invest.,30:76.
Kikkawa, Y., Yoneda, K., Smith, F., Packard, B., and Suzuki, K.,
    1975, The type II epithelial cells of the lung. II.Chemical
    composition and phospholipid synthesis, Lab. Invest., 32:295.
Meban, C., 1972, Localization of phosphatidic acid phosphatase
    activity in granular pneumonocytes, J.Cell Biol., 53:249.
Naimark, A., 1973, Cellular dynamics and lipid metabolism in the
    lung, Fed. Proc., 32:1967.
Schneeberger-Keeley, E.E., and Karnovsky, M.J., 1968, The
    ultrastructural basis of alveolar-capillary membrane
    permeability to peroxidase used as a tracer, J.Cell Biol.,
    37:781.
Smith, U., and Ryan, J.W., 1973, Electron microscopy of endothelial
    and epithelial components of the lungs: correlations of
    structure and function, Fed. Proc., 32: 1957.
Weibel, E.R., 1963, Morphometry of the Human Lung, Berlin, Springer.
Weibel, E.R., 1971, The mystery of 'non-nucleated plates' in the
    alveolar epithelium of the lung explained, Acta Anatomica,
    78:425.
Weibel, E.R., and Knight, B.W., 1964, A morphometric study on the
    thickness of the pulmonary air-blood barrier, J.Cell Biol.,
    21:367.

D I S C U S S I O N

LECTURER: Corrin   CHAIRMAN: Cumming

CUMMING: Thank you very much, Brian, a lucid demonstration.
              This paper is now open for questions and comments.

HEATH: I would like to ask about the shape of the
              endothelial cells of the pulmonary capillaries.
              Something which is becoming apparent at Liverpool
              recently is that we have always considered
              endothelial cells as endothelial cells - no
              discussion. But is has become apparent in the last
              few months that the shape of the endothelial cells of
              the pulmonary trunk are quite different from those of
              the aorta, and they are also different from those of
              the pulmonary veins. Furthermore when you disturb
              the haemodynamics of the pulmonary circulation the
              endothelial cells of the pulmonary trunk will change
              in shape to approximate to those of the aorta. What
              I am wondering is: do you have any information as to
              the shape and size of the endothelial cells of the
              capillaries in the alveolar capillary wall? And, do
              you know anything about their alteration on size and
              shape in various pathological conditions?

CORRIN: I certainly know that in many pathological conditions
              the endothelial cells thicken, their basement
              membrane thickens too, and often seems to be
              reduplicated, many layers thick, and it has been
              shown that this indicates that the endothelial cells
              have died and regenerated. When an endothelial cell
              regenerates it is not content with the basement
              membrane of its predecessor, but lays down a new one.
              So, we have thickening of the capillary endothelium,
              and I think what you are particularly referring to
              are those patterns of the endothelial cells which you
              have studied in the pulmonary trunk and the aorta and
              perhaps I hope you will show later in the week, but
              it is much more difficult to view the endothelial
              surface of a narrow capillary with a microscope as
              you have done with the pulmonary trunk and the aorta.
              the original chemistry and role of the hypophase in
              alveoli? And, do you think it has a relation to the
              proteinases and possibly the hyaline membrane?

CORRIN: Yes, I concentrated, perhaps wrongly, in dealing with alveolar lipoproteinosis on the lipid component. As its name indicates, there is also a protein component, and it has been shown by immuno-electrophoresis that the protein component derives very much from plasma. I should perhaps have mentioned that and stressed that there is a double component to alveolar lipoproteinosis. One from the blood and the other from the lipid secreted by the type II cells. It is likely then that the hypophase of the normal surfactant layer, the normal lining layer, probably derives from the blood. We have many pinocytotic vesicles, we do know sometimes that these carry fluid from the alveolar side of the interstitium, we have to speculate whether perhaps these also sometimes carry fluid in the opposite direction. I think it unlikely that bronchiolar cell secretion comes backward into the alveolus, and apart from the secretion of surfactant I think the only mechanism I can envisage for the hypophase coming into the alveolus is across the epithelium in some way, from interstitial fluid and also from blood plasma. The hyaline membrane has much fibrin, and much of this derives from the plasma, is a vascular exudate.

BIGNON: I was interested by the case you have shown of paraquat lung. We observed a case of paraquat lung ventilated by PEEP with an emphysema-like aspect, and I think it was different from the honeycomb lung, some areas were distended, with airspaces fully distended, but the air wall was thickened, the appearance was very different from what we see usually in lung fibrosis. This raises two comments: first, recently the group of Kimbell in the US reported that in the adult respiratory distress syndrome there is a lot of elastase in the pulmonary secretion, and we can imagine that elastase could be involved in such a disease. If you look at elastase emphysema you will see that at the beginning of the lesion you have such a lesion as you observed in your lung as in paraquat lung. I think that what you observed in such a lung can be related to destruction by enzyme. My question is: was your patient ventilated by high pressure?

CORRIN: I showed both light microscopy and electron microscopy of paraquat poisoning. All the electron microscopy was experimental, it was animal lung and there was no ventilation or treatment with oxygen.

The point I was really trying to bring out today was that this epithelial necrosis which I illustrated at the electron microscopic level with paraquat is not specific to paraquat. It is the identical effect, I think, of high concentrations of O_2 and I think that the picture of oxygen poisoning in the patient who had the complex situation of shock followed by high oxygen therapy is virtually identical with hyaline membrane disease. So, we are stressing the non-specific pattern of acute alveolar injury, which precedes diffuse alveolar damage.

CANDURA: I also wanted to ask some clarification on paraquat lung. In the Lancet, a few months ago, a letter was published from an Italian author who described the case of an individual who had fallen into an irrigation canal, and after 24 or 48 hours, I do not remember exactly, died with dyspnoeic symptoms which required assisted ventilation with intermittent positive pressure. The author of the letter wondered about the possible reason for his death and I suggested that it might have been due to paraquat ingestion. The answer I got a few days ago is that unfortunately the autopsy was not done, so my hypothesis could not be confirmed. I wanted to ask by what mechanism paraquat acts, but I think the answer has already been given to Bignon's question. Thank you.

CORRIN: I would like to comment on the patient who fell in the irrigation ditch. I would think that it is very unlikely that paraquat played a role, because when this is sprayed on the crops the paraquat which hits the soil is immediately inactivated, and I would envisage no danger in that respect. Strangely, all the damage from paraquat comes by ingestion rather than inhalation. The people spraying the crops, as far as I know, have not suffered any damage, and experimental animals which have been exposed to an aerosol have either died acutely or they have recovered.

CANDURA: But he drank the water in the ditch.

CORRIN: I would think by the time that any paraquat reached the ditch it would have been inactivated by passing through the soil.

CUMMING: There is one possibility, that if the man spraying the crops with paraquat emptied the remainder into the ditch, then it may have been still toxic. And then, there has to be some explanation for acute death due to dyspnoea by falling in a ditch.

CORRIN: Perhaps he drowned.

CHEVALIER: From your experiment with paraquat, do you have any idea of what could be the triggering mechanism for the transformation of Type II into Type I cells?

CORRIN: Yes, I do not think it is anything specific to paraquat, I would envisage that it is very similar to skin regeneration: if we lose the epidermis the neighbouring epidermal cells seem to sense that they have no neighbour and proliferate to fill the gap. I think that it is probably exactly the same to re-cover the alveolar surface; the cells which have not undergone necrosis proliferate and slide in. Perhaps there is some contact inhibition in the normal epithelium which is released when a few cells die.

SMITH: First of all in regard to this man who fell in the ditch, if I may just make a short comment. I would agree entirely that it seems highly unlikely that it could have been paraquat that killed him, particularly if it was washed off the fields then it would be inactivated; even if it were poured in its diluted form in which it is used as a spray, it is very much diluted from the original concentration and I would think that enormous quantities would have to be ingested. So I think it is very unlikely. Could I also make the observation, with regard to Corrin's paper, that in my experience, in animals particularly, you cannot get regeneration of the Type II pneumocyte following damage, even in our human cases I have seen that this is not a prominent feature at all. And this seems to be because the Type II and Type I pneumocytes are completely obliterated, they are both destroyed and there is no reservoir of cells left for them to regenerate, the alveoli remain, if you like, in a denuded state. A question I would like to ask is: you mentioned paraquat and you mentioned oxygen poisoning: I was wondering to what extent you think that oxygen poisoning in cases of paraquat intoxication may modify or even have a major role to play in the development of the fibrosis which occurs. As we have seen, both produce pathology of their own.

CORRIN: Which is very similar. But of course in the animal
               model no oxygen is given and paraquat can cause
               fibrosis by itself, so it is not essential to have
               the two. The mechanism by which paraquat poisons the
               cells is probably related to its low redox potential
               and the liberation of free oxygen radicals. For this
               reason superoxide dysmutase has been tried, as
               perhaps unsuccessful, treatment. Perhaps its action
               is by the production of superoxide radicals, perhaps
               this is why oxygen would only make it worse.

CUMMING: I would like to make a comment from the chair at that
               point. Virchow called the autopsy room the temple of
               truth and gave the pathologists a great sense of
               certainty in their views, but you, Brian suggested
               that ARDS was based upon superoxide secretion, and
               that was it. How certain are you that this is true?

CORRIN: I suggested that in the case of shock lung. I am
               fairly confident that in shock lung one can see, if
               one examines carefully the interalveolar septa, many
               polymorphs within the capillaries. There is a
               sequestration of polymorphs in the lung, the
               peripheral blood leukocytes diminish. According to
               workers in Minnesota the polymorphs aggregate and
               they arrive as microemboli and it is these workers
               who envisage them as being activated by sepsis or by
               traumatic shock, and we know that activation of
               polymorphs induces much superoxide radical production
               and that there is also lysosomal enzyme release, and
               for these reasons I envisage the release of these
               substances as the damaging agent.

CUMMING: Not unlikely.

DENISON: You may know that it has suggested that,
               paradoxically, in hypoxic conditions, tissues are
               less able to handle spontaneously generated
               superoxide and so the mechanisms of hypoxic damage
               and hyperoxic damage may be very similar both being
               due to a relative excess of superoxide. I wondered
               if you wanted to comment on that the whether you
               could make any distinction between the effects of
               hypoxia and hyperoxia, histologically in the lung.

CORRIN: In my talk I showed the patient who was poisoned with
               oxygen, with hyaline membrane formation and I showed
               a slide of an electron micrograph of animals exposed
               to 15% O_2, and again we saw necrosis of the alveolar
               epithelium and a bare basement membrane covered with

hyaline membrane. So, pathologically, the changes can be very similar.

FAGAN: I am rather sorry to see you classifying hyaline membrane disease on a simplistic sort of classification such that it all looks the same, because there are certain very major differences between infantile hyaline membrane disease and adult shock lung. There is absolutely no alveolar damage in uncomplicated cases of infantile hyaline membrane disease. The exudation is solely at the junctions of the duct systems, and there is no diffuse alveolar damage. The only time you get diffuse alveolar damage in the infantile form is when you have started giving hyperoxia and IPPV and things like that. A second and very important point is the content of fibrin. There is no fibrin in an uncomplicated case of hyaline membrane disease, except from very small flakes that you can see on EM, whereas of course the hyperoxic membranes and the other ones you showed are almost pure fibrin. So, I wondered if I could invite you to be a little more subtle about your classification.

CORRIN: I use this slide illustrating a vicious circle, it was white on black with the neonatal causes shown in red on the left and the adult in green on the starboard side. Certainly, to begin with, it is a state of prematurity aggravated perhaps by hypoxia during birth, but ultimately, and in the end state of the disease, we have fibrin and alveolar necrosis, perhaps secondary to the hypercapnia. I have been given 40 minutes to describe the pathological reactions of the alveolar epithelium, and I would agree that in all these many causes there are subtle differences, but it would be impossible to point these out in such a short time. I was, perhaps, elective, rather than being simplistic about this and point out the common features which I think make the subject more easily understood by our clinical and physiological colleagues, and if we wish to emphasize the subtle pathological differences, with no disrespect to them, I think they would be lost and perhaps uninterested before the long discourse was finished.

FAGAN: I have nothing but total sympathy for that answer.

CUMMING: I think we are coming to the end of the session. If I am permitted two small observations, it appears in

the temple of truth there are many hand maidens and
secondly the word simplistic, I think has come to
mean a simple explanation of which one disapproves.
We shall break for coffee.

THE EFFECTS OF IRRITATION ON THE STRUCTURE OF

BRONCHIAL EPITHELIUM

Peter K. Jeffery

Cardiothoracic Institute
Brompton Hospital
London, SW3 6HP

Atmospheric pollution, and in particular personal pollution by tobacco smoke, is a major cause of chronic inflammation of the bronchial mucosa. In man there is an increase in the number of mucus-secreting cells both in surface epithelium and in submucosal glands (i.e. gland hypertrophy) the consequence of which is the production of excessive bronchial mucus and its expectoration as sputum, (Reid 1954). There may also be impairment of mucociliary clearance and an associated persistent cough. It is by these clinical criteria that chronic bronchitis is defined, exacerbations of cough and sputum persisting for more than 3 months in any 2 consecutive years (WHO definition).

In laboratory animals these bronchitic changes may develop spontaneously as a result of infection (Jones, Baskerville & Reid 1975) or may be induced experimentally by short term inhalation of a variety of irritants including tobacco smoke (Lamb & Reid 1969), sulphur dioxide (Reid 1963, Lamb & Reid 1968), elastase (Christensen et al 1977), nitrogen dioxide (Freeman & Haydon 1964) or following administration of isoprenaline (Sturgess & Reid 1973 or methacholine (Kleinerman et al 1976). These bronchitic changes may thus be produced by infectiion, irritation or by administration of drugs in the absence of infection. We have chosen the specific pathogen free (SPF) rat, exposed to either tobacco smoke or sulphur dioxide, as the animal model for study. By observation of the changes in such an experimental animal model free of disease much can be learnt of the early pathogenesis of hypersecretory disease and it is with these changes that the following discussion is concerned.

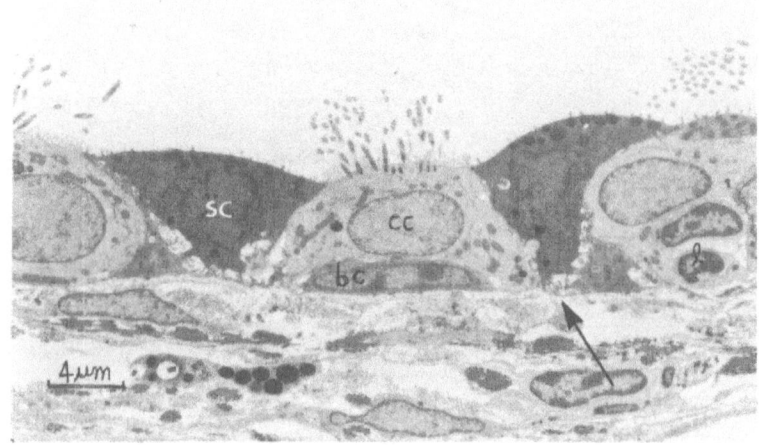

Fig.1.An electron micrograph of main bronchial epithelium from a rat
      not exposed to cigarette smoke. The epithelium is about 12μm
      thick. Serous cells (se) have electron-dense granules. Ciliated
      (cc) and basal (bc) cells and epithelial basement membrane
      (arrow). Glutaraldehyde + osmium tetroxide: uranyl acetate +
      lead citrate X.

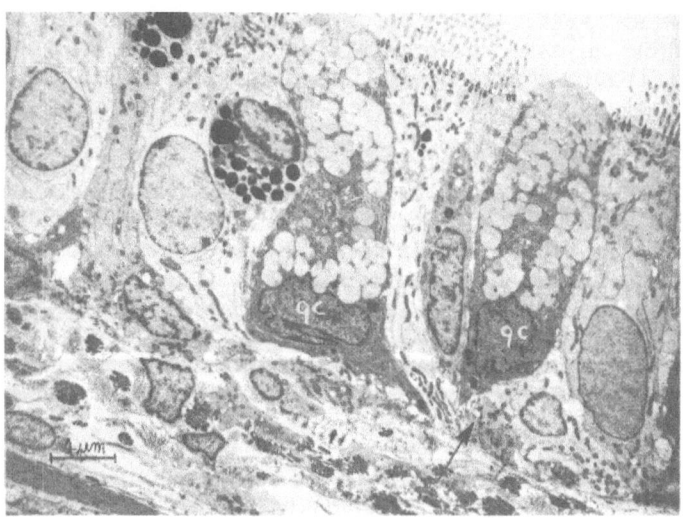

Fig.2.Main bronchial epithelium from a rat exposed to tobacco smoke
      for two weeks showing the increased epithelial thickness with
      goblet cells (gc) containing large numbers of electron-lucent
      granules which are confluent. Basement membrane (arrow).
      Glutaraldehyde + osmium tetroxide: uranyl acetate + lead citrate.
      Same magnification as Fig. 1.

Tobacco smoke

When SPF rats are exposed to an atmosphere of tobacco smoke (i.e. passive exposure) generated from 25 cigarettes per day (4 hours daily for up to 6 weeks) epithelial mucus-secreting cells are increased in number at all levels of the bronchial tree. The effect is dose related and greatest in the main bronchi and hilar airways, (Jones, Bolduc & Reid 1972). There is also a concurrent increase in epithelial thickness, cell division and in the concentration of cells per unit length epithelium.

Jeffery & Reid (1977 & 1980) have shown that the increase in epithelial thickness is not due to stratification of cells but to cell hypertrophy. Ciliated cells become taller, there is an increase in the amount of cell cytoplasm, in mitochondrial length and of surface microvilli also. Cilia are, however, normal for structure and density.

Secretory cell number increases and within each cell there is an increase in the volume of mucus. The small electron-dense secretory granules of the serous cell (fig.1.) are replaced by increased numbers of large electron-lucent granules typical of the mucous cell (fig.2). Cells of intermediate morphology are found (fig.3) suggestive of serous cell transformation: thus serous cell metaplasia contributes to the increased numbers of mucous cells seen with irritation. Light microscopic histochemistry shows there is an associated shift in the predominant type of intracellular mucus, from neutral to acidic; particularly of the resistant sialylated and sulphated types. This change may be important in disease as glycoproteins are a major determinant of mucus rheology and an increased sputum viscosity has been shown to correlate well with an increase in constituent sialic acid (Charman et al. 1974).

One of the early changes seen after tobacco smoke is an increase in the number of dividing cells. Normally a cell in division (mitosis) is rarely seen at any level of the respiratory tract (Bolduc & Reid 1976). With irritation, the number of dividing cells rises to a peak at 24 hours after exposure to the smoke, thereafter falling to a level just above that seen in controls (Wells & Lamerton 1975). There is no further rise inspite of continued exposure unless there is a break in the exposure routine after which there is a similar, though less pronounced rise as that seen initially (Bolduc 1976). Normally most of dividing cell population is basal, these being the 'stem' cells from which other basal and more superficial (secretory and ciliated) cells arise. With tobacco smoke there is a rise in the number of dividing basal cells but also many superficial cells divide. One of the interesting and recent findings is that differentiated mucous cells with secretory granules can also divide (fig.4)(Jeffery 1973, Ayers -

personal communication). Thus some of the increase in mucous cell
number may also be due to hyperplasia of an existing albeit small
mucous cell population. The extent to which serous cell metaplasia,
basal cell differentiation and mucous cell hyperplasia contribute to
the increase in mucous cell number seen after irritation is
currently being investigated.

The numbers of migratory cells found normally are not altered
significantly but neutrophil recruitment is often a feature (Kilburn
et al 1975) and this may contribute to the epithelial damage seen in
more chronic exposures. Lumsden & Lamb (personal communication) have
studied bronchioli of human lung resections for carcinoma and
carcinoids from smokers and non-smokers and report that the number
of epithelial mast cells which appear to have degranulated, is
highest in smokers. Thus the irritant-induced release of cellular
mediators is likely to be of importance in the inflammatory response
to irritation by tobacco smoke, particularly in regard the alteration
of epithelial permeability described (see Simani et al 1974).

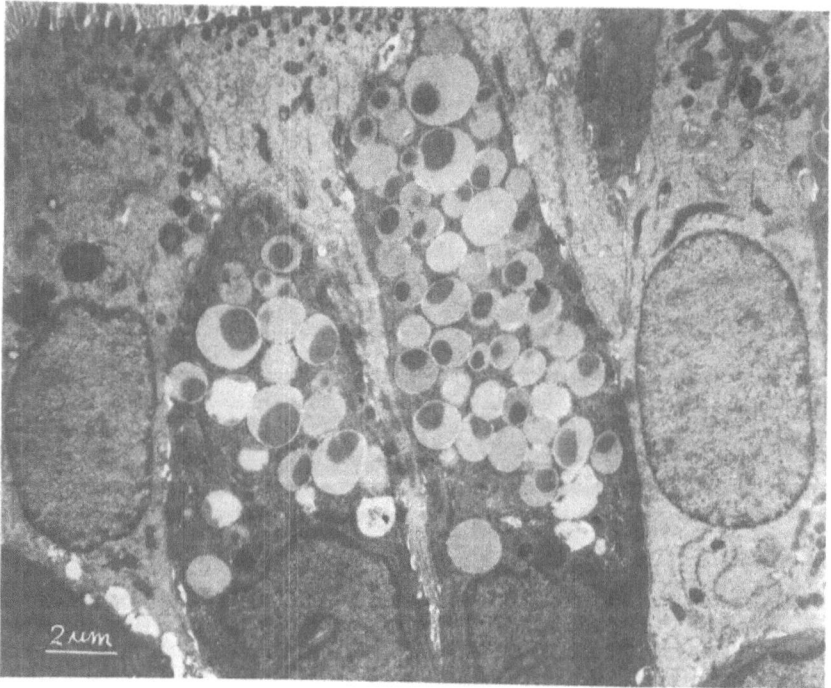

Fig.3. Electron micrograph of a cell showing features of both serous
       and mucous cells. Electron lucent granules have a dense core
       suggestive of serous cell transformation. From an animal given
       2 weeks tobacco smoke X

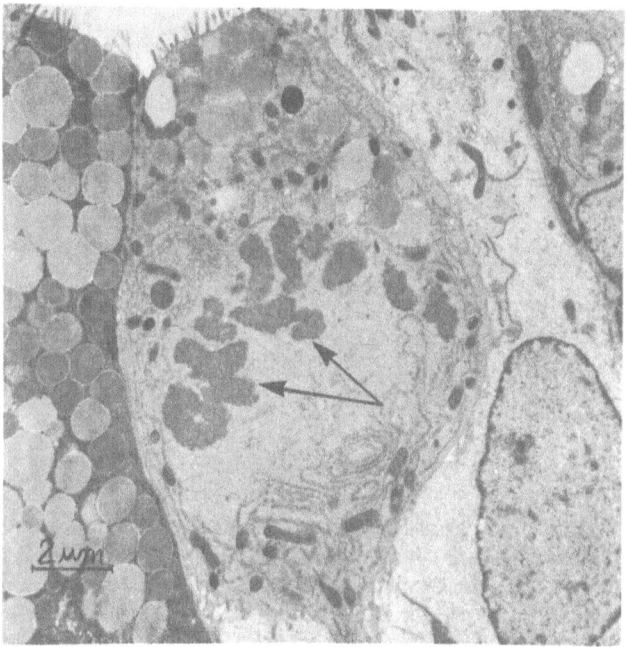

Fig.4. A fully differentiated mucous cell dividing in response to
       isoprenaline administration. The electron micrograph shows
       the loss of nuclear membrane , the sectioned chromosomes
       (arrow) and secretory granules at the cell apex.
       Glutaraldehyde & osmium tetroxide: uranyl acetate + lead
       citrate X

While distal broncholi are little affected with the doses used
here, smoke does reach this distal site and its products are
phagocytosed by the more distally located alveolar macrophages
(fig.6). Subsequent alteration of macrophage function or
proteolytic enzyme release by them may thus lead to alveolar wall
damage also.

Sulphur dioxide

Tobacco smoke is a complex organic irritant and it is difficult
to ascribe the induced changes to any one constituent, the effect
probably due to synergism of several. In contrast sulphur dioxide
is a simpler irritant, highly water soluble and which has been an
important atmospheric pollutant for many years. The concentration
which caused so many deaths during the worst smogs of London was of
the order of 2 ppm.

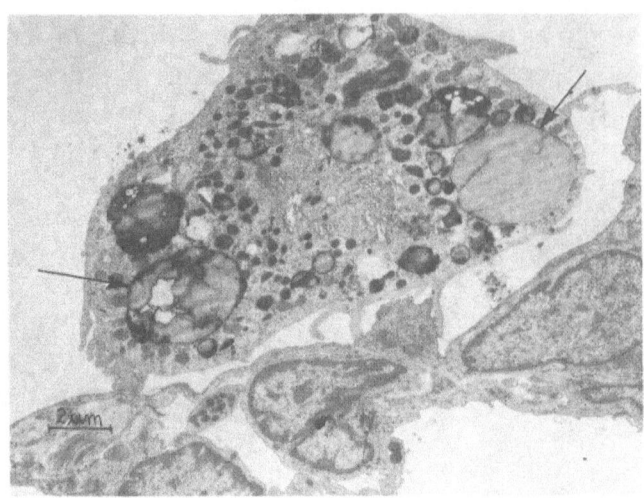

Fig. 5. Electron micrograph of an alveolar macrophage from an
animal given 2 weeks of cigarette smoke by inhalation. The
cell contains a number of large inclusions ("tar bodies") of
moderate electron-density (arrow). Smaller electron-dense
lysosomes are present also. Glutaraldehyde and osmium
tetroxide: uranyl acetate & lead citrate X

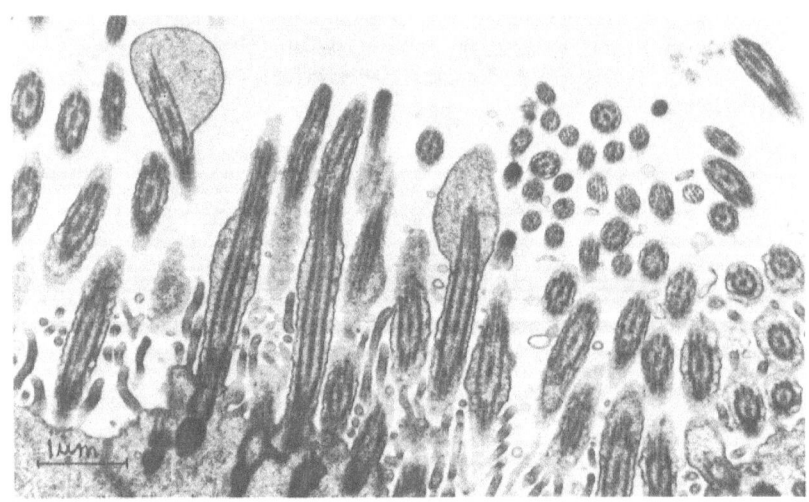

Fig. 6. Damaged cilia in rat distal airway after 3 weeks of sulphur
dioxide. There is bleeding of the ciliary shaft. Glutaraldehyde + osmium tetroxide: uranyl acetate + lead citrate X

Rats seem, however, particularly resistant to its irritant effect and in our experiments 400 ppm were given to rats for about 4 hours per day for up to 6 weeks. At this concentration there is early ulceration of the lining of the most proximal airways with a loss of both mucous and ciliated cells. Subsequent healing, associated with a rise in the mitotic index, leads to mucous cell hyperplasia. In contrast distal airways (less than 0.4 mm in diameter) do not show ulceration nor a significant increase in the mitotic index during the 6 weeks of exposure (Lamb & Reid 1968). Yet by the third week of exposure mucous cells appear and by 6 weeks they have increased in number to values significantly above those found in controls. In the absence of ulceration these findings indicate mucous cells may develop by cell transformation at this distal site. Our electron microscopic studies support this suggestion: cells with ultrastructural features of both Clara and mucous cells are found (Jeffery, Widdicombe & Reid, 1976). A similar transformation has been found in animals treated with repeated doses of Isoprenaline also (Jeffery & Reid, in preparation). Thus it appears that the serous to mucous cell transformation seen following tobacco smoke in the larger airways may be paralleled by a Clara to mucous cell transformation in the distal airway. However, unlike tobacco smoke, sulphur dioxide damages the ciliary escalator also with ciliary loss seen proximally and blebbing of the ciliary shaft distally (fig.6).

Histochemical changes

The intracellular mucus of epithelial cells shows a characteristic pattern of distribution along the bronchial tree for a given species. As a generalized response to irritation there is a shift in the predominant type (i.e. neutral or acidic) of mucus produced: the tendency is to an increase in the proportion of acidic mucus (acidity due to sulphate or sialic acid groups on the glyco-protein molecule) (Lamb & Reid 1968; Jones et al 1973). This shift is also seen following infection (Jones et al 1975) and administration of neuromimetic drugs (Sturgess & Reid 1973; Jones & Reid 1979).

Reversibility of irritant-induced changes

Are these irritant-induced changes reversible? There is now evidence that they are. The time taken for mucous cell number to return to normal following a 2 week exposure to tobacco smoke (i.e. recovery period) has recently been investigated (Jones, 1978; Rogers & Jeffery, in preparation). Our studies show that tracheal mucous cell number has returned to normal again 9 days following cessation of smoke: that of proximal and distal lung takes somewhat longer being 42 and 21 days respectively.

The recovery period following 6 weeks exposure to sulphur dioxide is longer: Lamb & Reid (1968) found elevated mucous cell numbers even 5 weeks after cessation of SO_2 exposure.

Baskerville (1976) has shown that the Isoprenalin-induced increase in airway mucous cell number returns to normal approximately 8 weeks following termination of drug administration. It seems likely that a reasonably quick return to normal reflects short term and mild exposure to an irritant. Further experiments are needed to see if cessation after chronic and more severe exposure requires proportionally longer recovery periods. Interestingly elastase seems to be an exception as Christensen et al., (1977) found that epithelial secretory cell number progressively increased following a single intratracheal dose of elastase.

Anti-inflammatory Agents

The effects of two anti-inflammatory agents on the bronchial response to tobacco smoke have been investigated with some success. Phenymethyloxadiazole (PMO) when included in the tobacco as 2% by weight, inhibits the tobacco smoke-induced mucous cell increase in the trachea but not that seen in the lung (Jones & Reid, 1978). There is partial inhibition of the increase in mitotic index, epithelial thickness and of the changes to certain cell organelles (Jeffery & Reid, 1980). Indomethacin given daily by intraperitonial injection either before (2 mg/kg) or, both before and after daily exposure to tobacco smoke, inhibits the mucous cell hyperplasia seen at all airway levels of the lung but not that of the trachea (Greig, Ayers & Jeffery, 1980) (fig.7). These studies indicate that the bronchial response to irritation can be modified. However, neither of the above anti-inflammatory agents inhibited the shift in the type of intracellular mucus produced.

The means by which mucous cell hyperplasia is inhibited - whether by depression of the mitotic response, metaplasia or both - is currently under investigation.

Summary

The role of atmospheric pollution, and in particular personal pollution by tobacco smoke, in the aetiology of bronchitis in man and the experimental animal has been discussed. These changes may be produced in the laboratory animal by a variety of agents including tobacco smoke, sulphur and nitrogen dioxides, elastase, isoprenaline and methacholine. Following tobacco smoke(subchronic) there is an increase in the number of mucus-secreting cells, in epithelial thickness, in cell division and concentration. Sulphur dioxide (400 ppm) causes epithelial ulceration proximally following which there is mucous cell hyperplasia: distally there is similar

TABLE I

Percentage increase (due to tobacco smoke
inhibited by Indomethacin

	TRACHEA	UPPER AXIAL	LOWER AXIAL	UPPER LATERAL	LOWER LATERAL
Indomethacin 2 mg/kg	18%	40%	36%	57%	66%
4 mg/kg	22%	71%	68%	79%	100%

Fig. 7. Percentage inhibition of mucous cell hyperplasia.

hyperplasia without prior ulceration. There is evidence that both cell (serous and Clara) metaplasia and mucous cell hyperplasia contribute to the increased numbers of mucus-secreting cells. The changes are reversible with time after cessation of the irritant and the initial response may be modified by concurrent administration of anti-inflammatory agents.

The author acknowledges the support given by both the Cystic Fibrosis Trust and the Medical Research Council for the studies presented here. Also I am grateful to the following publishers for allowing reproduction of figures 1 & 2 - Wiley & Sons; figure 3 - Marcel Dekker and figures 5 & 7 - Journal of Pathology.

References

Bolduc, P., 1976, Cell turnover of the bronchial and alveolar
    lining in the rat lung in various types of hypertrophy.
    Ph.D. thesis. Univ. of London.

Bolduc, P., & Reid, L., 1976, Mitotic index of the bronchial and
    alveolar lining of the normal rat lung, Am.Rev.resp.Dis,
    114:1121.

Charman, J., Lopez-Vidriero, M., Keal, E., Reid, L., 1974, The
    physical and chemical properties of bronchial secretion, Brit.
    J. Dis. Chest, 68:215.

Christensen, T.G., Korthy, A.L., Snider, G.L., & Hayes, J.A., 1977,
    Irreversible bronchial goblet cell metaplasia in hamsters with
    elastase-induced panacinar emphysema , J. Clin. Invest,59:396.

Freeman, G., & Haydon, G.B., 1964, Emphysema fter low level exposure
    to NO_2, Arch. Environ.Hlth.,8:125.

Greig, N., Ayers, M., & Jeffery, P.K., 1980, The effect of
    Indomethacin on the response of bronchial epithelium to tobacco
    smoke, J. Pathol., In press.

Jeffery, P.K., and Reid, L., 1977, The respiratory mucous membrane.
    In Respiratory Defense Mechanisms (ed. J.D.Brain et al(Vol.5
    of Lung Biology in Health and Disease (ed.C.Lenfant) Dekker
    pp.193-246.

Jeffery, P.K., & Reid, L., 1980, The effect of tobacco smoke, with
    or without phenylmethyloxadiazole (PMO), on rat bronchial
    epithelium: a light and electron microscopic study, J. Pathol.
    in press.

Jeffery, P.K., Widdicombe, J.G., & Reid, L., 1976, Anatomical and
    physiological features of irritation of the bronchial tree.
    In Air Pollution and the Lung (ed. E.F.Aharonson et al) Wiley
    pp. 253-267.

Jones, R., 1978, The glycoproteins of secretory cells in airway
    epithelium. In Respiratory Tract Mucus, Ciba Found. Symp.
    54 (new series) (ed. R. Porter) Elsevier pp. 175-188.

Jones, R., Baskerville, A., & Reid, L., 1975, Histochemical
    identification of glycoproteins in pig bronchial epithelium:
    (a) normal and (b) hypertrophied from enzootic pneumonia,
    J. Pathol, 116:1.

Jones, R., Bolduc, P., & Reid, L, 1972, Protection of rat bronchial
    epithelium against tobacco smoke, Br. Med.J.,2:142.

Jones, R., Bolduc, P., & Reid, L, 1973, Goblet cell glycoprotein and
    tracheal gland hypertrophy in rat airways: the effect of
    tobacco smoke with or without the anti-inflammatory agent
    phenylmethyloxadiazole, Brit.J.exp.Pathol, 54:229.

Jones, R., & Reid, L, 1978, Secretory cell hyperplasia and
    modification of intracellular glycoprotein in rat airways
    induced by short periods of exposure to tobacco smoke, and the
    effect of the anti-inflammatory agent phenylmethyloxadiazole,
    Lab. Invest. 39:41.

Jones, R., & Reid, L, 1979, β-agonists and secretory cell number
    and intracellular glycoprotein in airway epithelium, Am.J.Path,
    95:407.

Kilburn, K.H., Wayland, N.Mc., & Thurston, R.J., 1975, Cellular
    effects of cigarette smoke and hamster airways, Chest 67/suppl.
    54S - 55S (17th Aspen Lung Conference).
Kleinerman, J., Sorensen, J., Rynbrandt, D., 1976, Chronic bronchitis
    in the cat produced by chronic methacholine administration(abstr)
    Am.J.Pathol., 82:45a.
Lamb, D., & Reid, L., 1968, Mitotic rates, goblet cell increase and
    histochemical changes in mucus in rat bronchial epithelium
    during exposure to sulphur dioxide, J.Path. & Bact., 96:97.
Lamb, D., & Reid, L., 1969, Goblet cell increase in rat bronchial
    epithelium after exposure to cigarette and cigar tobacco smoke,
    Br.Med.J. 1:33.
Reid, L., 1954, Pathology of chronic bronchitis, Lancet 1:275.
Reid, L., 1963, An experimental study of hypersecretion of mucus in
    the bronchial tree, Br.J.exp.Pathol., 44:437.
Simani, A.S., Inone, S., & Hogg, J.C., 1974, Penetration of the
    respiratory epithelium of guinea pigs following exposure to
    cigarette smoke, Lab. Invest., 31:75.
Sturgess, J., & Reid, L., 1973, The effect of isoprenaline and
    pilocorpine on (a) bronchial mucus-secreting tissue and (b)
    pancreas, salivary glands, heart, thymus, liver and spleen.
    Br.J.exp.Pathol., 54:388.
Wells,A.B., & Lamerton, L.F., 1975, Regenerative response of the
    rat tracheal epithelium after acute exposure to tobacco smoke:
    a quantitative study, J.Nat.Cancer Inst., 55:887.

DISCUSSION

LECTURER: Jeffery   CHAIRMAN: Cumming

DENISON: Going to your last summary slide, which reminded us
                that the PMO that is present in the cigarettes
                protected the trachea and not the distal airways,
                could this be simply because you failed to get the
                PMO any further than the trachea? What evidence have
                you that it got any further?

JEFFERY: There is no evidence that PMO got any further than
                the trachea, and indeed we think that this is the
                explanation of these results, PMO being given by an
                inhalation route and therefore it would be expected
                that it would affect the larger airways, more so than
                the distal. However, indomethacin given by
                intraperitoneal route, it would be presumed, would
                affect more significantly those airways in the lung
                than in the trachea, which is supplied mainly by
                systemic circulation. So I think this is right. It
                has been suggested that we ought to try the reverse,
                put indomethacin in the cigarette. However, I do not
                think that indomethacin would come over the 800
                degrees C at the tip and still be effective. PMO on
                the other hand appears to be a very stable compound,
                and this has been researched in America, and it
                appears to be effective even when it traverses that
                high temperature.

PRODI: Just a comment. I would expect that if PMO is put
                inside the tobacco it would come out together with
                the cigarette smoke. So, behaving as particulate
                matter exactly in the same manner, it would reach the
                areas where the tobacco smoke is deposited.

JEFFERY: It would presumably depend on the size of the
                particle generated.

PRODI: Is there any reason that the size distribution of PMO
                would be different from cigarette smoke?

CUMMING: Yes.

PRODI: Why?

CUMMING: It is a very poly disperse smoke, and the aggregation depends on the pH and many things, so, PMO being neutral I would expect it to aggregate perhaps in smaller particles than the larger aggregates of tars.

PRODI: At the temperature conditions of the cigarette I would expect that PMO would be in the form of vapour.

CUMMING: Probably.

PRODI: So, either at once or later on, this would end up by condensing. I would not really expect any serious size fractionations between the various components.

CUMMING: What is the boiling point of PMO? Anybody know?

JEFFERY: I do not know. One of the suggestions of course that was made is that perhaps PMO might in some way be modifying the cigarette smoke itself and therefore reducing its inflammatory action on the tissues. However, earlier works by Dalham and his colleagues in Sweden showed that if you gave the parent compound, oxalamine citrate, or indeed PMO by the intraperitoneal route it also has an anti-ciliostatic effect on the trachea. So it would appear that it is working at the tissue level rather than by modification of the smoke. We did do some rather simple studies to see if the pH of the smoke was altered by PMO, and it was not.

WIDDICOMBE: May I ask two questions, please. First of all, do you see any likelihood that the drugs you were using, PMO and indomethacin, may have had behavioural actions on the rats, so that when they were exposed to the cigarette smoke they either coughed less, in the case of PMO, or have a smaller respiratory response or different respiratory response, so that either the uptake or the deposition or both, of the cigarette smoke, was different?

JEFFERY: In fact we did not look at respiratory rate, it is quite difficult, they never stay still for long enough. But one behavioural thing we did notice, which perhaps does not support your suggestion, is that the animals given PMO with the cigarette smoke appear to exhibit longer and more vigorous cleaning activities than did the animals given cigarette smoke alone. In other words, not only could we see it, but their fur appeared to be cleaner than those animals given tobacco smoke alone.

WIDDICOMBE: What is the mechanism of that? Have you any ideas?

CUMMING: That is more difficult.

JEFFERY: Indeed it is.

WIDDICOMBE: My second question concerns my ignorance of cell
                biology. When you get labelled thymidine in the
                goblet cells, as your pictures very elegantly showed,
                is there any chance that this is being transferred
                from dividing basal cells?

JEFFERY: I purposely kept away from some of the more complex
                aspects of the experiments. We in fact looked at two
                groups of animals. In the first group of animals we
                killed them one hour after giving them the tritiated
                thymidine. So if an animal was to be killed on day 1
                or day 3 or day 14, they were given their thymidine
                one hour before they were killed. All cells that
                they appear labelled are in the S phase and these are
                the results which I have shown you today. In the
                second group of animals we gave the thymidine at the
                beginning of the experiment and then they received no
                more thymidine. Then with killing, what we are
                looking at is the passage of thymidine, which we
                would in this case expect to be in the basal cell
                compartment, moving to a more superficial cell
                compartment. But those results I have not shown you
                today. I showed you the results of our A group.

WIDDICOMBE: I would like to hear a little bit about the results
                of group B.

JEFFERY: I think I would rather not comment on the B group at
                this stage.

CORRIN: Have you offered any of your smoking friends a
                cigarette containing 2% PMO? Is it acceptable to a
                smoker or does it spoil his pleasure?

JEFFERY: I have not done the experiment myself, but Tor Dalham
                again, in Sweden, is rather notorious for his dinner
                parties and I would not like to comment on the
                effects of this, but in a pot on the table were
                cigarettes, and he knew which ones contained PMO and
                which ones did not, and some of his guests did smoke
                cigarettes with PMO included, and when they were
                later asked had they noticed the difference in the
                flavour, they said that they detected no difference.

FABBRI: I would like to know if the excess of acid
                   glycoproteins, during chronic exposure to smoke may
                   determine an inhibition of the activity of the cilial
                   epithelium, besides the direct action of smoke.

JEFFERY: I think that the question is relating to the effect
                   of acid glycoproteins on ciliary motility. The
                   answer is: we do not know. In fact, we are
                   interested in the effects on the shift of viscosity,
                   which I might say is characteristic of the chronic
                   bronchitics as well. There is a good correlation
                   between the increase in sialic acid in the
                   glycoprotein and viscosity. However, I cannot say
                   whether the increase in viscosity would affect
                   ciliary motility or, indeed, how it would influence
                   mucociliary clearance. We know that there is an
                   optimum level for mucociliary clearance in terms of
                   mucous viscosity; if it is too thin, such as in
                   bronchorrhea, particles are not moved. If it is too
                   thick, particles are not moved. But the effects on
                   ciliary motility directly from that change, I cannot
                   say.

PAOLETTI: You know the difficulty of correlating the
                   experimental effects of air pollution in the
                   laboratory and the effects under normal conditions,
                   and you make your studies at very high exposures of
                   SO_2. I want to know if you made some studies at a
                   lower level of SO_2 and, if not, what are the problems
                   of absorption.

JEFFERY: While working with John Widdicombe at St. George's
                   Hospital we did try a lower concentration, in the
                   range of 50 to 100 ppm; in that case we did not get
                   epithelial ulceration in the trachea, and there was
                   an increase in goblet cell number, but it was not
                   significant. We saw no significant changes in the
                   distal airways, which is interesting, but I think in
                   the light of John's comments from his talk it might
                   be expected that very little got down to the trachea.
                   There is a paper by Dalham and Strandberg in which
                   they looked at the absorption of SO_2 by the nares of
                   the rabbit, and they put 200 ppm in the inhaled air.
                   When they measured the concentration in the trachea
                   of the rabbit they found it was of the order of 2
                   ppm. So this, I think, emphasizes what John
                   Widdicombe has been saying, the efficiency of the
                   nares in these lower animals.

BIGNON: Did you check if there was some leukocyte
               recruitement in these animals at the level of the
               airways?

JEFFERY: Yes, there was leukocyte recruitment, polymorph
               recruitment, that is neutrophil recruitment,
               especially into the intrapulmonary airways and into
               the epithelium, and this coincides with the results
               of Kilburn and colleagues who have looked at the
               effects of carbon dust and cigarette smoke in
               experimental situations. There was polymorph
               recruitment.

CANDURA: I do not know whether to put this question to you, or
               to Cohen, or to someone else who can give me an
               answer. I am thinking of this; based on the
               conceivable hypothesis that hyperproduction of mucus
               can actually be a defence mechanism against an
               invading agent which can penetrate the respiratory
               system and produce effects even on a distant target,
               would it then be right to modify tobacco so as to
               limit this hyperproduction, or not?

JEFFERY: I think it is a very interesting point, and I would
               not disagree with you that the production of excess
               mucus is protective in its nature. As we have seen
               from Turner Warwick's talk, the normal immunological
               response is initially protective in nature. However,
               as you pointed out, an excessive response will go
               beyond the protective function and will of itself
               cause further disease, there is a fine balance. What
               perhaps encourages me a little about this work is
               that indomethacin as you saw from my histogram did
               not totally inhibit the goblet cell response to
               tobacco smoke. It was significant, but formed an
               intermediate stage. Perhaps what we ought to be
               thinking about is partial but not total, inhibition,
               because we would not wish to withdraw that aspect
               which would be protective to the individual.

CUMMING: I will be forgiven a comment on excess mucus
               secretion. It seems a general view that it is
               possible to say what is excess mucus secretion, but I
               think that if one asked the question: what is the
               normal level of mucus secretion and what is the
               observed level of mucus secretion, you would find
               that the question is unanswerable. So I think one
               has to be very careful before saying that mucus is
               secreted in excess. We can certainly say that there
               is an increased sputum production, and if we confine

ourselves to that will be no difficulty. The problem
often facing the morbid anatomist is that he sees a
larger number of glands, but of course you see a
larger number of glands in the myxoedematous thyroid,
and no one would say that was hypersecreting. So, we
do have a problem about the definition of what is
excess mucus secretion. Would you like to comment on
that, Peter?

JEFFERY: I think the most acceptable definition of excess
mucus secretion that we have at the moment is that it
is in excess when it becomes expectorated as sputum.
But I will make a further comment, I think you raised
an interesting aspect. What I have been showing you
is an increase in the number of cells within the
intracellular mucus. We know very little about the
rate of discharge of that mucus, and I think we need
to do a lot more on that aspect.

CHEVALIER: As far as another type of smoking is concerned, do
you have any evidence from your laboratory or other
laboratories that after THC inhalation there is such
an irritative response of the bronchial glands? THC
is tetrahydrocannabinol.

JEFFERY: Forgive me for my ignorance, I know nothing of
hashish! I do not know of any studies of this
nature, with hash or marihuana. But there are
studies in the literature. I cannot comment.

CUMMING: Ladies and gentlemen, the time allocated for this
session has come to an end. If you are not going to
the beach I can prolong it for ten minutes more, or I
terminate it now.

INTERACTION BETWEEN MINERAL FIBRES AND PULMONARY CELLS

Jean Bignon, Marie-Claude Jaurand and Patrick Sebastien

Service de Pneumologie, ERA CNRS 845, INSERM U 139

40, Ave de Verdun, 94010 Creteil cedex

It is now well documented from human and animal data that asbestos exposure is associated with the development of two kinds of diseases : 1/ Fibrosis, not only of lung parenchyma (asbestosis) but also of pleura (pleural plaques) ; 2/ Cancer, mostly bronchogenic carcinoma and pleural or peritoneal mesothe-lioma, and also laryngeal cancer, gastro-intestinal tumors and perhaps cancer of other site (1, 2). Initially, such diseases were observed only in asbestos workers, some of them such as mesothelioma or lung cancer after a long latency period of 20 to 40 years (3). At present, with the increasing consumption and accumulation of asbestos dusts in our environment, the asbestos-related diseases increase steadily in incidence and are also observed in association with moderate para-occupational (domestic) or environmental exposures (3,4). More recently, other fibres, either natural (zeolite) or man made fibres (glass fibres) were shown to have the same potential hazards ! (5)

The respiratory system being the major target organ for inhaled fibrous dusts, several points concerning the pathogenetic effects of fibres will be discussed : penetration and deposition inside the airways ; alveolar biotransformation ; clearance and translocation to other sites, particularly to pleura ; biological activity at the level of tissular, cellular and subcellular effectors of the respiratory system. These various points have been investigated by different experimental studies on animals and on cells isolated from the respiratory tract (alveolar macrophages (AM)), tracheal epithelium, lung fibroblasts, meso-thelial cells). Moreover, in order to better understand the

biological and molecular events, basic studies have been carried
out using simplified models, such as red blood cells, natural
(ghosts) or synthetic (liposomes) plasma membranes and proteins
and lipids (6) ; all these experiments were designed to investigate
different type of fibres with well defined physico-chemical
parameters (7).

1. FIBROUS DUSTS

The word *fibre* is used for elongated particles with a
length diameter ratio (aspect ratio) equal or superior to 3.
Actually, biologically reactive inorganic fibres have usually
an aspect ratio equal or superior to 10, with a diameter generally
less than 3 μm (8).

1.1. Asbestos is a generic term for a variety of hydrated
silicate minerals which have a common attribute, namely the
ability to be separated into thin and relatively soft and
flexible fibres. They are silicates made by a framework of
tetraedric SiO_4.

a/ Chrysotile is a magnesium hydrated silicate formed by sheets
of SiO_4 and $Mg(OH)_2$ (brucite layer). Ultimate chrysotile have
a hollow cylindrical form. The outer diameter of such fibrils
is in the range of 30 nm. Chrysotile fibres are in fact formed
by the association of numerous such ultimate fibrils disposed
concentrically in a direction parallel to the fibre axis
(Fig. 1, 2). It is well known that chrysotile fibres have
powerful sorptive properties, able to fix organics (benzo
3-4 pyrene), gas (SO_2) and metals (9).

b/ The basic crystal form of amphiboles is a double silica
chain (Si_4O_{11}). The chains are paired "back to back" with a
layer of hydrated cations : magnesium, fer, calcium, sodium.
The four commercial varieties of amphiboles are : amosite,
crocidolite, anthophyllite, tremolite. Individual amphibole
fibres are generally straight with a diameter from 0.13 μm for
crocidolite to 0.5 μm for other amphiboles (Fig. 2).

1.2. Other asbestiform fibres having the same shape and
size characteristics as amphiboles have been recently described
(8). Some of them are already associated with health hazards ;
it is the case of erionite-zeolite fibres found in association
with environmental mesothelioma in a rural area of Turkey (10).
Fibrous clays (attapulgite or palygorskite, sepiolite) are
largely used commercially because of their sorptive properties ;
recent data might indicate a potential hazard in man after
inhalation of such fibres (11).

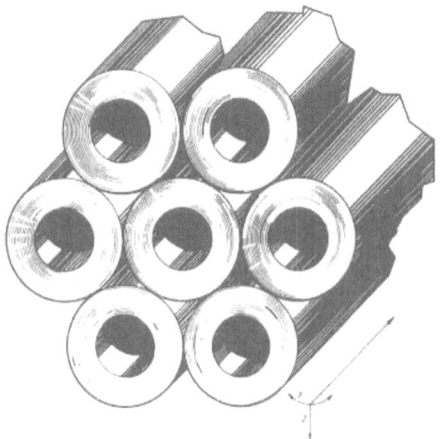

Figure 1   Diagramm showing the hollow tube
pattern of chrysotile fibres
(from Brouet et al., 46).

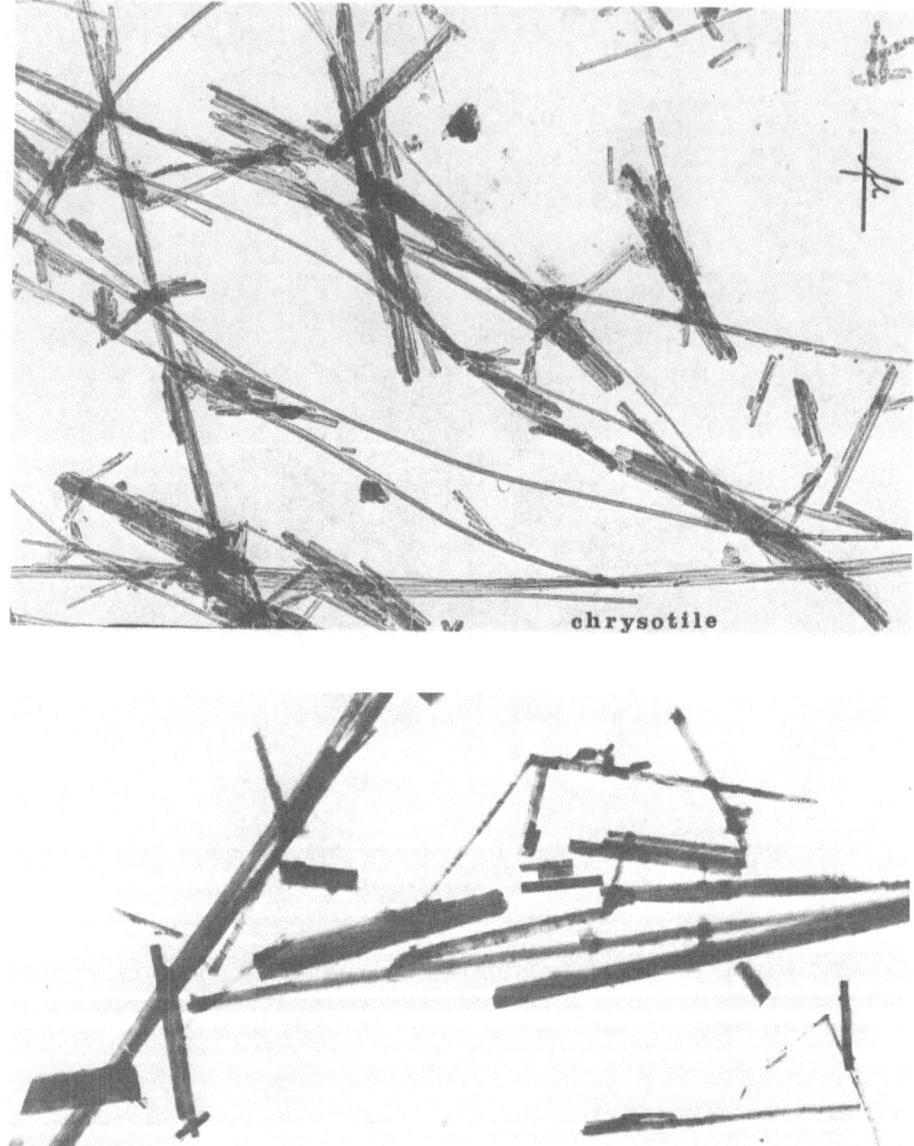

Figure 2   Electron micrography of chrysotile and amphibole type
           fibres

2. LUNG PENETRATION AND DEPOSITION OF FIBRES

Different models have been published to describe deposition of inhaled compact particles in the human respiratory system (12). The aerodynamic behaviour of fibres and the characteristics which influence their deposition, however, differ from those of compact particles (13). Four deposition mechanisms are involved : inertial impaction, settlement under gravity, direct interception and brownian diffusion. In a model of a regular dichotomous branching pattern for the respiratory system, the deposition of fibrous dusts will differ according to the site of the airways, the regimen of airflow and the shape and size of fibres (14).

In conductive airways, besides inertial impaction and settlement under gravity fibres deposition is related to direct interception, particularly for long fibres in a turbulent airflow regimen. In a laminar airflow which is the case in peripheral airways, long fibres if they are thin and straight will show a tendency to alignement with air streamlines and a capacity to penetrating deeply into the alveolar airspaces. By contrast, curly fibres such as chrysotile asbestos should be intercepted at the site of airway branchings (Fig. 3). In alveolar spaces, fibres deposite either by brownian diffusion or by interception in the case of long fibres.

Mathematical models predicting the deposition of fibres have been described (14) ; they fit more or less well with the metrologic data obtained from human lungs ; so, in lung parenchyma, 90 % of amphiboles and 70 % of chrysotile fibres are less than 5 μm in length (Fig. 4). However, after an heavy asbestos exposure, the mean length increase steadily, with 15 at 20 % of fibres with a length more than 8 μm (Table 1). The mean diameter of parenchyma fibres was from 0.09 to 0.16 μm ; when mutiplying this actual diameter by 3, we obtain an aerodynamic diameter less than 1 mm. Nevertheless, this mathematical theory on fibre deposition does not explain why the numerical concentration of fibres is higher in the peripheral (subpleural) areas of the lung, in humans (15) as well as in rats (16).

3. BIOTRANSFORMATION OF FIBRES

Several in vitro experiments have demonstrated that asbestos chrysotile is very unstable in acid medium ; thus, the magnesium of the brucite layer is rapidly leached by different acids : chlorhydric acid, oxalic acid and other organic acids (17). This leaching is associated with a dramatic increase of specific surface area and with changes of cristallinity, chemistry and possibly other mineralogical properties. By contrast, amphiboles or glass fibres seem to be more stable in acid medium : only cations of the superificial layers are leached from amphibole fibres.

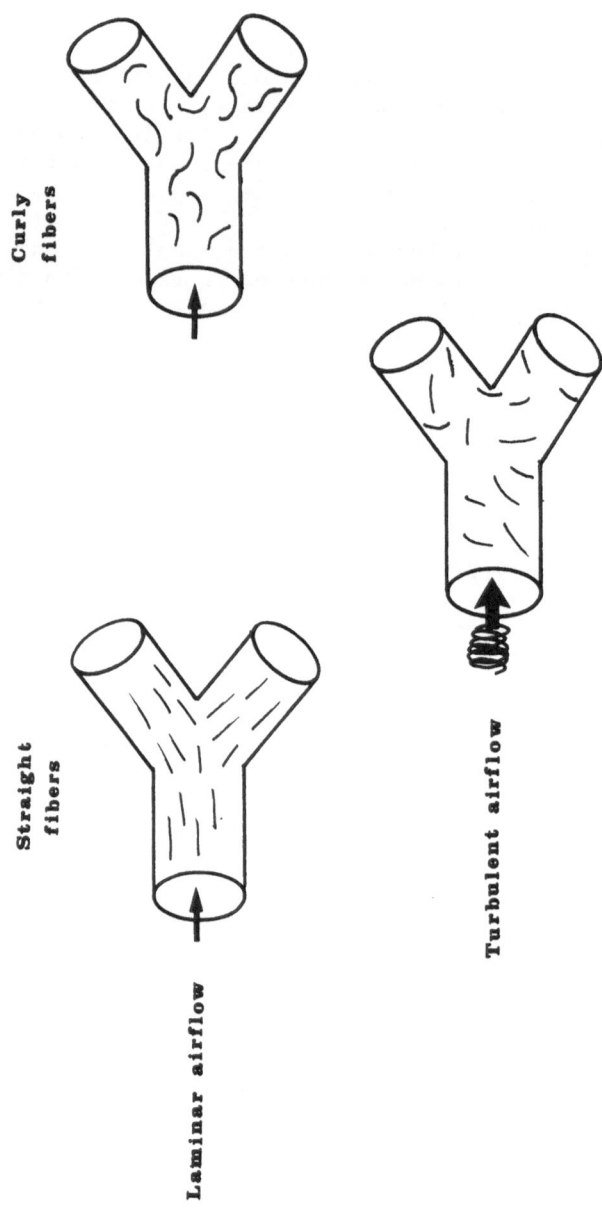

Curly fibers

Straight fibers

Laminar airflow

Turbulent airflow

Figure 3   Diagramm indicating the behaviour of fibres (straight or curly) in the streamline at the level of airway branching.

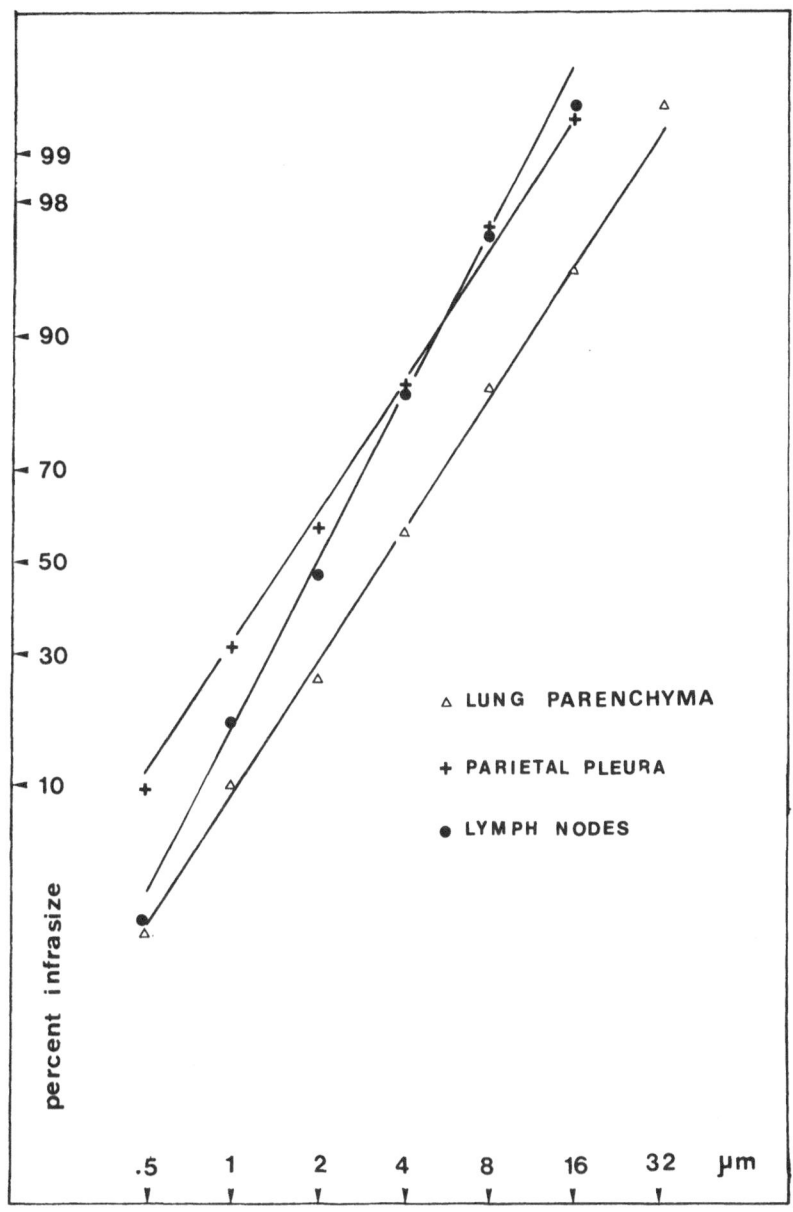

Figure 4 Fibre size distribution in human lung parenchyma,
          parietal pleura and lymph nodes

Such findings raised two questions about the behaviour of
fibres in vivo : are fibres residing in lung tissues leached ?
how this biological transformation modify the toxic properties
of fibres ?

A small proportion of fibres (from one out of 10 to 1000)
in retention in the lung is transformed into ferruginous bodies
(15). This phenomena is not specific to asbestos ; all inorganic
fibres, if long enough (glass fibres, zeolite fibres, carbon
fibres and so on), can also induce such transformation. Among
asbestos dusts, asbestos bodies are rather observed with long
respirable amphiboles fibres ; by contrast the short and thin
fibrils of chrysotile give very few asbestos bodies. It is now
demonstrated that such biotransformation takes place in the
phagosomes of alveolar or tissular macrophages, frequently
transformed in giant cells (18) (Fig. 5). Ferritin and haemosi-
derin granules sediment at the surface of fibres, possibly
attracted by electrostatic forces. Subsequently, this ferruginous deposit is cemented around the fibres by mucopolysaccharids
creating a ferroprotidic coating. This coating is often discontinuous, giving a beading aspect which suggests some heterogeneity in the surface chemistry or cristallinity of the fibres.
This coating seems to be associated with a decreased toxicity
of fibres (19).

Chrysotile fibres recovered from human lung parenchyma
have usally a modified chemistry, indicating a dramatic lost of
magnesium. This in vivo release of magnesium happens very
quickly, as soon as 24 hrs, when chrysotile fibres were phagocytosed by AM (20).

4. LUNG BURDEN FOR FIBRES

The lung burden for mineral fibres results from the difference
between the fraction of inhaled fibres deposited in the airspaces
and the fraction cleared or translocated from the lung parenchyma
(21) (Fig. 6). The majority (98 %) of fibres deposited at the
surface of the conductive airways is rapidly cleared by the
mucociliary escalator towards the gastro-intestinal tract or
the sputum . By contrast, fibres having reached the alveolar
spaces are engulfed by alveolar macrophages and cleared up
towards the terminal bronchioles or translocated across the
alveolar membrane in the pulmonary vascular and lymphatic
circulations.

Among asbestos, chrysotile fibres disappear rapidly from
the rat lung (22). This must be relevant to the fact that
asbestos chrysotile is less abundant than amphiboles in human
lung (23)(Table 1) ; this finding contrasts with the predominant
commercial utilisation of chrysotile (more than 90 %). The
mechanisms which make the chrysotile fibres to be rapidly

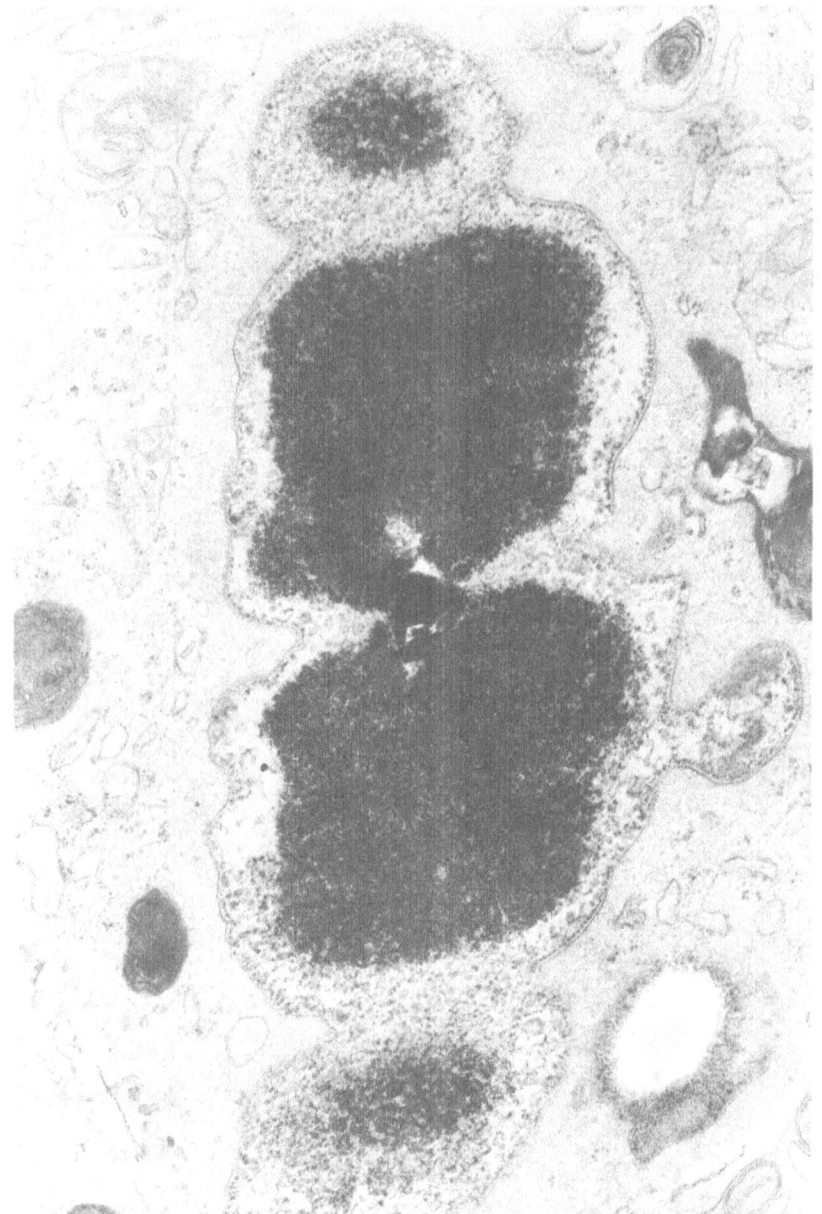

Figure 5   Micrography showing the bead pattern of the part of an asbestos body enclosed in a vesicle. The fibre is in the center of the ferruginous core.

Table 1. Number and size of fibres in human lung in different groups
of past asbestos exposure

PARAMETERS	GROUP 1 heavy exposure	GROUP 2 moderate exposure	GROUP 3 environmental exposure
Number per cc of parenchyma (light microscopy) (LM)	383×10^6	2×10^6	0.3×10^6
Number per cc of parenchyma (TEM)	33×10^9	1×10^9	1×10^9
All fibres (TEM) (µg per cc parenchyma)	4200	60	35
Amphibole type fibres (%)	68	45	19
Mean length (TEM) µm	5.5	4.8	2.5
Mean diameter (TEM) µm	0.16	0.10	0.09
Fibres (TEM) longer than 8 µm (%)	20	18	3

LM : light microscopy   TEM : transmission electron microscopy cleared from the lung remain however unelucidated.

On the other hand, it has been demonstrated that fibres have a great ability to migrate far away from their initial deposition and that short fibres migrate more easily that long fibres. By contrast, long fibres are selectively retained and concentrated in lung parenchyma (16, 23). The discovery of asbestos fibres in human parietal pleura suggests that fibres can migrate to the pleura. However, only small fibres, mostly ultimate fibrils of chrysotile, are found in this location. The mechanism by which fibres can be transported from the lung to the parietal pleura is still subject of controversy : either migration across the pleural cavity or translocation by the lymphatic route (24) and/or by the systemic arterial circulation (25).

Thus, in summary from a practical point of view, several points have to be emphasized :
. Short and straight fibres can penetrate deeply into the lung ; but they are cleared and/or translocated more quickly than long respirable fibres. Only short fibres mostly chrysotile are found in the parietal pleura.
. Long fibres can produce ferruginous bodies and are concentrated in the parenchymal connective tissue.
. Alveolar clearance of fibres lasts during the whole life so that it is possible to find ferruginous bodies in sputum or in pulmonary lavage fluid even 20 to 30 years after the cessation of asbestos exposure. Thus, the numeration of coated fibres in these biological samples appeared good indices for evaluating the concentration of fibres in lung parenchyma, even a long time after exposure cessation.

5. LUNG TISSUE RESPONSES

5.1. Early inflammatory response. The initial events induced intraalveolarly by asbestos fibres are not well known. In a previous work (26), we showed that 68 hours after an intratracheal instillation of 10 mg of chrysotile fibres in the rabbit, there was a decrease number of alveolar macrophages in the bronchoalveolar lavage ; however, the viability of these AM was normal and their enzymatic content was increased. On the microscopic sections of the lung, there was an inflammatory reaction around the peripheral airways, with many monocytes, macrophages and polymorphonuclear leucocytes (PMN) sequestrated in the bronchiolar wall tissue. This leuco-attraction has been recently found intra-alveolarly by others after intratracheal instillation of asbestos. It was attributed to the release of a chemotactic factor by AM (27). In the lung washing fluid of patients with severe asbestosis, we also found an increase of PMN leucocytes (23). These preliminary data in animals and in

humans deserve more basic work in order to understand more
clearly the biological events which make fibres induce lung
fibrosis!

5.2. <u>Fibrogenesis</u>. Asbestosis corresponds to the fibrosis
of alveolar connective tissue, due to the retention of airborn
asbestos dusts. Clinically, radiologically, functionnally and
pathologically, advanced cases are identical to idiopathic
pulmonary fibrous (3). Recently, however several interrogations
have been raised about the early stages of asbestosis. It has
been shown that the disease can begin as a small airway disease,
with dusts and collagen concentrated in the wall of peripheral
airways. The practical interest of this finding is the possibility
to detect asbestosis earlier by the small airways functional
tests, before an advanced lung fibrosis is developed (28, 29).

The mechanisms by which fibres induce lung fibrosis are
not well known. Several important experimental results are now
acquired : long fibres and large doses are more fibrogenic for
the lung than short fibres, in humans as well as in animals
(30, 31). Moreover, chrysotile seem more fibrogenic than
amphiboles (30).

The cellular and molecular events leading to an increase
of the connective tissue synthesis are not yet fully understood ;
however, it might be possible that AM could be stimulated by
fibres to liberate a fibrogenic factor (32), as demonstrated
for quartz by Heppleston and Styles (33). As mentioned earlier
regarding the migration of fibres to pleura, we do not know the
mechanism underlying the fibrous plaques formation at the level
of the parietal pleura.

5.3. <u>Carcinogenesis</u>. The increase risk of lung cancer and
mesothelioma has been clearly demonstrated in occupationnaly
exposed people (1, 2, 3, 4). For medico-legal compensation, in
asbestos workers, lung cancer is considered to be related to
asbestos exposure only if it is associated with lung fibrosis.
However, actually we do not know the exact relationship between
lung and pleural cancer and asbestos associated fibrosis ;
indeed, we are still far from the certitude that cancers are
induced by subacute inflammatory reaction around fibres !

In the past years, the two stages concept of initiation-
promotion has been largely used in order to classify the carcinogenetic agents (34). Fibres seem to act mostly as promoters,
because they did not cause mutations in bacteria (35) neither
induced sister chromatid exchanges in mesothelial cells in
culture (36) and because they enhanced the effects of other
carcinogenetic substances such as tobacco smoke as demonstrated
by epidemiological studies (37) or ionizing radiation as

demonstrated by a preliminary study on the rats, where radon
222 inhaled prior to the intra-pleural injection of 2 mg of
chrysotile provided a very high incidence of mesothelioma (38).

6. CELL RESPONSES

Although a lot of experimental works have been carried out
every-where in the world, we still do not know the cellular
and molecular events responsible for fibrogenesis and cancerogenesis
induced by mineral fibres. Fibres have numerous physico-chemical
characteristics (shape, size, external surface area, surface
charge, chemistry), intimately associated and difficult to
analyze separately in relation to a biological effect !

The carcinogenicity induced by fibres can be compared to
that observed by Oppenheimer et al with plastic film (39) ;
foreign bodies such as long fibres or plastic films cause a
frustated phagocytosis which release enzymes and superoxide
radicals responsible for chronic inflammation and cancerogenesis
(40). As far as fibres are concerned, the role of length and
diameter in cancerogenesis has been emphasized by the experimental
work of Stanton et al (41, 42) : after an intrapleural implanta-
tion of synthetic glass fibres, with the same chemistry, the
probability of pleural sarcoma was 100 % when the fibres were
longer than 8 μm and had a diameter less than 0.25 μm. Actually,
this might be understood by taking into account the aspect ratio
or the external surface area of fibres ; indeed the thinnest
fibres which were the more carcinogenic in the Stanton et al
experiment had the largest surface area per unit of weight.

Other parameters must also play a role, as indicated by
the modifications of the biological reactivity (haemolysis,
release of enzymes by AM) when the surface charge (43) and
surface chemistry (44, 45) of fibres were modified by acid
treatments. Thus, in our experience (44) UICC chrysotile incubated in vitro with AM induced only the release of lysosomal
enzymes (β galactosidase), whereas oxalic acid leached UICC
chrysotile (>85 % Mg released) was cytotoxic inducing the release
of lacticodehydrogenase and β galactosidase.

Recent experiments in our laboratory (44) and in U.K. (45)
have indicated that there was direct agreement between the degree
of inflammatory response in vitro and the tumor-promoting activities of chrysotile leached at various degrees. So, 100 % Mg
leached chrysotile which was cytotoxic in vitro for AM, gave
almost no tumor after an intrapleural injection of a 20 mg dose
in the rat. By contrast, mesothelioma were produced by untreated
UICC chrysotile in 43 % of rats, by crocidolite in 49 % and by
glass fibres in 14 %. Similar in vitro studies are now in progress

for testing other types of fibres after acid treatment ;
preliminary results indicate that amphiboles such as crocidolite
or amosite had an in vitro response opposite to that of chrysotile :
unleached amphibole type fibres induced the release of both
types of enzymes, while leached fibres induced only the release
of lysosomal enzymes!

These data relating to the basic interaction between
fibres and cells are still too preliminary to allow a definite
classification of fibres according to their noxious effects to
the human health. It is possible that fibres acting as tumor-
promoters induce or enhance the release of inflammatory materials
from intact cells and that this release is directly related to
their cocarcinogenic properties ; so the mechanism understanding
of fibre cancerogenesis mechanisms is perhaps within reach ;
then there is no doubt that considerable progress will be made
and will help the political decisions concerning asbestos and
fibre health related problems.

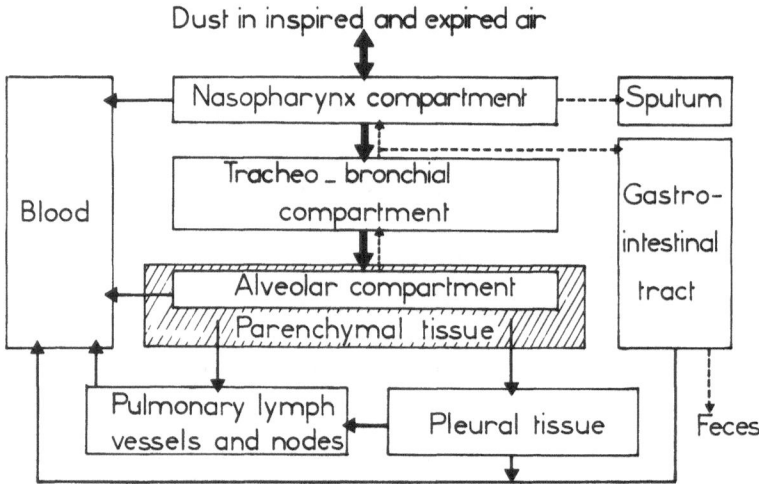

Figure 6 General scheme for deposition, clearance, translocation
          and retention fibres, derived from the ICRP lung model
          (heavy arrows : deposition ; light dotted arrows :
          clearance pathway ; light arrows : translocation
          pathways). (from Bignon et al, 23).

REFERENCES

1. IARC Monographs on the evaluation of carcinogenic risk of
    chemicals to man. Asbestos. vol 14, International
    Agency for Research on Cancer, Lyon (1977).
2. Health hazards of asbestos exposure. I.J. Selikoff and E.C.
    Hammond eds., Ann. N.Y. Acad. Sci., 330 (1979).
3. M.R. Becklake, Asbestos related diseases of the lung and
    other organs : their epidemiological and implications
    for clinical practice. Amer. Rev. Resp. Dis., 114 :
    187 (1976).
4. Public Health risks of exposure to asbestos. Commission of
    the European Communities. Pergamon Press, Luxembourg,
    (1977).
5. J. Bignon, Introduction. Les contaminants fibreux. Rev.
    Fr. Mal. Resp., 7 : 675 (1979).
6. M.C. Jaurand, J. Bignon, L. Magne, A. Renier and J. Lafuma,
    Interaction des fibres avec les globules rouges et les
    macrophages alvéolaires in vitro. Rev. Fr. Mal. Resp.,
    7 : 717 (1979).
7. J. Bignon, M.C. Jaurand and L. Magne, Effets cytotoxiques
    des asbestes en relation avec leurs propriétés physico-
    chimiques. Etude in vitro sur globules rouges, macro-
    phages alvéolaires et cellules mésothéliales. Contrat
    CCE 264 77 6 ENV F (1979).
8. C.C. Gravatt, P.D. Lafleur and K.F. J. Heinrich, Workshop on
    asbestos : definitions and measurement methods. NBS
    Special Publication, 506 (1978).
9. J. Goni, J. Bignon and G. Bonnaud, Les asbestes. Leurs mena-
    ces sur l'environnement urbain et sur la santé. Revue
    Nuisances et Environnement, Janvier-Février-Mars (1974).
10. M. Attivinili and Y.I. Baris, Malignant mesotheliomas in a
    small village in the Anatolian region of Turkey : an
    epidemiologic study. JNCI, 63 : 17 (1979).
11. J. Bignon, P. Sebastien, A. Gaudichet and M.C. Jaurand,
    Biological effects of attapulgite. Biological effects
    of mineral fibres. IARC Conference, Lyon, 25-27 Septem-
    bre (1979) (in the press).
12. Task Group on lung dynamics. Deposition and retention models
    for internal dosimetry of the human respiratory tract.
    Health Phys., 12 : 173 (1966).
13. V. Timbrell, The inhalation of fibres. " Pneumoconiosis"
    Proceedings of the International Conference, Johannes-
    burg. H.A. Shapiro ed. Oxford University Press (1970).
14. R.L. Harris and V. Timbrell, The influence of fibre shape in
    lung deposition. Mathematical estimates. "Inhaled
    Particles", IV, W.H. Walton ed. Pergamon Press Oxford
    (1977) , p. 75.

15. P. Sebastien, A. Fondimare, J. Bignon, G. Monchaux, J. Des-
     bordes and G. Bonnaud, Topographic distribution of asbes-
     tos fibres in human lung in relation to occupational and
     non-occupational exposure. "Inhaled Particles", IV,
     W.H. Walton ed., Pergamon Press Oxford (1977), p. 435.
16. A. Morgan, J.C. Evans and A. Holmes, Deposition and clearance
     of inhaled fibrous minerals in the rat. Studies using
     radioactive tracer techniques. "Inhaled Particles", IV,
     W.H. Walton ed., Pergamon Press Oxford (1977), p. 259.
17. J. Goni, J.H. Thomassin, M.C. Jaurand and J.C. Touray,
     Photoelectron spectroscopy analysis of asbestos dissolu-
     tion in acidic media of biological interest. "Origin and
     distribution of the elements". L.H. Ahrens ed. Pergamon
     Press, Oxford & New York (1979), p. 807.
18. J.M.G. Davis, Further observations on the ultrastructure and
     chemistry of the formation of asbestos bodies. Exp.
     Mol. Pathol., 13 : 346 (1970).
19. J. Bignon, Effets biologiques des polluants particulaires
     inorganiques. XVII Congrès National de la Tuberculose
     et des maladies respiratoires, Clermont Ferrand,
     Masson ed. Paris (1974), p. 207.
20. M.C. Jaurand, J. Bignon, P. Sebastien and J. Goni, Leaching
     of chrysotile asbestos in human lungs. Environ. Res.,
     14 : 245 (1977).
21. J. Bignon, P. Sebastien and M. Bientz, Review of some factors
     relevant to the assessment of exposure to asbestos
     dusts. "The Use of biological specimens for the assess-
     ment of human exposure to environmental pollutants".
     A. Berlin, A.H. Wolff & Y. Hasegawa eds. Martinus
     Nijhoff Publishers, The Hague, Boston, London (1979).
22. J.C. Wagner, G. Berry, J.W. Skidmore and V. Timbrell, The
     effects of the inhalation of asbestos in rats. Br. J.
     Cancer, 29 : 252 (1974).
23. J. Bignon, P. Sebastien and A. Gaudichet, Measurement of
     asbestos retention in human respiratory system related
     to health effects. "Workshop on asbestos : definitions
     and measurement methods". C. C. Gravatt, P.D. Lafleur
     & K.F.J. Heinrich eds., NBS Special Publication (1978).
24. E. Taskinen, K. Ahlman and M. Wiikeri, A current hypothesis
     of the lymphatic transport of inspired dust to the
     parietal pleura. Chest, 64 : 193 (1973).
25. K. Kanazawa, F.J.C. Roe and T. Yamamoto, Milky spots (taches
     laiteuses) as structures which trap asbestos in meso-
     thelial layers and their significance in the pathoge-
     nesis of mesothelial neoplasia. Int. J. Cancer, 23 :
     858 (1979).
26. M.C. Jaurand, J. Bignon, A. Gaudichet, L. Magne and A. Oblin,
     Biological effects of chrysotile after SO_2 sorption. II
     Effects on alveolar macrophages and red blood cells.
     Environ. Res., 17 : 216 (1978).

27. C. Schoenberger, G. Hunningrake, J. Gadek and R.G. Crystal, Role of alveolar macrophages in asbestosis : modulation of neutrophil migration to the lung following asbestos exposure. Amer. Rev. Resp. Dis., 121 : 257a (1980).

28. L. Di Menza, F. Ruff, J. Bignon, G. Bonnaud and G. Brouet, Obstruction des voies aériennes périphériques au cours de l'exposition professionnelle à l'amiante. Ann. Anat. Pathol., 21 : 261 (1976).

29. D.R. Gracey, M.B. Divertie and A.L. Brown, The blood air barrier in pulmonary asbestosis study of a case by electron microscopy. Chest, 63 : 46 (1973).

30. J.M.G. Davis, S.T. Beckett, R.E. Bolton, P. Collings and A.P. Middleton, Mass and number of fibres in the pathogenesis of asbestos-related lung disease in rats. Br. J. Cancer, 37 : 673 (1978).

31. G.W. Wright and M. Kuschner, The influence of varying lengths of glass and asbestos fibres on tissue response in Guinea Pigs. "Inhaled Particles", IV, W.H. Walton ed., Pergamon Press Oxford (1977).

32. A.C. Allison, I.A. Clark and P. Davies, Cellular interactions in fibrogenesis. Ann. Rheum., 36 : 8 (1977).

33. A.G. Heppleston and J.A. Styles, Activity of a macrophage factor in collagen formation by silica. Nature, 214 : 521 (1967).

34. I. Berenblum, The mechanisms of carcinogenesis. A study of the significance of cocarcinogenic action and related phenomena. Cancer Res., 1 : 807 (1941).

35. M. Chamberlain and E.M. Tarmy, Asbestos and glass fibres in bacterial mutation tests. Mutation Res., 43 : 159 (1977).

36. H. Kaplan, A. Renier, M.C. Jaurand and J. Bignon, Sister chromatid exchanges in mesothelial cells cultured with chrysotile fibres. "The in vitro effects of mineral dusts". Cardiff 3-7 Septembre 1979 (in the press).

37. E.C. Hammond, I.J. Selikoff and H. Seidman, Asbestos exposure cigarette smoking and death rates. Ann. N.Y. Acad. Sci., 330 : 473 (1979).

38. J. Lafuma, A. Hirsch, G. Monchaux, M. Morin, J.L. Poncy, R. Masse and J. Bignon, Mesothelia induced by intra-pleural injection of different types of fibres in the rat. Synergestic effect of other carcinogens. "Biological effects of mineral fibres", IARC Conference Lyon, 25-27 Septembre 1979 (in the press).

39. B.S. Oppenheimer, E.T. Oppenheimer, I. Danishefsky, A.P. Stout and F.R. Eirich, Further studies of polymers as carcinogenic agents in animals. Cancer Res., 15 : 333 (1955).

40. I.M. Goldstein, Effects of phorbol esters on polymorphonuclear
     leukocyte functions in vitro carcinogenesis. "Mechanism
     of tumor. Promotion and cocarcinogenesis" vol 2, T.J.
     Slaga, A. Sivak and R.K. Boutwell eds, Raven Press,
     New York (1978).

41. M.F. Stanton, M. Layard, A. Tegeris, E. Miller, M. May and
     E. Kent, Carcinogenicity of fibrous glass : pleural
     response in the rat in relation to fiber dimension.
     J. Natl. Cancer.Inst., 58 : 587 (1977).

42. M.F. Stanton and M. Layard, The carcinogenicity of fibrous
     minerals. "Workshop on Asbestos : defintitions and
     measurement methods". C.C. Gravatt, P.D. Lafleur &
     K.F.J. Heinrich eds., NBS Special Publication, (1978).

43. W.G. Light and E.T. Wei, Surface charge and asbestos toxicity.
     Nature, 265 : 537 (1977).

44. M.C. Jaurand, J. Bignon, L. Magne, A. Renier and J. Lafuma,
     Interaction des fibres avec les globules rouges et
     les macrophages alvéolaires in vitro. Rev. Fr. Mal.
     Resp., 7 : 717 (1979).

45. A. Morgan, P. Davies, J.C. Wagner, G. Berry and A. Holmes,
     The biological effects of magnesium leached chrysotile
     asbestos. Br. J. Exp. Pathol., 58 : 465 (1977).

46. G. Brouet, J. Bignon, G. Bonnaud and J. Goni, Incidence sur
     la santé de la pollution atmosphérique par l'asbeste ou
     autres particules fibreuses. Rev. Tuberc. Pneumol.,
     35 : 461 (1971).

DISCUSSION

LECTURER: Bignon   CHAIRMAN: Cumming

CUMMING: Can I begin by asking: you apparently have difficulty
                    in detecting chrysotile by the light microscopic
                    technique. Am I to understand that it is because
                    there is an increased clearance of chrysotile? Or a
                    diminished deposition?

BIGNON: It is difficult to detect chrysotile because of the
                    diminution of the size of chrysotile. The ultimate
                    fibre is about 300 Augstrom in diameter, and cannot
                    be seen by light microscopy. But using the electron
                    microscopy we see a lot of chrysotile. Nevertheless,
                    in the human lung, the percentage of chrysotile is
                    usually less than amphiboles.

CHEVALIER: When you injected the fibres into the trachea what
                    was the precise procedure?

BIGNON: The animals were sacrificed at 70 hours, and we make
                    either a washing in different animals and study the
                    alveolar fluid and various clinical modifications and
                    we make the histological variffication.

CHEVALIER: I would like to know also what was the vehicle of the
                    fibres.

BIGNON: It was associated with phospholipids. We discovered
                    that it was easier to see and to separate the fibrils
                    of chrysotile for experiments in animals by this
                    technique.

COHEN: It is perhaps rather a naive question, but do all
                    workers or all people who are exposed to similar
                    amounts of asbestos suffer similarly? And, if not,
                    what is responsible for the differences in their
                    responses?

BIGNON: My data relate to heavy asbestos exposure, it is very
                    imprecise, because it is based on a retrospective
                    questionnaire. As I said at the end of my talk I am
                    not sure that the number of fibres inside the
                    alveolar spaces is absolutely related to the number
                    of fibres inside the parenchymal tissue. So, there
                    is no correlation between our data and the biological
                    data in some of these cases. It could be due to the

fact that the biological events in the alveolar spaces in these people with long exposures is less related than the biological events inside the parenchymal tissue. Did I answer your question?

CUMMING: In two subjects exposed to the same dose load in the atmosphere, say one million particles per cubic metre, is there any difference in the deposition between these two, or the way they handle the fibres.

COHEN: Deposition and clearance.

BIGNON: I cannot answer this question because you and I are not animals, and we do not know the exact dose, and it is very difficult to know what dose has been inhaled.

COHEN: Perhaps I will ask a slightly easier question. The enzymes that you monitored, several showed some effect, but the beta-glucoronidase showed very little effect. Is it just a less sensitive assay, or is there any other reason?

BIGNON: I cannot answer. I have no explanation for this difference between the results in humans and the result in vitro, because it seems a less sensitive assay.

CUMMING: I think you have answered the question, it is a less sensitive assay.

BIGNON: In human studies, yes. There is a different story concerning the macrophage enzymes because recent data showed the interaction of fibres with peritoneal macrophages, this showed that the synthesis of elastase was inhibited or reduced by incubation with asbestos fibres.

CORRIN: In calculating the total asbestos content of the lung you made this 5,000. The calculation assumes that the distribution is even throughout the lung. Is that fair?

BIGNON: I said that it was an assumption, but there is a very large heterogeneity. We have measured different locations and there is a very important heterogeneity, but for the present purpose I wanted to know what percentage of fibres was inside the alveolar space compared to the total amount of fibres inside the lung, and as our data are more or less

confirmed by the experimental data of Morgan I think
they are relevant to some actual facts.

CUMMING: Especially bearing in mind that these are two orders
of magnitude of difference.

BIGNON: Yes

CORRIN: May I ask another question. I have difficulty in
correlating two features which I picked out from your
talk. You got good discrimination between controls
and exposed people, especially by light microscopy,
counting the ferruginous bodies. And yet the
chrysotile fibres are difficult to pick out by light
microscopy.

BIGNON: I know. My good discrimination was based on
ferruginous bodies, because if you have ferruginous
bodies you can say that they come from the alveolar
space. But if in the laboratory one does the
electron microscopy of fibres and you have some
chrysotile fibres on the grids you would say:
perhaps it comes from the air of the laboratory. You
are not sure that it comes from the alveolar spaces.
And in control subjects it is a very difficult
technique for lower levels of contamination. I said
that if you check only the ferruginous bodies the
method is very discriminative.

CORRIN: Both forms of asbestos.

BIGNON: No, only for amphibole, because the chrysotile has
very few ferruginous bodies.

CORRIN: Today, increasingly, the exposure would be, I think,
to chrysotile. The amphiboles are losing favour with
the manufacturers.

BIGNON: Yes, but in the industry, perhaps in Quebec, it will
be necessary to do that, because they are mostly
exposed to chrysotile. But in our industries of the
EEC countries usually they use an association of
mostly chrysotile plus amphiboles. In that case we
can think that amphiboles might be interpreted as a
marker. And I think it is a simplified way of
managing these data in a particular point of view.

CUMMING: One must bear in mind that most countries now are
bringing in legislation to prevent amphibole from
being used and moving towards chrysotile. So,

perhaps in the future we shall find these difficulties becoming more prevalent. We might check this question about the non-isometric distribution of bodies in the lung that Corrin mentioned by turning to this interesting point you made that every alveolus has his own guardian macrophage. And as this turns out in the numbers we might ask the question: is it true, in the experience of microscopists that there is only one guardian macrophage per alveolus? Or have people seen ten? Because if this was os it would cast some doubts on the assumption of the figures we have made. Anybody know? A difficult thing to do, I suppose.

BIGNON: If you examine the lung of a smoker you will see more than one macrophage, about three per alveolus. But usually the count we obtain when washing the lung of a small animal, that is ten consecutive washings, you can assume that you get about one macrophage per alveolus.

CUMMING: That raises a very interesting biological question, doesn't it?

BIGNON: Yes.

CUMMING: How can it be that each alveolus has just one macrophage guarding it? And when it is coughed up is replaced by another single macrophage. It postulates a very interesting control mechanism for the population of macrophages.

BOGNON: I think there is another point coming from this observation. It was said that this single macrophage was washed out in the fluid, but it is a long distance, ten times the macrophage diameter to transfer across the alveolar space.

CUMMING: Yes, just one alveolar space. It has to go around the perimeter of course, not across which is further still.

BIGNON: Yes.

CUMMING: Are there any other points of discussion? If not, thank you for your presentation. We will move on now to Milic-Emili, who will talk about the mechanics in abnormal environments.

EFFECT OF ACCELERATION AND WEIGHTLESSNESS ON LUNG MECHANICS

J. Milic-Emili

Meakins-Christie Laboratories
McGill University
Montreal, Canada H3A 2B4

It has been long recognized that earth gravity (1G) plays an important role in determining the static mechanical properties of the respiratory system. Indeed, when changing from upright posture to recumbency, there is a substantial change in the functional residual capacity (FRC).[1] As with relaxed respiratory muscles, the FRC is determined by the balance between the inward recoil of the lung and the outward recoil of the chest wall, the postural changes in FRC could be due to changes in the static pressure-volume (P-V) curve of either the lung or chest wall, or both. In normal man, however, the static P-V curve of the lung does not change appreciably with body posture.[2] Accordingly, the effect of earth gravity appears to affect mainly the mechanical properties of the chest wall, more specifically the abdomen.[1] In the upright position gravity acts in the inspiratory direction on the abdomen and in the expiratory direction on the rib cage; the effect of the abdomen is greater at small than at high lung volumes because at large volumes the height of the abdomen is smaller and its wall stiffer. In the supine position the gravitational effect changes little with volume and the action of gravity on both abdomen and rib cage is expiratory. Hence, the reduction in FRC between the upright and supine posture is due almost entirely to the abdomen.

Centrifugal acceleration (increased G) can profoundly affect the subdivisions of lung volume, the changes depending both on the magnitude of the acceleration and its direction relative to the body.[3] For the weightless state (zero G) no measurements of subdivisions of lung volume are available, although some simple

343

lung function studies (e.g. vital capacity) have been reported.[4]
These were carried out during brief periods (less than 1 min) of
parabolic flight in an aircraft, and the results indicate that the
vital capacity does not change appreciably when going from 1 to
zero G.

Several attempts have been made to measure the static P-V
curve of the lung during exposure to acceleration. Bondurant[5]
reported significant falls in lung compliance during exposure of
seated and supine subjects to accelerations of +3 to +5 G. Bryan
et al[6] found no change during head-to-foot acceleration (G_z) up to
+ 2 G_z, except at low lung volumes. More recently, Glaister et al[7]
applied transversal acceleration (G_x) to prone subjects and found
a progressive reduction in lung compliance with increasing acceleration. They concluded, however, that "acceleration has no influence
on the mechanical properties of the individual lung units".[3,7]
More specifically, the changes in static P-V curve of the lung
observed during acceleration may simply be the result of an increased gradient in pleural surface pressure (see below) with increased acceleration, while the intrinsic elastic properties of
the lung may remain unchanged. This notion is supported by
Sutherland et al.[8] The latter authors also predicted that in the
gravity-free state the static P-V curve of the lung will differ
from that obtained at normal earth gravity only at low lung volumes
(near residual volume). At higher lung volumes (between FRC and
full inspiration) the P-V curve should be the same at 1G as at
zero G.

One of the problems in measuring the static P-V curve of the
lung is in controlling the position of the esophageal balloon. In
fact, it has been long recognized that there is a vertical gradient
in pleural surface pressure[9,10] which is also found along the
esophagus.[2,6,8] With increasing acceleration this vertical gradient
increases such that, at constant lung volume, the pleural surface
pressure becomes more negative in the upper lung regions and less
negative in the dependent lung zones.[6]

Apart from its effect on the overall static P-V curve of the
lung discussed above, this gravity-dependent gradient in pleural
surface pressure has also a most profound effect on the regional
distribution of gas.

The first clear evidence that in the intact thorax there is
vertical gradient in alveolar size was provided in 1966 from measurements of the distribution of regional lung volumes with radioactive xenon.[11,12] Fig.1 illustrates the average (± SE) static
distribution of gas at functional residual capacity (FRC) and at
residual volume (RV) in eight seated normal young men. Regional
volumes are expressed as a percentage of the volume of each region

at total lung capacity (TLC), i.e., as a percentage of the regional
TLC (TLC_r, the subscript r denoting a regional variable). Because
at TLC in normal subjects the volume of the lung units (alveoli) is
probably nearly uniform throughout the lung, regional volumes expressed as a percentage of TLC_r are also an expression of the relative size of the alveoli. The volume per alveolus, expressed as
a percentage of the alveolar volume at TLC (TLC_{alc}), is shown on
the upper abscissa of Fig.1.

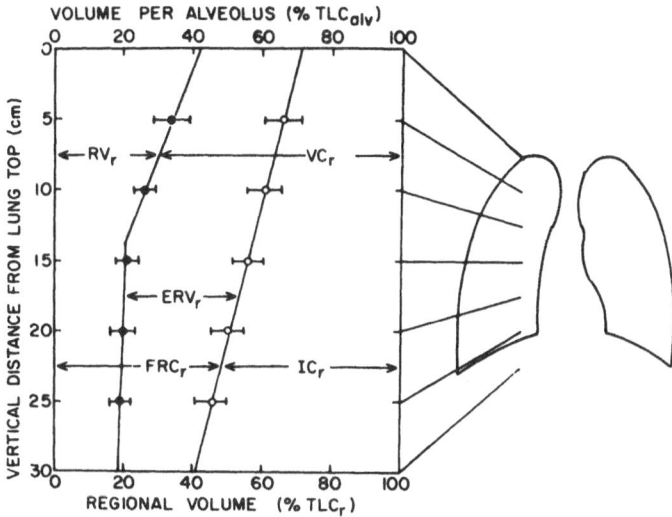

Fig.1.   Regional subdivisions of lung volume in seated men.
         Filled and open circles represent average results obtained
         on eight healthy men (ages 33-39) at RV and FRC, respec-
         tively. Bars indicate 2 SE. RV_r, regional residual
         volume; VC_r, regional vital capacity; ERV_r, regional
         expiratory reserve volume; FRC_r, regional functional
         residual capacity; IC_r, regional inspiratory capacity
         (After ref. 8).

     As indicated in Fig.1, at both RV and FRC the apical lung units
are more expanded than the basal ones. On the other hand, no significant differences in expansion were found at constant lung height
between left and right lungs.[11] This behaviour, which was also
found in supine, prone and lateral decubitus[12] appears to be
due primarily to the presence of a gravity-dependent vertical

gradient in pleural surface pressure.[9-14]

    In the earliest studies of regional distribution of gas within
the normal lung, a simple mechanical model (Fig. 2) has been tenta-
tively proposed for explaining the uneven static distribution of
gas determined with radioactive xenon.[8,11,12] This model was based
on the following assumptions:(a) that the vertical gradient of
pleural surface pressure is fixed (i.e., it is the same at all
degrees of overall lung inflation) and is due entirely by distortion
of lung tissue by its own weight rather than to the effect of gravity
on the chest wall; (b) that the intrinsic elastic properties of
the normal human lung are uniform; and (c) that small (peripheral)
airways closure occurs at low lung volumes (i.e., near RV).

    Since then, Agostoni and his co-workers[9,10] have elegantly
shown that in various experimental animals (e.g., dogs and rabbits)
the topography of pleural surface pressure depends not only on the
lung weight but also on the shape, weight and mechanical properties
of the chest wall. Recently, however, evidence has been provided
supporting the original serendipitous notion that in man the
gradient in pleural surface pressure is normally due chiefly to
the weight of the lung itself.[15,16] Nevertheless, by performing
selective muscle efforts (e.g., voluntary contraction of the diaphragm alone) the topographical distribution of the pleural surface
pressure can be substantially modified in man (for refs. see 17).

    Whether or not in normal man the vertical gradient in pleural
surface pressure is indeed the same at all degrees of lung inflation
is still not precisely known. On the basis of the available evidence, however, it appears that the assumption of a lung volume-
independent gradient is still tanable, at least under most normal
physiological conditions. As stated above, this is no longer the
case during selective contraction of different respiratory muscles[17]
and perhaps also in anesthetized-paralyzed man.[18]

    There is good evidence that the vertical gradient of pleural
surface pressure increases during acceleration.[3,6] As a result,
the regional differences in alveolar expansion become more pronounced with increased G, as shown in Fig.3 which depicts the
relationships between regional and overall lung volumes obtained
by Bryan et al[6] in a seated young subject at normal earth gravity
($+ 1 G_z$) and during increased head-to-foot acceleration ($+ 2 G_z$).

    On the basis of their results, Bryan et al[6] suggested that
in the weightless state (zero G) the alveolar expansion should be
uniform throughout the lung, i.e., at zero G the relationship
between regional and overall lung volume should for all regions
fall along the line of identity (broken line) in Fig.3. Recent
studies by Michels et al[16] support this notion. Indeed, in

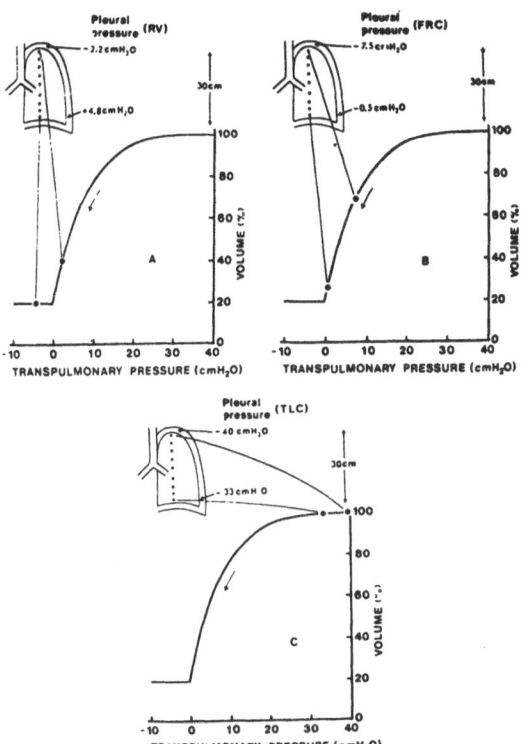

Fig. 2. Effect of vertical gradient of pleural surface pressure on
static distribution of gas within the lung in seated man.
It is assumed that the intrinsic static mechanical properties of the lung are uniform, as indicated by a unique
volume-pressure curve. Values of pleural surface pressure
at the apex and base at the three lung volumes (RV, FRC
and TLC), shown in upper part of each panel, were computed
according to Milic-Emili et al,[11] who assumed that the gradient does not change with lung volume. In C, at full inspiration (TLC), all lung regions are expanded virtually
uniformly, in spite of the differences in pleural pressure
down the lung. At both RV (A) and FRC (B), the pleural
pressure gradient causes the upper regions to be considerably more expanded than the lower zones. Note that at RV
the small airways in the most dependent lung zones are
probably closed and the gas in the alveoli subtended by
them is "trapped (From ref. 13).

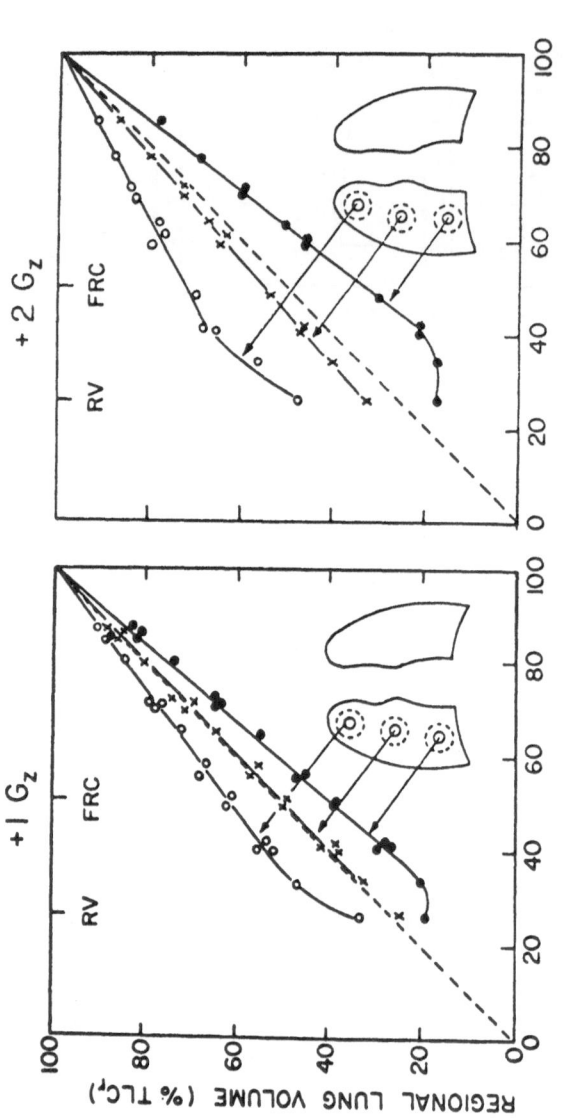

Fig. 3. Ordinates: Regional lung volume, expressed as percentage of regional total lung capacity (% TLC$_r$); Abscissae: overall lung volume, expressed as percentage of total lung capacity (% TLC). The broken lines (slope = 1) indicate regional percentile expansion equal to overall lung expansion. The vertical distance from lung top to the center of the scintillation counters with which regional volumes were measured was 4.5 cm (open circles), 13 cm (crosses) and .23.5 cm (filled circles). Data from a young seated subject at normal earth gravity (+ 1 G$_z$) and increased acceleration (+ 2 G$_z$) on a human centrifuge.(from ref. 6).

experiments carried out under zero G condition in a parabolically flying aircraft, they found that in young normal subjects the "phase IV" disappeared when a bolus of argon was inhaled at 1 or more G and exhaled at zero G.

As the presence of phase IV (or closing volume) implies regional differences in lung emptying, the results of Michels et al[16] imply that in the weightless state (a) the vertical gradient in pleural surface pressure is abolished (hence the regional emptying of the lung is uniform) and (b) that in man the intrinsic elastic properties are uniform. In this connection it should be noted that recently Berend et al[19] have provided evidence that in normal humans (unlike rabbits, cats and dogs) the static pressure-volume characteristics of excised upper and lower lobes are identical. While at a gross regional level the elastic properties of the normal human lung appear relatively uniform, substantial intraregional inhomogeneities are present[17] and cause intraregional inhomogeneties in ventilation distribution even in weightlessness.[16]

In 1975, Engel et al[20] have provided the most convincing evidence to date for airway closure at low lung volumes in healthy man at normal gravity (earth gravity). The results of Michels et al[16] indicate that at zero G there is no airways closure in young normal individuals. One would expect, however, that in older subjects, as well as in some patients with pulmonary disease (e.g., chronic obstructive lung disease) small airways closure to be present even in the condition of weightlessness. In this connection it should be noted that closing volume (phase IV) and closing capacity increase with increasing acceleration,[3,21] as expected on the basis of the fact that acceleration causes an increased vertical gradient of pleural surface pressure.

With the forthcoming experiments in spacelabs there will be ample opportunity to expand our knowledge of the regional lung function in the weightless state. At the same time, further research during increased acceleration on the human centrifuge seems to be required in order to better understand the effect of increased gravitational forces on regional lung function in man.

REFERENCES

1. E. Agostoni, and J. Mead, Statics of the respiratory system, in: "Handbook of Physiology," Section 3, Respiration, Vol.1 Am. Physiol. Soc. Washington (1964).

2. J. Milic-Emili, J. Mead, and J.M. Turner, Topography of eso-phageal pressure as a function of posture in man, J. Appl. Physiol. 19: 212 (1964).

3. D. H. Glaister, Effect of acceleration, in: "Regional Differ-
    ences in the Lung" J.B. West, ed, Academic Press, New York
    (1977).

4. D. B. Michels, and J. B. West, Distribution of pulmonary venti-
    lation and perfusion during short periods of weightlessness,
    J. Appl. Physiol.: Respirat. Environ. Exercise Physiol. 45:
    987 (1978).

5. S. Boundurant, Effect of acceleration on pulmonary compliance,
    Fed. Proc. 17: 18 (1958).

6. A. C. Bryan, J. Milic-Emili, and D. Pengelly, Effect of gravity
    on the distribution of pulmonary ventilation, J. Appl. Physiol.
    21: 778 (1966).

7. D. H. Glaister, M. R. Ironmonger, and B. J. Lisher, The effect
    of transversely applied acceleration on lung mechanics in
    man. Rep. No. 1340. Flying Personnel Research Committee.
    M.O.D. (Air Force Dept.), London, England, 1975.

8. P. W. Sutherland, K. Katsura, and J. Milic-Emili, Previous
    volume history of the lung and regional distribution of gas,
    J. Appl. Physiol. 25: 566 (1968).

9. E. Agostoni, Mechanics of the pleural space, Physiol. Rev.
    52: 75 (1972).

10. E. Agostoni, Transpulmonary pressure, in: "Regional Differences
    in the Lung" J. B. West, ed, Academic Press,New York (1977).

11. J. Milic-Emili, J. A. M. Henderson, M. B. Dolovich, D. Trop,
    and K. Kaneko, Regional distribution of inspired gas in the
    lung, J. Appl. Physiol. 21: 749 (1966).

12. K. Kaneko, J. Milic-Emili, M. B. Dolovich, A. Dawson, and
    D. V. Bates, Regional distribution of ventilation and perfus-
    ion as a function of body position, J. Appl. Physiol. 21:
    767 (1966).

13. J. Milic-Emili, Pulmonary Statics, in: "MTP International
    Review of Science;"Respiratory Physiology, J. G. Widdicombe,
    ed, Butterworths, London (1974).

14. J. Milic-Emili, Ventilation, in: "Regional Differences in the
    Lung" J. B. West, ed, Academic Press, New York (1977).

15. R. Greene, J.M. B. Hughes, M.F. Sudlow, and J. Milic-Emili, Regional lung volumes during water immersion to the xiphoid in seated man. J. Appl. Physiol. 36: 734 (1977).

16. D. B. Michels, P. J. Friedman, and J. B. West, Radiographic comparison of human lung shape during normal gravity and weightlessness. J. Appl. Physiol. Respirat.: Environ. Exercise Physiol. 42: 851 (1979).

17. L. A. Engel, and P. T. Macklem, Gas mixing and distribution in the lung. in: "International Review of Physiology,"Vol. 14, Respiratory Physiology II, J. G. Widdicombe, ed, University Park Press, Baltimore (1977).

18. K. Rehder, H. M. Marsh, J. R. Rodarte,and R. E.Hyatt, Airway closure, Anesthesiology 47: 40 (1977).

19. N. Berend, C. Skoog, D.W. Galangher, and W. M. Thurlbeck, Lobar pressure-volume characteristics of excised human lungs, Physiologist 22: 9 (1979).

20. L. A. Engel, A. Grassino, and N. R. Anthonisen, Demonstration of airway closure in man, J. Appl. Physiol. 38: 1117 (1975).

21. N. R. Anthonisen, Closing Volume, in: "Regional Differences in the Lung" J. B. West, ed, Academic Press, New York (1977).

DISCUSSION

LECTURER: Milic-Emili   CHAIRMAN: Cumming

BONSIGNORE: If the gravitational force increases we have a
                 situation where pleural negativity increases at the
                 apex whilst there is positive pressure at the base.
                 For the apex, the situation is the same as in deep
                 inhalation and for the base it is the same as in
                 complete exhalation. How does pressure behave below
                 the diaphragm under these conditions, because I
                 suppose that the increase in gravitational force
                 leads to an increase of negativity below the
                 diaphragm and this increase has no influence on the
                 pleural gradient, which is affected only by the
                 weight of the lung.

MILIC-EMILI: This is a very good question. What happens is that
                 the form of the thoracic wall changes because if the
                 pleural pressure at the level of the diaphragm
                 changes, the form of the diaphragamatic cupola also
                 changes and so on. The problem is this: will the
                 diaphragm still remain flacid with acceleration. In
                 other words will the pressure at the end of an
                 expiration still remain zero, or does the diaphragm
                 distend and the transdiaphragmatic pressure change?
                 This is one of the complex problems and the message
                 that I wanted to convey, these are studies which
                 should be carried on using more modern techniques to
                 see what is the form of the rib cage, the lungs, and
                 so on. What I said was very schematic and whilst you
                 should not believe all I said, it was more or less
                 true.

DENISON: When people are accelerated head to foot and the
                 forces act downwards so that the blood becomes
                 heavier and fails to reach the top of the lung, so
                 the specific gravity of the lung will be less at the
                 top, which is blood-free, than the bottom which is
                 heavy with blood. This means that the more you are
                 accelerated, then the more should the specific
                 gravity or the pressure gradient between two upper
                 parts of the lung approach that predicted from the
                 blood-free weight of the lung - does it do this?

MILIC-EMILI: We tried to do such measurements, not with the
                 centrifuge because there are limitations there, but

we tried to see if we could affect the gradient of pleural pressure by breathing against a positive pressure, so that you would put the apex of the lung in zone 1 and all the blood went to the lower part, and we were not able to see any difference. So, I cannot answer that question, but the interesting part, with head to foot acceleration there are big problems of ventilation-perfusion, because most of the blood is going to go to the lower portions and perhaps you get airways closure in the lower parts of the lung. But whereas for acceleration we know roughly the distribution of ventilation for quasi static slow inspiration, as determined by the distribution of compliance no measurements have been made of ventilation in accelerated subjects at normal flow rates. The only study of the effect of flow rate on the distribution of ventilation has been done under conditions of normal gravity.

DENISON: My second question. You spoke, and elegantly showed that the properties of the lung at the top and the bottom are essentially similar, but of course if you take the lungs out of the chest you can observe that they have many sharp corners at the edges of lobes. Supposing that we now cut a lobe transversly, do you really think that there is isotropic distribution of forces even at the sharp edges?

MILIC-EMILI: First of all workers in Winnipeg have shown that in man there are no differences in the elastic properties of upper and lower lobes. A good question requires a good answer, a knowledgeable answer: I do not know!

THE EFFECT OF PRESSURE ON THE LUNGS

David Denison

Brompton Hospital
London, U.K.

Mainly for fun, this talk became an exercise in communicating by diagrams. Now its luck is being pushed further by being printed in essentially the same form. It attempts to cover a field that may be unfamiliar to some people, and begins with a man standing in a tall but empty bath.

As soon as water begins to come above the feet of this man his shape changes. His tissues are rehung on his skeleton as the effects of gravity are removed and blood and tissue fluids are pushed upwards.

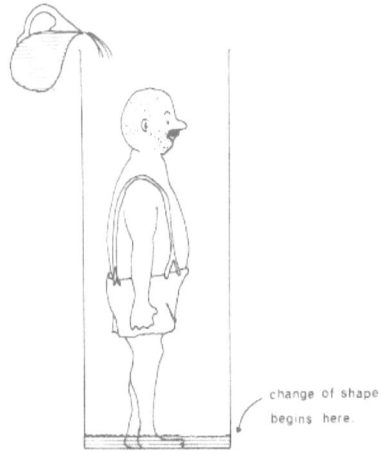

change of shape begins here.

By the time the water has reached his neck there has been a complete change in the shape of his lungs, as pressure on the abdomen forces the diaphragm up and the greater external pressure on the base than the apex of his chest pushes the lower ribs in.

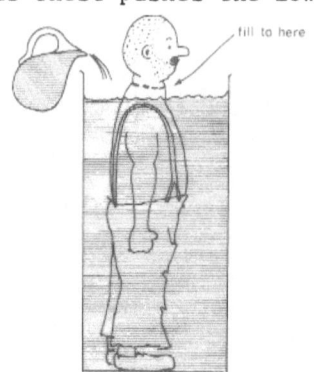

There is also an extra half liter of blood in the lungs, heart, and pleura. This excites stretch receptors in the atria and great veins. The signal is interpreted as an excessive volume of circulating fluid. That leads to increased urine secretion (Gauer-Henry reflex. Immersion has also displaced more blood into his head and other parts of the upper systemic circulation. The lower parts are relatively free of blood, but local blood <u>flow</u> is probably unchanged due to regional "autoregulation."

If, after filling the bath to the line around his neck, the man blew against a manometer, we could record the changes in his mouth pressure and body volume and use Boyle's law to calculate the <u>absolute</u> volume of his lungs without running into the expense, drift, and imprecision associated with whole-body plethysmographs that are filled with air.

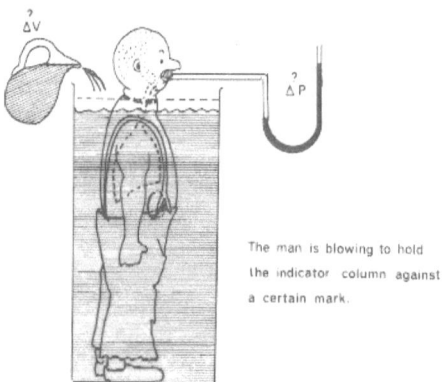

However, although it can be determined accurately, the measured volume is of limited interest because it is smaller than the corresponding volume in air and depends, as will be shown, on the precise state of immersion.

At the stage shown below (left), changes in the shape of the man
are largely complete. His systemic circulation is weightless and
will preserve the same distribution of blood in any posture. His pulmonary circulation is engorged but still influenced by gravity
because his lungs contain air. Since his lungs are air-filled, their
shape, and the distribution of blood within them, will still vary
with posture. They will continue to change shape as more water is
added to the bath. Any changes in the rest of the body will be
secondary to those in the chest.

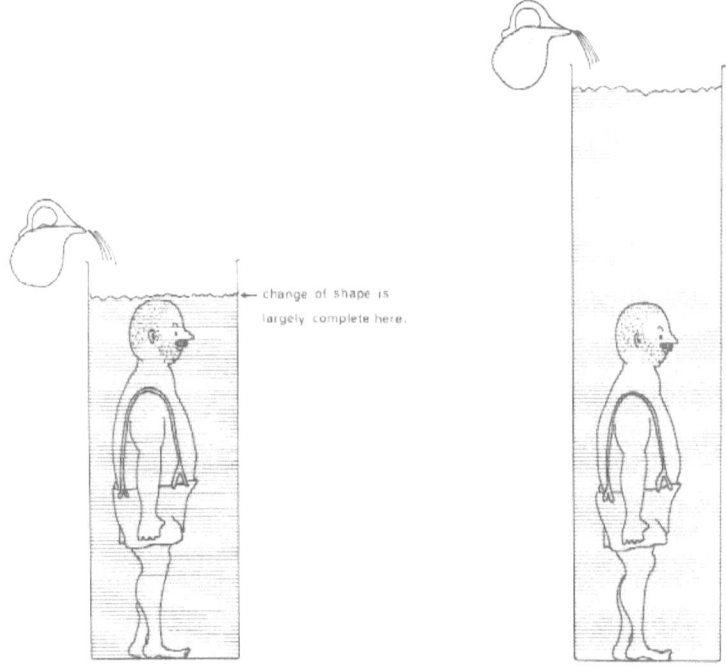

In water, ambient pressure increases linearly with depth by
1 atmosphere/10 meters. So, for example, at a depth of 10 meters the
man is compressed by a pressure of 2 atmospheres (one of air plus
one of water).

The chest wall is a floppy structure that, at most, withstands
a pressure of 1 to 2 meters of water, so it continues to change its
shape as the external pressure rises. Because its wall is floppy,
the pressure inside the chest is always very close to that outside.

However, the relationship between the common gas pressure inside the lung and the graded hydrostatic pressure of the water outside is not simple since the chest wall has an irregular shape. As a rough guide: in a man standing erect at FRC, the pressure inside the chest is the same as the external pressure at a level 19 cm caudal to the sternal notch. It corresponds to this low anatomical position due to the larger surface area that is presented by the base than the apex of the chest.

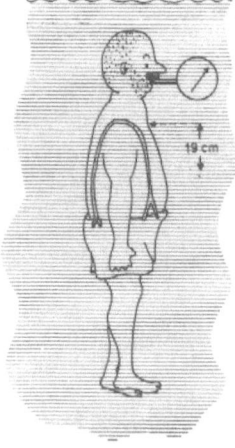

When a man lies supine underwater, the pressure inside his lung at FRC is roughly the same as the external pressure at a level 6 cm posterior to the sternal notch.

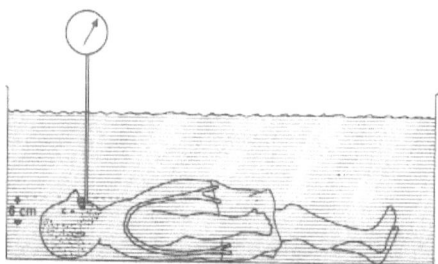

In a man in any attitude, at FRC, the intersection of the planes 19 cm caudal and 6 cm posterior to the sternal notch, in the midline, is the centroid of pressure, or eupneic point, of his chest, or is at least a good guide to it.

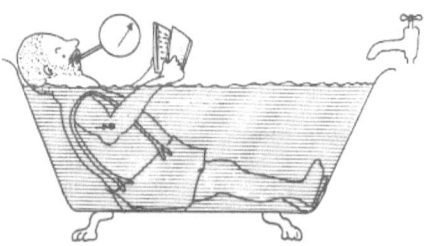

Lung compression is one factor limiting the depth of simple
dives. Thus, at 30 m ambient pressure is 4 ATA (1 of air + 3 of
water). In a dive to this depth TLC is compressed to RV. Some
people supposed "breath-hold" divers would not be able to go any
deeper.

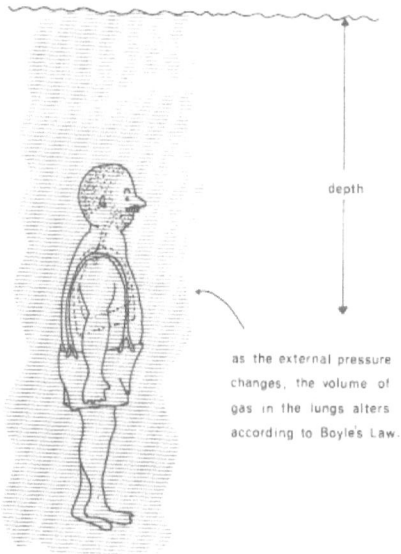

as the external pressure
changes, the volume of
gas in the lungs alters
according to Boyle's Law.

However, the depth limit to breath-hold dives is surprising.
In 1974 Enzo Maiorca of Italy surfaced unconscious from a depth of
87 m. At these depths ($P_B \equiv 10$ ATA), lung gas is compressed to a
tenth of TLC and about a liter of blood is forced into the chest
in compensation.

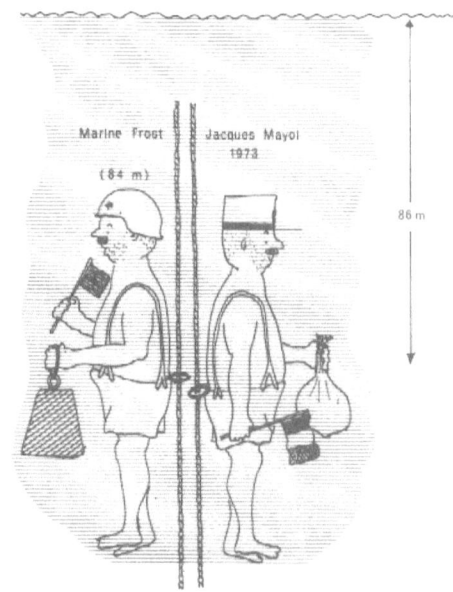

Another problem in breath-hold diving is that of buoyancy. At
TLC an adult man is about 2.5 kg positively buoyant, which is enough
to keep his head out of water. At RV he is some 2.5 kg negatively

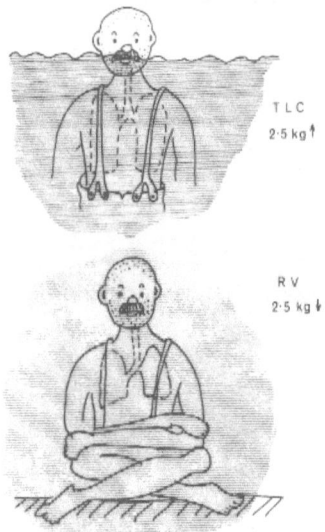

buoyant and can sit at the bottom of a pool with ease. These facts
have some interesting consequences. Putting the wrong 2.5 kg above
water is risky...

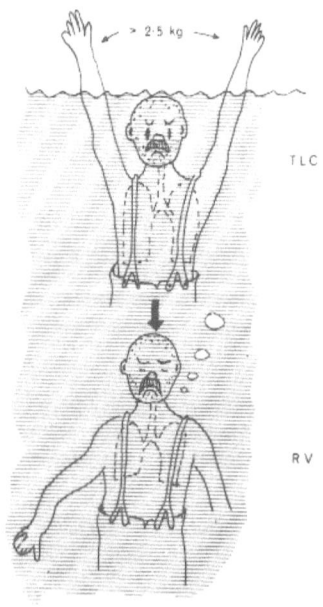

... and shouting for help is unwise.

We usually breath-hold for about 1 minute at most and break due to CO_2 accumulation. Hyperventilation can prolong the hold to 3 or even 4 minutes in a fully rested man. The break is then due to hypoxia. Any exertion shortens these times greatly.

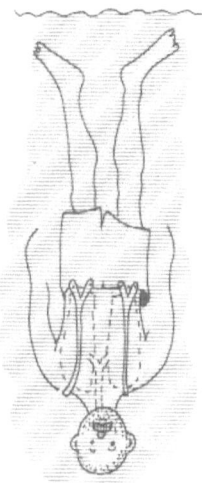

A good swimmer can generate some 5 kg of thrust, so half his maximum power output is spent simply overcoming buoyancy as he leaves the surface at TLC. As he descends and his lungs get smaller it will be easier to swim. The rise in lung pressure will force soluble CO_2 back into the circulation but will raise his alveolar PO_2 misleadingly. At some stage, a combination of hypoxic and hypercapnic signals will persuade the man that it is time to come up.

It is at just this point that the man is in most trouble:

- First, since his lungs are compressed, he must spend about half his aerobic capacity overcoming body weight to leave the bottom at all.

- Second, as he ascends, ambient pressure falls. That greatly speeds the decline in alveolar and arterial PO_2.

- Third, the decompression also causes a marked fall in alveolar and arterial PCO_2. This leads to cerebral vasoconstriction that aggravates the brain's hypoxia.

These events are most marked when the fractional change in pressure is greatest, as the water surface is approached.

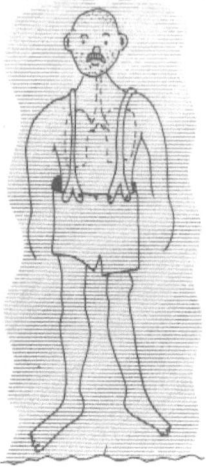

If the man has misjudged the time to ascend he may become unconscious on the way up and drown. This sequence, which is accelerated by previous hyperventilation, is a common cause of death in breath-hold divers and in children straining to do "lengths" underwater in a swimming pool.

Breath-hold divers may risk decompression sickness, but only if they make repeated dives. Young Japanese ama (Cashido) dive briefly, but often, to 8-10 meters without harm, but the older ladies (Funado), who dive deeper for greater rewards, show small punched-out areas of necrosis on the articular surfaces of bones. These probably are beds of "end" arteries that have been blocked by bubbles of nitrogen coming out of solution on ascents late in a series of dives. Polynesian pearl divers, who go even deeper for even greater rewards, have very short working lives. Many perish from hypoxia, but it is also thought that many die of "the bends," as this disorder is called.

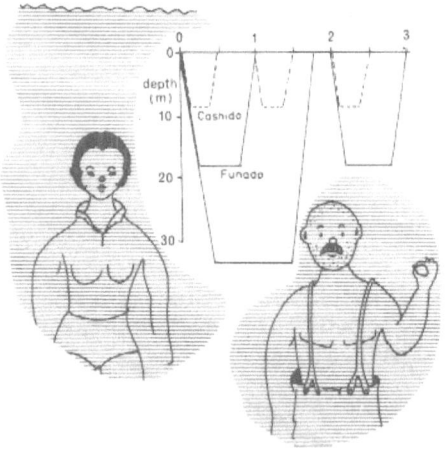

All marine mammals, i.e., toothed whales, baleen whales, true seals, eared seals, manatees, dugongs, walruses, and sea otters, regardless of their diverse evolutionary origins, dive much deeper, for much longer, and much more frequently, without coming to harm. They can do this because their lungs differ from our own.

All terrestrial mamals, even beavers, hippopotamuses, and river otters, have lungs of the same sort, with floppy peripheral airways. On compression, these collapse before the alveoli they serve are empty, trapping compressed air in contact with pulmonary capillary blood.

By contrast, all marine mammals, including the sea otter that evolved most recently, have peripheral air passages which are reinforced with cartilage or muscle all the way to the mouth of the alveoli. When these lungs are compressed during a breath-hold dive, alveolar gas is displaced into the airways away from pulmonary blood. This striking example of convergent evolution protects these creatures from oxygen toxicity, nitrogen narcosis, and the bends, but does not explain why they can stay underwater for so long.

It is possible to stay underwater for some time by breathing air at atmospheric pressure through a tube. This maneuver drops the pressure within the chest to that of the air being breathed, but leaves the outside pressures unchanged. As a result the lungs are compressed, more blood is displaced into the chest, and inspiration is opposed.

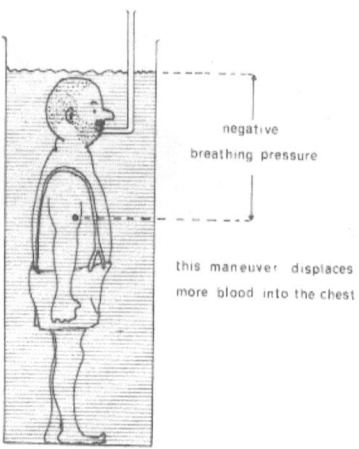

There are many stories of people eluding pursuit by breathing through tubes. However, ignoring the fascinating problems of added dead-space which arise, because of chest fatigue the depth limit to snorkel breathing is about 1.5 meters.

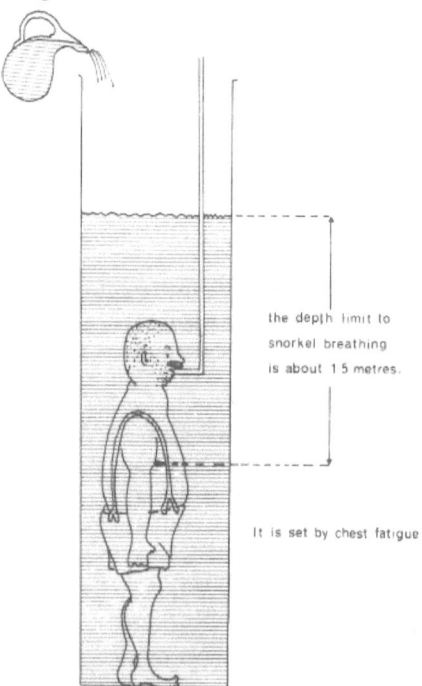

If a man in a diving helmet is given an appropriate gas supply at ambient pressure, he can invade the top 0.3 to 0.4 km of the sea. If he had an armored diving suit or submarine, he could go much deeper, even to 10 km, but if trapped he would have no chance of escape unless he could be transferred at atmospheric pressure to a vessel alongside. Luckily the top 0.3 km is in several ways the richest part of the sea.

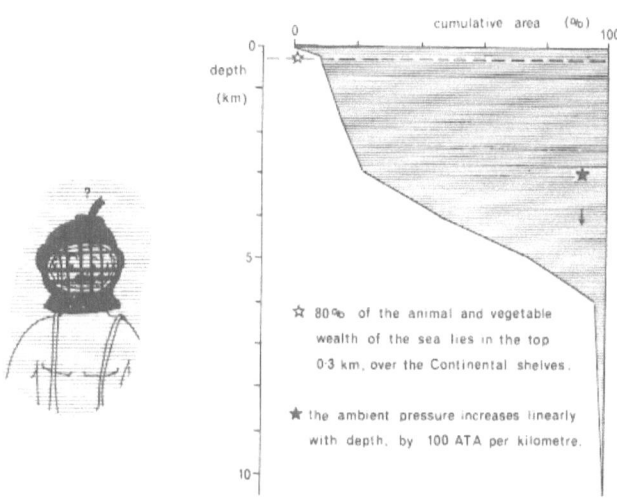

☆ 80% of the animal and vegetable wealth of the sea lies in the top 0·3 km, over the Continental shelves.

★ the ambient pressure increases linearly with depth, by 100 ATA per kilometre.

Three chemical factors limit the man's ability to even explore this upper part. First, there is oxygen poisoning, due largely to the toxic effects of the superoxide ion. This can occur at depths as shallow as 10 meters. The same is true of a hazard mentioned earlier - decompression sickness. The inert gas narcoses are probably due to distortion of nerve membranes by the molecules of gas dissolved within them.

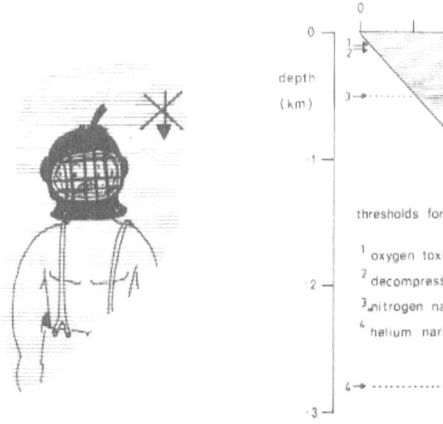

thresholds for:

[1] oxygen toxicity
[2] decompression sickness
[3] nitrogen narcosis
[4] helium narcosis

Although the proper choice of gas will allow a man in a diving helmet to <u>descend</u> to 300 meters or more, the gas that dissolves in his tissues will come out of solution on ascent. As a rough guide, it will form bubbles and cause some clinical problems if he makes a free ascent after staying at depth for longer times than the graph below shows.

Decompression sickness limits the man's ability to <u>ascend</u>, and is the major obstacle to exploring the sea.

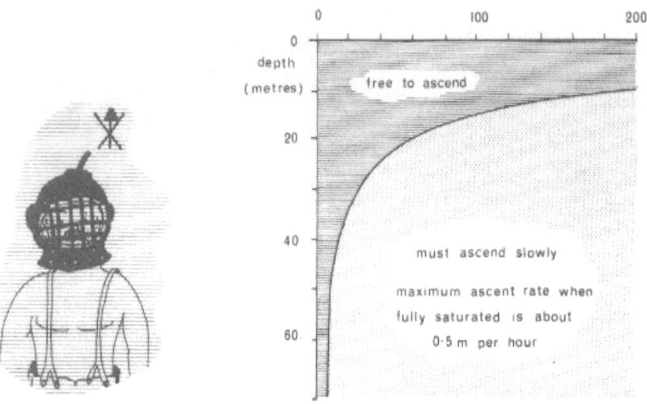

The other great danger for divers and escaping submariners is the risk of lung rupture on free ascent. When full, the lung has a bursting pressure equivalent to 2 meters of water.

The man at the left, with a 6 liter TLC, is escaping from a submarine at a depth of 30 meters. He leaves at FRC (3 liters) planning to hold his breath to the top. His lungs will be full when he reaches 10 meters and will burst at 8 meters.

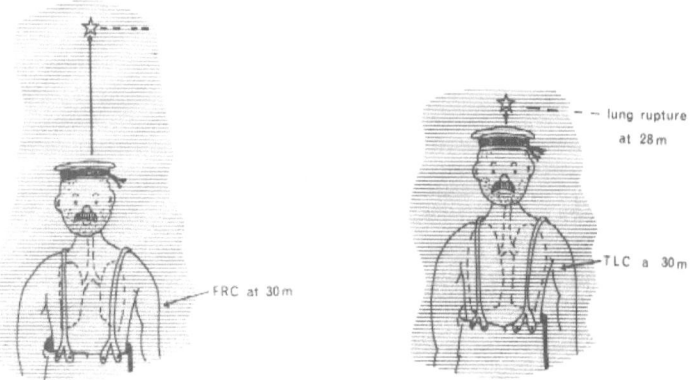

The man at the right has decided to hold his breath at TLC. His lungs will burst after an ascent of 2 meters. But he can escape safely by breathing out slowly and evenly all the way to the top.

At this stage we can leave the sea and think briefly about the pressure of the air which surrounds us. Empedocles, a Sicilian living in Athens about 450 B.C., made a historic discovery while playing with a "clepsydra" or wine-thief. This household gadget was used as a burette to transfer wine from a barrel to a tumbler. It was also used as a water-clock to stop lawyers from speaking too discursively in court.

Empedocles noted that if he put his finger over the top of the clepsydra before sinking it in the barrel, he could prevent wine from entering. He argued that something must be inside already, and so he showed that air existed. This well-documented experiment, which was consciously repeated many years later by Hero of Alexandria and then described by Lucretius, was particularly important as the first demonstration of the physical existence of something that could be conceived but not perceived.

Many years later again, about 1640 A.D., Galileo read Lucretius and repeated Empedocles' experiment. He then invented a pump capable of developing a pressure of 4 atmospheres and used it to fill a copper flask with compressed air. The flask was sealed by a leather "mitral" valve. Galileo then weighed the flask, finishing the balancing with grains of sand. He then vented most of the air by making the valve incompetent and found he had to remove some sand to reestablish balance. In this manner he showed that air had weight.

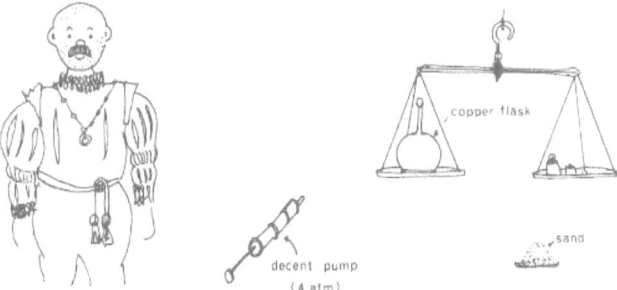

As a direct consequence, Galileo's pupil Torricelli developed his pudding-basin manometer to weigh the atmosphere, and the young French mathematician Pascal and his brother-in-law Florin Perier used Torricelli's manometers to demonstrate that the weight (or pressure) of the atmosphere decreased with altitude.

In fact, barometric pressure decays almost exponentially with altitude, halving every 5.5 kilometers. Since the fractional composition of the atmosphere remains constant to very high altitudes indeed, ambient oxygen pressure also decays exponentially with altitude and can be described by essentially the same curve.

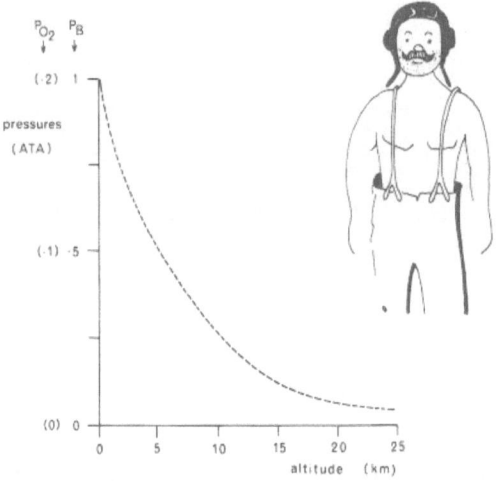

This fall in ambient and thus alveolar oxygen pressure interferes with many subtle bodily processes long before any measurable fall in oxygen uptake occurs. Small changes in night vision can be detected at an altitude of 2 km. Only slightly higher, a definite deterioration in learning ability and slowly developing hyperventilation appear, when alveolar PO_2 is about 65 mm Hg.

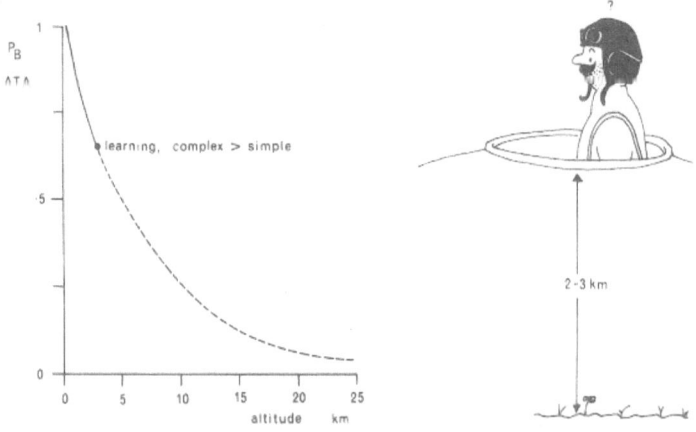

One kilometer higher, alveolar PO_2 has fallen to 50 mm Hg or so, and performance of complex but familiar tasks will begin to degenerate. The subject will not appreciate this, however, since insight is one of the first characteristics that is lost.

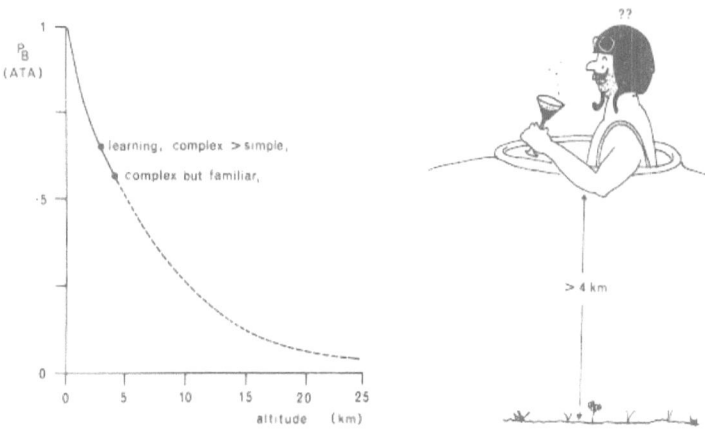

Even higher, judgment will become dangerously impaired. Performance of simple well-learnt tasks deteriorates as alveolar PO_2 falls to 43 mm Hg, and consciousness is lost when it drops below 38 mm Hg.

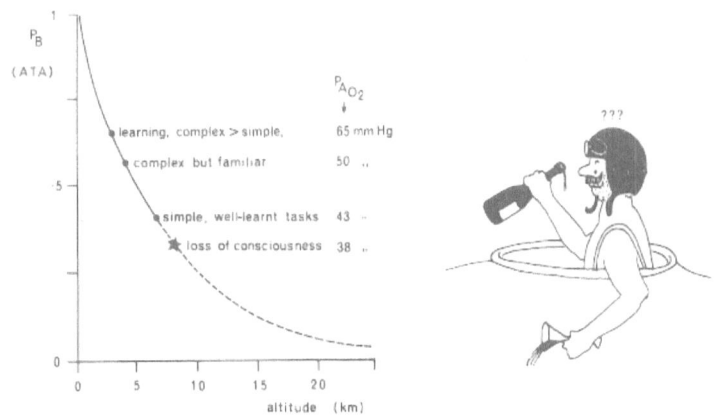

The threshold dose of hypoxia for loss of consciousness is an alveolar PO_2 below 38 mm Hg for more than some 150 mm Hg sec. The higher the altitude above 8 km, the more rapidly is that dose reached and the shorter the "time of useful consciousness" becomes, as the graph below shows.

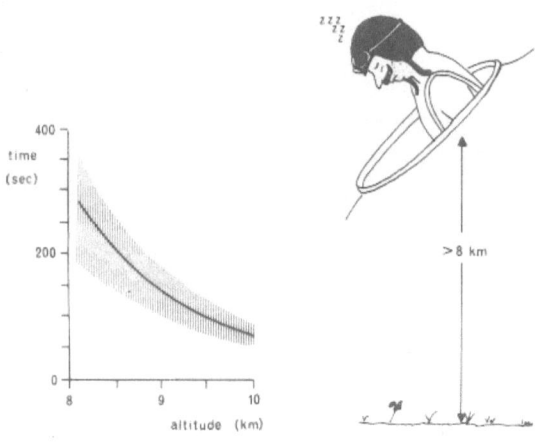

Giving the pilot oxygen will allow him to ascend considerably further, but a threshold to impaired performance is reached at an altitude of 12.2 km, where barometric pressure is 141 mm Hg and, despite modest hyperventilation, alveolar PO_2 has again fallen to 65 mm Hg.

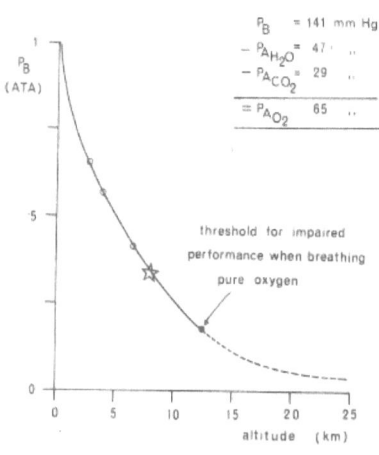

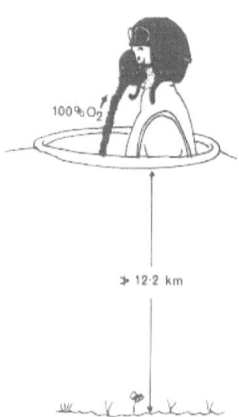

Because he is now breathing pure oxygen, the full force of any fall in barometric pressure is seen as a fall in alveolar PO_2, and the pilot only has to go slightly higher to lose consciousness entirely.

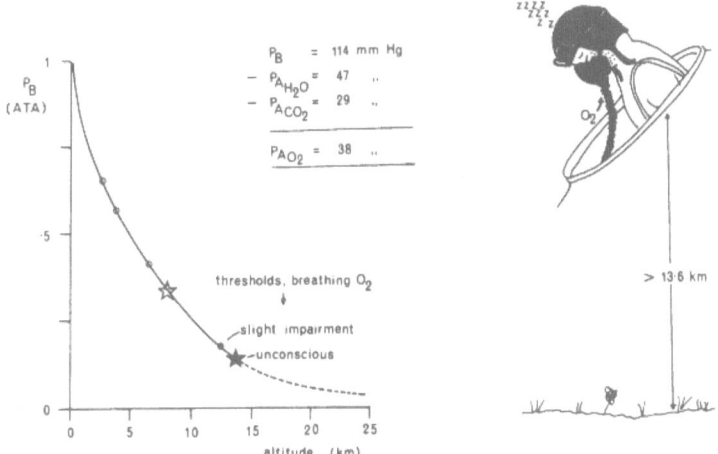

It is practical to preserve consciousness by giving oxygen through an oronasal mask at a positive breathing pressure. Up to 30 mm Hg can be tolerated for half an hour.

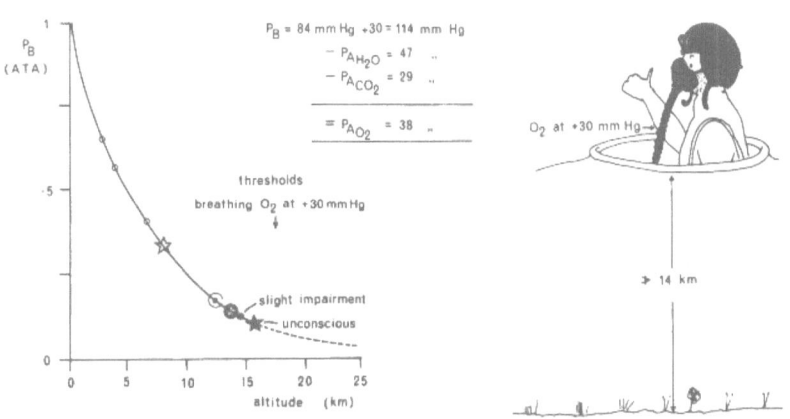

However, as in diving, pressure-breathing introduces fresh physiological problems. This system fails
after breathing at +30 mm Hg for 30 min, because of
chest fatigue.

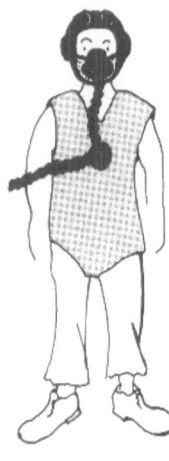

One solution is to surround the "chest wall"
with an equal and opposite pressure, which must extend over the hernial orifices. The oronasal mask
will seal to +70 mm Hg, but the system fails when
one liter of blood has been displaced into the legs
(half of which passes through the vessel walls into
the tissues).

This displacement of fluid can be overcome by extending the
counterpressure to the legs, sparing the knees to preserve their
mobility. The combination takes a man to an altitude of 20 km, but
then leads to painful distension of the neck and axillae and
petechial bleeding in the parts that are unprotected.

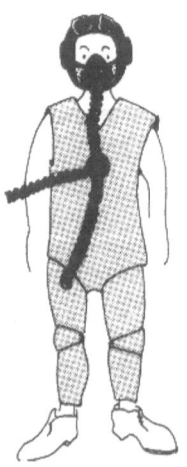

These complications are prevented by giving the man pressurized sleeves and headgear, which will give him temporary protection to a height of 25 km.

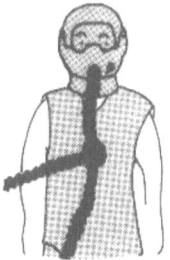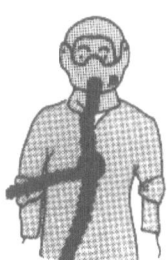

At this stage you might hope that the topic was exhausted, but one important aspect remains.

Normally, aviators are protected at high altitudes by pressurizing the aircraft cabin in general rather than the respiratory tract and selected body surfaces alone.

 Freight and passenger aircraft, which do not need to be very maneuverable, have high internal pressures and are designed to lose pressure over many seconds if the structure fails.

 Fighter aircraft, which must be lightweight and aerobatic, have low internal pressures. If there is a structural failure they decompress very rapidly indeed, say in one to two seconds.

Rapid decompression of either sort carries an obvious risk of lung rupture, but also abruptly puts the man in a more hostile situation. Since it will take a while for any protective system to come into operation, he will inevitably be exposed to a pulse of brief profound hypoxia. The duration and severity of the pulse will depend on the range and rate of the decompression, and on the speed of response of the protective devices.

The severity of the pulse of hypoxia also depends on the composition of the gas that was breathed just before the cabin failure, since any nitrogen in the alveoli has to be washed out before the

full benefit of breathing oxygen is realized. This is illustrated
in the graph below and is the final example of some of the many
effects of applying changes in pressure to the lung.

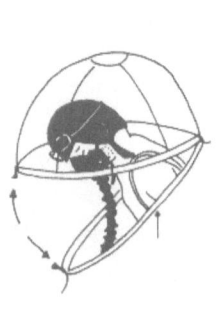
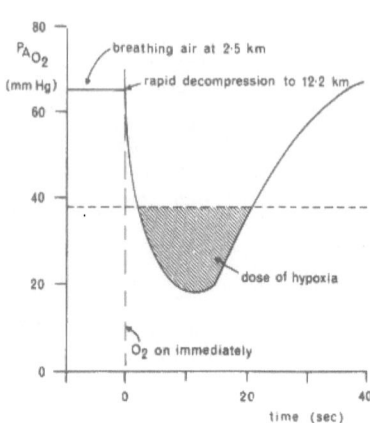

REFERENCES

Bennett, P.B., and Elliott, D.H. (1975) (eds), The Physiology and
    Medicine of Diving and Compressed Air Work. London, Baillière
    Tindall, 566 pp.

Denison, D.M., and Kooyman, G.L. (1973), The structure and function
    of the small airways in pinniped and sea otter lungs. Respir.
    Physiol., 17: 1-10.

Denison, D.M. (1981), High altitudes and hypoxia, in: Edholm, O.G.,
    and Weiner, J.S. (eds), The Principles and Practice of Human
    Physiology. London, Academic Press, pp. 241-307.

Ernsting, J. (1966), Some Effects of Raised Intra-pulmonary
    Pressure in Man. Slough, Technivision (for NATO).

Ernsting, J. (1978), Prevention of hypoxia: acceptable compromises.
    Aviat. Space Environ. Med., 49: 495-501.

Rahn, H., and Yokoyama, T. (1965) (eds), Physiology of Breathhold
    Diving and the Ama of Japan, Washington, D.C., NAS-NRC,
    369 pp.

Varène, P., and Valiron, M.O. (1980), Effects biologique des gaz
    inertes. Bull. europ. physiopath. resp., 16: 79-109.

D I S C U S S I O N ˎ

LECTURER: Denison   CHAIRMAN: Cumming

MILIC-EMILI: Can I ask you about these seals. First of all do
                they trap gas and then lose it via collateral
                ventilation, or do they just get straight
                atelectasis?

DENISON: They do not trap gas at all. If you take a seal or
                dolphin lung and press on it gently it will empty
                completely. And we attribute this to the reinforced
                structure of the airways. When they dive they
                usually dive in expiration, they have ribs that are
                modified, having only one articulation on the spine,
                they have more false fibs, they have fewer ribs
                altogether and the chest is much more compliant. A
                blue whale has only four ribs and if you watch them
                underwater you can see them blow out before they
                begin to dive. They just do not need it. The
                remarkable thing, I think, is the re-expansion of the
                lungs that occurs when they come to the surface. But
                a very beautiful demonstration of the rigid nature of
                their airways is given by the flow-volume curve, that
                is something Cohen has shown in a pet seal which he
                called Houdini. You know that our flow-volume curve
                would look like this; he trained this pet seal to do
                a forced flow-volume curve and it looked like that,
                because of course, like us, the seal is maintaining a
                constant pressure but his airways were not
                collapsing.

SANT'AMBROGIO: Is the reason why diving mammals do not develop
                decompression signs because they are able to shift
                all the air from the air exchanging zone into the
                dead space, I mean the rigid, non gas exchanging
                area.

DENISON: Exactly so. And if you measure the arterial nitrogen
                tension in a seal during a dive to 300 m we would
                expect the nitrogen tension to rise to about 27
                stmospheres, something like that. In fact it rises
                to about 3 atmospheres, and then falls continuously
                during the dive, even though the seal is at 30
                atmospheres. It just requires a pressure of 3
                atmospheres to close the airways and thereafter the
                nitrogen in the blood is leaking away into the

tissues. When the seal comes to the surface it stays at that pressure until it takes its first breath.

CUMMING: I should perhaps reinforce this important concept among those of you who like swimming and diving. It is quite easy to get down to 20 m, and it gets progressively easier; but when you get to the equilibrium point you continue to sink and you may never rise again, especially if you hyperventilate before you begin the manoeuver.

DENISON: I did a lot of diving when I was in the Arctic and the Antarctic, where you need three wet suits, and of course your buoyancy chain and I have got down to about 300 feet, luckily with two friends, because I was just falling like a stone, I dare not take my weight off because there was 15 ft. of ice above me, and if I kept it on I would go down for ever. It shows the advantage of diving with friends.

CUMMING: Perhaps that is the most important lesson: do not dive alone.

NEWMAN TAYLOR: Can we just stick with your diving seal for a moment? You told us how he avoids decompression sickness, what I am not clear about is how he manages to stay down without requiring oxygen and with the accumulation of CO_2 that is going to occur.

DENISON: No-one is clear about this either. What is known is that the moment that a seal dives it gets a massive constriction of all of its arterioles except for the heart and the brain. The heart and brain have an aerobic metabolism, and the seal will come to the surface when his arterial blood is beginning to desaturate severely. It is clear that they can anticipate how long it is going to take to come back to that particular hole in the ice, because they can swim long distances away, so they must sense it before it is absolutely necessary. What we do not understand is how the rest of their body works during that time, when they are actively swimming in that total absence of oxygen.

NEWMAN TAYLOR: After he comes up, what about the acidotic load which he presumably accumulated?

DENISON: These are enormous. Seals and dolphins have a very large blood volume relative to their useful body mass, and they have a special valve in the diaphragm which only allows the peripheral blood to enter the heart slowly. If it entered suddenly they would die. You can sometimes disorganise seals by anaesthetising them when they lose this reflex.

WIDDICOMBE: How is the valve controlled?

DENISON: I do not know.

WIDDICOMBE: It is smooth muscle in the diaphragm, is it not?

DENISON: As I understand it, it is an exaggeration of the crura of the diaphragm around their massive venae cavae. For two reasons they have a massive blood volume: one is so that the blood can come into the chest, and the other is so that it can accommodate some of the peripheral change. If you look at the vena cava of a walrus it is bifurcated, and it is very big.

CUMMING: No other questions? Thank you very much David.

WATER VAPOUR HANDLING IN THE AIRWAYS

Gordon Cumming & Warren Warwick

The Midhurst Medical Research Institute
Midhurst, West Sussex

The preservation within the lungs of their appropriate milieu interieur requires that ambient air be cleansed of particulate and noxious gaseous matter and that clean air be delivered to the alveoli warmed to body temperature and fully saturated with water vapour. This process must continue under the most adverse of ambient conditions, ranging between air at $-45^{\circ}C$ and almost devoid of water vapour as in polar regions, to air which is inspired above body temperature as in the world's deserts.

This process should be carried out economically, both in terms of energy and conservation of water. In addition, body temperature must remain constant. Thus the elegant solution adopted by the camel is denied to man. The camel, inspiring gas at a higher temperature than his body temperature, merely stores the excess heat and permits the body temperature to rise since the setting of the sun will reverse the process of heat gain - fitting him for further exposure on the morrow. To maintain the function of his brain (with its bad temper) evolution has devised a heat exchanger in his long nose in which water is evaporated during inspiration, via the rete mirabilis to cool the internal carotid arterial blood and thus maintain the appropriate biochemical conditions in the brain. This cameo of the heat handling mechanisms of the camel serves to illustrate the scientific principles which are used equally well, albeit with slightly differing objectives, in the human.

The principle involved concerns the changes in heat content, the enthalpy, of the air in the respiratory passage. The enthalpy of the air is negative when heat must be added to the air and positive when the air has a heat content greater than air at BTPS.

Negative enthalpy decreases when gas temperature is increased
depending upon the specific heat of the gas, which in turn is
determined by its molecular properties, and the accompanying change
of state of water to vapour in the lungs. The change of enthalpy
determined by the latent heat of vaporisation of water is 0.58
cals/mg.

Since enthalpy changes of both types occur during inspiration,
it is useful to understand both the magnitude of the changes
involved and the relative contributions from each source. These
relationships are depicted in Fig. 1. At an alveolar temperature of
37°C one litre of gas contains 44 mg. of water and enough water must
be added to each litre of inspirate to reach this value. This
latter figure is expressed in terms of relative humidity. The
curved lines indicate relative humidity from zero to 100% at any
temperature indicated on the abscissa. The addition of specific
heat is indicated by the slopes of the straight lines labelled
'enthalpy of air'.

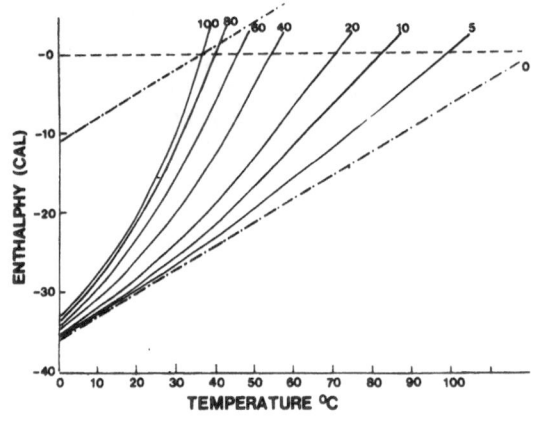

Figure 1

The transfer of heat and water vapour is accompanied by a
cooling of parts of the respiratory passageways. The residual
coolness permits these airways to recapture during expiration some
of the heat and water given up to the air during inspiration.

However, not all of the heat and mass of water can be recaptured for
the conducting airways maintain an average temperature despite the
continuous input of heat from blood flow through their walls during
both inspiration and expiration.

Therefore, for each part of the conducting airways that participates in heat and water transfer, and for the conducting airways as a whole, the enthalpy changes can be summarised by the general equation:

$$QE - QI + QB = 0 \quad \text{or} \quad QB + QE = QI \qquad (1)$$

where

QB =   the enthalpy change to the conducting airway walls from blood flow.

QI =   the enthalpy given up to the air passing through the conducting airways during inspiration.

QE =   the enthalpy given back to the conducting airways by the expired air.

The enthalpy of the air is defined by the equation:

$$E_T = E_H + E_{H_2O} = E_H + C_{H_2O} \qquad (2)$$

E_T =   total enthalpy (calories per litre of air)

E_H =   enthalpy due to the temperature of the air (calories/litre of air)

E_{H_2O} =   enthalpy due to the content of water vapour in the air (calories/litre of air).

C_{H_2O} =   mg of water per litre of air.

The water content of air has a calorific equivalent such that 0.58 calories of heat are required to vaporise 1 mg of water.

The enthalpy of alveolar air will be used as the reference enthalpy and has been defined as zero for convenience of calculations. Since usual ambient air requires the addition of both heat and water, the enthalpy of ambient air is less than that of alveolar air and is therefore negative.

The use of Fig. 1 is best illuminated by an example. Suppose one litre of ambient air is inspired at a temperature of $10^{\circ}C$ and a relative humidity of 40%, conditions which might exist on an autumn day. The air enthalpy line indicates a negative enthalpy of -8.1 cal/l, whilst the 40% humidity line indicates a negative enthalpy of -31.3 cal/l. Thus we see that the enthalpy of change of state is four times that of specific heat.

As a second example consider ambient conditions of $0^{o}C$ and zero humidity. Raising the temperature from $0^{o}C$ to $37^{o}C$ (a huge change) produces an enthalpy change from -36.9 cal/1 at zero to -25.4 cal/1 at 37° a net change of 11.5 cal/1. However, a change in humidity from zero at 37° to 100% at $37^{o}C$ produces a change of -25.4 cal/1, again change of state 2.5 times as great as specific heat. These two examples show that water vapour saturation dominates the heat exchange picture, temperature change being of secondary importance.

We have seen here that two mechanisms of enthalpy change contribute to heat flux and have some idea of the energy involved. The next questions are deceptively simple - do these two processes take place simultaneously? If so, are they at all times in equilibrium? If one litre of dry air is inspired through the nose and its temperature rises by $1^{o}C$ from $20^{o}C$ to $21^{o}C$, what happens to the relative humidity? Since the nose is thought to be an efficient warmer and humidifier, does the inspirate pass from $20^{o}C$ dry to $21^{o}C$ fully saturated, or does it remain zero saturated, or is there some intermediate situation? A priori considerations would lead us to the latter conclusion.

The anatomy of the nasal passages is well adapted to the process of air warming. For adequate heat transfer to occur, there must be a large surface area of contact between gas and mucosa, a sufficient supply of heat to the mucosal surface (from its blood supply), and a sufficient time of contact between gas and mucosa (determined by the velocity of gas flow and an adequate mixing device for the flowing gas turbulence within it).

The surface area of the nasopharynx in the human has not been closely studied, but estimates in the literature suggests a value of about 160 cm^{2} for the nose and 100 cm^{2} for the pharynx. The excellence of nasal mucosal blood supply is not in doubt, and it has been suggested that this supply, with its autonomic innervation, plays an important role in the maintenance of the thermal equilibrium of the body as a whole. The efficiency of both blood supply and its control mechanism is evidenced by the outpouring of nasal secretion that occurs in rhinitis.

The time of contact of air with the mucosa is more difficult to quantify. The gap between the turbinates is about 2 mm and the length of the gap may be about 50 mm so that the area of cross section of the air passage in each nostril is about 1.0 cm^{2}.

Inspiratory flow rate averages 200 mls^{-1} during quiet nasal breathing; the peak rate is greater than this. Consequently the linear velocity of gas flow over the turbinates is about 100 cm s^{-1} and the dwell time of any gas molecule is of the order of 0.2 s. Turbulent flow is readily induced at such high velocities and the shape of the turbinates both increases surface area for heat exchange and maximises mixing by turbulence.

Measurement of water vapour pressure

Direct measurement of the partial pressure of water vapour became possible with the development of the mass spectrometer coupled to an appropriate sample inlet system.

Water has a mass/charge ratio of 18 and this peak is readily detected by a mass spectrometer. Problems, however, arise in the characteristics of the inlet system. Since water vapour from expired air will naturally condense in any inlet system at room temperature, the system must be heated. Furthermore, the analyer itself must also be heated to above 37°C. The most useful temperature of the sample line whilst measuring water vapour is about 80°C. By this means a plot of expired water vapour partial pressure against expired volume for a single breath may be constructed and such a plot is shown in Fig. 2 for vital capacity inspirates at two different saturations.

Many authors have measured end expired water vapour partial pressure during mouth breathing, with concordant results suggesting a value of 38-40 mmHg. Since full saturation at 37°C is associated with a partial pressure of 47 mmHg, it is apparent that expired gas is not fully saturated at body temperature. The measurements of expired temperature previously mentioned indicate that expired gas temperatures of 37°C have been commonly measured when the ambient air temperature is about 23°C. These two facts, as well as a priori considerations permit the conclusion that expired gas is fully saturated at 32°C - with the resultant partial pressure of 38 - 40 mmHg. The a priori argument is as follows: since gas leaving the alveoli at 47 mmHg and 37°C leaves the nose at a lower temperature, then the cooling observed must be associated with condensation of vapour to liquid water; under such conditions the gas must remain saturated with water vapour.

From this simple observation that end expired gas is fully saturated with water vapour at 32°C, it is possible to infer the events which occur during inspiration in the area of uncertainty within the airways. The first question which poses itself is – what is the mechanism which cools the expired gas, and furthermore cools it for every expirate? Since the a priori consideration above suggests that the cooling of the airways must occur before expiration begins, it must occur during inspiration. If we recall that the airways are called upon to add between 11 and 19 mg. of water per litre during inspiration, the origin of the cooling mechanism and its site become apparent.

Inspection of Fig. 1 indicates that the negative enthalpy of gas at 32.5°C (oropharyngeal temperature) with an 80% saturation is 10.6 cal/l and this heat must be transferred from the airways to the inspired gas resulting in surface cooling. During inspiration, therefore, there exists a temperature gradient from 23°C (ambient) to 37°C (alveolar) passing through 32.5°C (oropharyngeal). The precise site and extent of the gradient along the airways cannot be inferred from the present observation. During expiration over this cooled surface there is both thermal retention (as the surface is warmed by alveolar gas) and fluid retention (as water condensing on the surface yields up its enthalpy of change of state).

THE SHAPE OF EXPIRED WATER VAPOUR CURVE

The lower curve of Fig. 2 depicts the events during a vital capacity expirate following an inspirate of similar size from ambient at 23°C and 40% r.h. in terms of partial pressure of water vapour. Initially, the mass spectrometer samples air which has just entered the mouth and leaves it immediately, so indicating the ambient water vapour pressure. The curve rises rapidly to level out at about one litre of expirate and continue at a similar level (about 40 mmHg) until the end of breath. All curves have a similar form when ambient air is breathed.

Figure 2

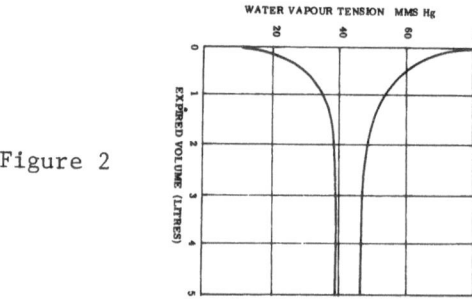

From our knowledge of the events within the airways, it is now possible to interpret this curve. Since the airways are cooled and gas is therefore fully saturated, the curve indicates a temperature less than 37°C and the temperature will rise during expiration as more heat is added to the system from the alveolar gas. It is also clear that the process of heating should continue until air at 37°C leaves the mouth. In fact this does occur, and even when residual volume has been reached the gas temperature is still below body core temperature.

Another piece of experimental evidence is offered by the upper curve of Fig. 2. This curve has been produced by inspiring saturated air at a higher temperature than that of the body (about 50°C), so that the P_{H2O} is about 80 mmHg. The heat content of such an inspirate is high and during inspiration the cooled surface is rapidly warmed so that the subsequent expirate travels up a temperature gradient which is nearly 50°C initially, and 37°C latterly. Under such conditions alveolar humidity is demonstrated without difficulty and the partial pressure of 47 mmHg is observed.

From these two curves it is possible to conclude that there exists between the oropharynx and the alveoli a gradient of mucosal temperatures, and that this gradient persists, though less steeply, during expiration when normal ambient air is breathed; but it can be abolished and even reversed by an appropriate input of heat from inspired air.

What is not clear is the anatomical site of this gradient, since it could be in the first few centimetres of the trachea so that at the carina full conditioning is achieved as reported by some earlier workers. Direct observation and definition of this gradient has not yet proved technically possible, and the information available is based on calculation and inference.

Since the nose adds about 75% of the required enthalpy during quiet nasal breathing and oropharynx only 27% during mouth breathing, it is clear that the completion of the humidification process must take place in the airways. The ability of the nasal passages to add heat and water vapour to inspired air are respectively 2.69 cals s^{-1} and 0.40 mg s^{-1}, so that it is evident that the blood supply for these processes is very different, and that the ability to add 73% of the enthalpy during mouth breathing depends upon some other characteristic of the airways. This reserve capacity for the adding of enthalpy depends primarily on the greatly increased lateral surface of the airways resulting from the rapidly increasing branching of these structures.

At some point in the branching system the incoming air will be fully warmed and saturated and the enthalpy difference becomes zero. It may be possible to find this point by calculation so that the relative contribution of nose, mouth and airways can be perceived.

In order to compute the contribution made by the airways it is necessary to define the condition of the air as it passes down the trachea during nose breathing and during mouth breathing. This can be done in the form of a table.

TABLE 1

Enthalpy States in the Airways

	Ambient Air 23°C, 40% rh.	Oropharynx nose breathing	Pharynx mouth breathing
Total enthalpy (cals)	-24.63	-6.65 (21%)	-17.95 (58%)
Enthalpy of specific heat	- 4.02	-1.28	- 2.74
Enthalpy of latent heat	-20.61	- 5.37	-15.21

The negative enthalpy represents that quantity which must be added by the bronchi during quiet breathing at rest. Nose breathing requires 21% to be added and mouth breathing 58%, both mainly from further water evaporation from the bronchi.

From this Table it may be seen that during nose breathing only 21% of enthalpy needs to be added by the bronchi whilst during mouth breathing 58% is called for. In a patient with a tracheostomy the whole enthalpy must be added by the airways.

Thus the bronchial tree makes an important contribution in water vapour handling, and constitutes the reverse heat exchanger permitting adaptation to adverse environments.

DISCUSSION

LECTURER: Cumming   CHAIRMAN: Bonsignore

DENISON: You began very kindly by giving us an equation which
                is so simple that even you could understand it,
                therefore we too have been able to. Would you like
                to look at it again and show us where the equal sign
                is?

CUMMING: That is right, thank you.

RICHARDSON: You said early on in your talk that the nose was not
                an important air conditioner, and you could prove
                that by showing that when a patient was given a
                tracheostomy he still continues to live. I would
                like to suggest that that is not so, because almost
                always, when a patient is given a tracheostomy, in
                the first few days he develops a bronchial infection,
                bronchitis. After that, however, he does seem to
                make some adjustment and can live very well with the
                tracheostomy. So, I would suggest that the air
                conditioning system of the airways does include the
                nose, but it also includes other systems and it is
                best to look at it as a system which is held up by
                belt, braces, and probably strings as well.

CUMMING: I entirely agree. If I said that the nose is not an
                important air conditioner I misled you, as I said in
                quiet breathing it provides 8.5% of the total air
                conditioning during that situation. What I meant to
                imply, as you rightly say, is that the nose is the
                first step in the conditioning process. I think I
                put quantitative terms on what the nose can do,
                because I talked of the enthalpy transfer rates, if
                you remember it is 3.4 calories and 2.8 mg of water,
                so without making a value judgement that is the
                quantitative capacity of the nose effect. During
                quiet breathing through the nose, excellent, you
                require to recruit maybe only the first two
                generations or orders of the bronchi. But since you
                have been using the maximum rate in the nose already,
                any further demand for conditioning has to be met by
                the braces, the belt being already on the last
                buckle.

WIDDICOMBE: Can I ask you a little bit more about tracheostomised patients following Paul's comment. Of course most of them are flat on their back and do not exercise, some of them are ambulant and at least in theory could exercise but probably do not very much. What do you think would happen if they were to exercise vigorously?

CUMMING: It is quite straightforward, if you assume that you are knocking out 260 cm^2 of lateral surface with good blood supply, then you have to supply that below the tracheostomy, and we could look at where was 260 cm^2, maybe just two orders of bronchi. If they exercise, with every increment in ventilatory volume you must increment the amount of lateral surface to produce the enthalpy change. It is a simple matter to compute, in a tracheostomy, to what level the interface would go. Could I have the last slide, please? It is possible to look at the gradient of temperature down the airways and to plot it using the model of the branching system and the equations of heat exchange that I have talked about. Then to show what the gradient in any position is under any given condition, and I think we show that here. Here is my Case A, which is a very simple case, and there is the oropharynx and the gradient becomes zero at eleven orders. Thereby using six orders of bronchi. If you want to exclude the nasopharynx we have to include enough lateral surface from here, which will bring it down by only one order. So, the tracheostomised patient, during exercise, would recruit practically the same as a man breathing through his nose, because the addition of 260 to 2000 is not going to make a great deal of difference. Here is Case C which is an intermediate one, that goes down to order number 7. That is an actual plot computed of the water vapour. Here are the alveoli, and here is the mouth, and that is what you see going down. Of course, if you measure that in a mass spectrometer, as it comes out here, you will get the curve that I demonstrated. One can do a calculation to show how the computed slope fits with the experimental slope, and that is not a bad fit.

WIDDICOMBE: You raised the possibility that nervous receptors in the airways might respond to changes in conditioning. Most of the studies on those receptors in the airways suggest that they are not very sensitive to temperature or humidity and Richardson has looked at this and may want to comment. Of course they have a temperature coefficient, like all nervous receptors, but it is not very high. The possibility exists that although they may not respond to short-term acute changes in temperature or humidity, if these are maintained long enough to change the structure of the epithelium they could then change their sensitivity.

CUMMING: To change the structure takes a considerable time.

WIDDICOMBE: That is what I wanted to ask you: how long would it take if somebody is breathing very cold and dry air, for the mucosal structure to be changed?

CUMMING: I would have thought that it would not be a relevant explanation for exercise asthma. It would be too long, I guess, though I am no pathologist. Changing the temperature of the airways pushes down the level of coolness and you recruit far more lateral surface and far more sensors in that situation. The experimental evidence is not too bad, because if you take a person in whom you can generate exercise asthma by breathing on a treadmill regularly, and then you let him breathe with air at $37^{\circ}C$ saturated, with water vapour they may not exhibit exercise asthma. So it something to do with enthalpy. It may not be temperature, it may be rate of change, it can be anything, but it is certainly strongly related to enthalpy change. In a simplistic way I think it may actually be either temperature or, better still, change of temperature, or even rate of change of temperature. I do not know whether Paul thinks that is a possibility.

DENISON: If I understand you correctly, your calculations have assumed that there is no resistance to the flux of water molecules. But is is everyday's experience to wake up with a very dry mouth indeed. And I would think that it would be very difficult to find any water vapour molecules on the surface of my tongue this morning whatsoever. How would that affect the heat transfer down the respiratory tract?

CUMMING: I cannot answer the question. The question I need to answer to clarify that point is: what is the maximum water transfer rate in the mouth, under the conditions you specified? Because the members of my team never work with a dry mouth in the morning, for obvious reasons, so we were never able to measure it. I am very happy to have you come along but the rate of change is low anyway, we are talking of 0.4 mg/1, maybe that falls to 0. This makes scarcely any difference to what happens distally. It is very interesting: what happens up top is easily offset just by recruiting another generation of bronchi, and this is very important; you cannot overcome the capacity of the bronchi to condition air. There may be one circumstance in which it is possible to overcome it, and that is for someone who dies in a fire, in a conflageration and may actually inhale flame. I guess that might overcome the capacity of the bronchi. But with 10,000 cm^2 of warm water they make a very good heat exchanger.

SANT'AMBROGIO: Are those figures that you showed in one of your last slides for the lateral surface the cumulative figures?

CUMMING: Yes. That point, this is nought, this is Case A, is the point in the airways where enthalpy becomes 0. In other words, when the temperature is 37°C, 100% saturated.

SANT'AMBROGIO: Sorry, I am not asking that. What I am asking is: is the function showing a cumulative figure, after that generation or order?

CUMMING: Yes. That represents the water vapour partial pressure along the airways. It is cumulative surface area.

SANT'AMBROGIO: Another point that I think may be of some interest is that somebody observed the temperature of the brain stem in anaesthetised animals which were breathing through their upper airways, with tracheostomy, and with resting ventilation. They measured the difference in the order of 0.51^{o}C, which might make a difference.

CUMMING: A similar experiment is to measure the pulmonary artery and the pulmonary venous blood and to see what the total heat transfer in these two situations is. But no one has managed to do this experiment well.

DENISON: I cannot give you the name now, but I do know, and I can give you the reference of at least 12 people who have measured it, or failed to find significant temperature gradients across the pulmonary bed.

PRODI: Pressure is very important from the point of view of particle dynamics. During inspiration humidity reaches a very high value in a fairly short time; you showed that in expiration you may have a situation of super-saturation, so the particle growth by water condensation could be very important. I was just wondering how is the heat transfer to the last generations that are recruited for air conditioning. How much time do these airways require to heat up again and lose their power to cool again the air which is expired? Because I would expect that during the pulse some heat could be supplied to these generations and heat them up and of course they would lose the possibility to cool the air again. Is that important? Or what kind of a profile of supersaturation could be taken for the expired air?

CUMMING: With regard to the capacity of the bronchial blood supply to reheat the cool air, this can be investigated by taking in a vital capacity, and holding the breath for as long as possible, a minute if we do not hyperventilate, and measuring the profile during expiration yet again. We have done this on numerous occasions and so far failed to establish any difference. Water vapour is a noisy signal and we could not measure better than 1 torr, so it may be that it is within 1 torr, that is an area of uncertainty, but for practical purposes the bronchial blood supply is such that normal breathing does not give any substantial input of heat to the mucosal surface. The second point, with regard to supersaturation is an important one. If you inhale, to take a simple case, a monodisperse aerosol, which is hydrophobic, like a stearate, which is commonly used, it goes down and ignores the flux of water which is going on rapidly all around it and it behaves in the way we have seen. If now we take a particle like disodium chromoglycate which is hydrophilic, it goes down surrounded by water flux and what happens is that it is at ambient temperature when it starts, it may be cooler because we have got the Kelvin effect as it comes out through the small orifice, so the particles are cool, water vapour will condense on it, they will aggregate, so the flux of water vapour in the airways modifies markedly the behaviour of hydrophilic aerosols. That is a great problem, and it may be the answer to the question although, if you take isoprenaline sulphate in a narrow dispersion with a mean particle size of 1 micron, you take a puff and inspire it, 95% of that dose is taken up by the buccal mucosa and only 5% gets to the bronchi. The evidence that 95% goes to

the buccal mucosa is that if you measure the blood
concentration of the various congeners 95% are
sulphated.

FAGAN: I think Denison has given us a clear clinical
              description of 'alcoholic enthalpingitis'. What I
              would be most interested to know: do you think that
              that sort of lesion is due to drying or due to
              chilling? Because it would seem that this might be a
              potent inducer of acute inflammatory reaction, and
              this might be the most rapid method of adaptation
              towards stress, because it was certainly my
              impression that any form of acute inflammation in the
              tracheobronchial tree is a potent accelerator or
              producer of asthma, so this might perhaps be the
              linkage.

CUMMING: Before making a comment on that suggestion, I would
              want to know what is the oropharyngeal partial
              pressure of water vapour in a person who says his
              mouth is dry, because it may not be dry in terms of
              physical measurement; that it is dry to sight, to
              touch and to perception I do not deny, but I would
              first like to measure it. Let us assume however that
              it is low, let us assume it is ambient. All this
              tells me is that the capacity of the mouth to
              transfer water has been exceeded for a finite time,
              and hence the surface layer has been depleted. The
              rate of requirement of adding water although it is
              only 0.4 mg/l has been greater than the blood supply
              will bring to the surface, and I guess this is a
              reflection of blood supply rather than anything else.

DENISON: Going back to my previous question: do we understand
              that in the way in which water molecules leave the
              surface of the respiratory tract, can we simply
              assume that they are free to go wherever they want
              to, or are there mechanisms retaining them?
              Supposing that I take a beaker of water and put a
              thin film of paraffin on it, I can prevent the water
              vapour molecules from leaving.

CUMMING: I made the point very clearly: the determining
              function of water vapour flux is the surface
              conditions. In the calculations I have used I made
              an assumption: I have said that in the tract as a
              whole, nose, mouth and bronchi, there is a
              uni-molecular layer of water. The unimolecular layer
              cannot be rapidly augmented by the blood supply. It
              could be bimolecular, ten-molecular, that does not

make very much difference because you do not deplete the molecule over any large surface to any great extent, because the content of water, even one molecule thick, is quite high in such a large lateral surface. Now, to your point about resistance, the resistance is clearly defined physically: it says that a monomolecular layer of water exerts a vapour pressure which is in line with the laws of physics, and that its mixture with molecules of gas is also defined by the laws of physics, and that if you add a convective component to that, that is also explicable by the same laws. You might say that it may not be a monomolecular layer and I have to say yes you are perfectly right in saying that the surface is completely dry. It is very unusual, but I think you could make such a statement. That it were 50 molecules thick would make no difference, because the top layer would be wet. It depends on what you mean by 'wet', I suppose.

BONSIGNORE: Thank you, we pass now to the second lecture by Candura.

POLLUTION IN CONFINED ENVIRONMENTS

Francesco Candura

Istituto Medicina del Lavoro
Via Boezio, 24
27100 - PAVIA, ITALY

The particular danger of an incident of pollution in a confined environment has always been well-known, and, this subject is already present in the first known studies of environmental hygiene and are referred to in all the classic texts.

At one time, the subject of the pollution of confined working and living environments was directed mainly towards pollutants caused by the persons working in such environments (the excess of CO_2 and of water vapor, the so-called "anthroprotoxins", etc.). At the present time we have unfortunately - in addition to such pollutants - a vast array of others, to such an extent that even a simple introduction to the subject will necessarily be incomplete.

On the other hand if the empirical identification alone of the dangers of a specific work environment can permit the definitive removal of the affected subject, the specialist is asked for something more. In particular, the occupational doctor is required to identify the precise etiology, to demonstrate the "cause-and-effect" relationship between one or more pollutants in the work environment and the onset of the symptomats. This is not only because of the clinical implications, since a therapy which is aimed at the causes should be founded on logic, but also because of the medical-legal consequences, as the etiological diagnosis is the foundation for indemnity as well as for its ethical implications, since the absence of indemnity for occupational damage that is the result of the failure to establish an etiological diagnosis seems frankly immoral. I recall that my professor, Salvatore Maugeri, taught me that occupational medicine is a discipline which aims to establish causes.

All this makes the identification of air pollutants essential, and this maybe effected by an adequate study of the technological cycle. In fact, as may be seen in fig. 1, which attempts to classify the major disciplines on which occupational medicine is founded, technology proves to be at the basis not only of prevention but also of the cure itself.

Therefore, the accurate study of the various phases of the individual productive cycles may permit identification of the air pollutants under examination in each work environment. Such identification, if conducted systematically can lead to a real census of the risks, such a census being the basis of any serious attempt at prevention, as it is evident that in order to fight effectively it is first necessary to recognize the enemy.

Once the possible potential risks in each work environment have been determined – and in particular, the various possible air pollutants – it is necessary to verify their presence, by studying the possible fluctuations of concentration in time, and ultimately to test the degree of noxiousness by means of adequate studies of animals and, above all, by means of adequate epidemiological investigations (fig. 2). Epidemiological studies should be given preference, as naturalists have always taught that noxious agents in nature which maybe toxic for one species, may be food for another. This is the foundation for the so-called "biodegradability" of a given noxious substance. It is, therefore, preferable to devote ourselves to studying how this causes illness in man rather than in animals.

This procedure, which may be called "descending", is the most logical, and it must be integrated with a procedure which we might call "ascending" and which, starting from the clinical picture and attempts to identify the noxious agent, that is it attempts to identify the cause (in this case, the responsible air pollutant).

At any rate, when it is desirable to generalize the subject to "all" work environments and to "all" air pollutants – in confined atmospheres – reference should be made to a classification (Table 1).

This attempt at classification does not constitute an exception to the rule which says that all classifications are deficient: this classification is not only incomplete, but it is also inexact, since rather than classifying the air pollutants, it attempts to classify the

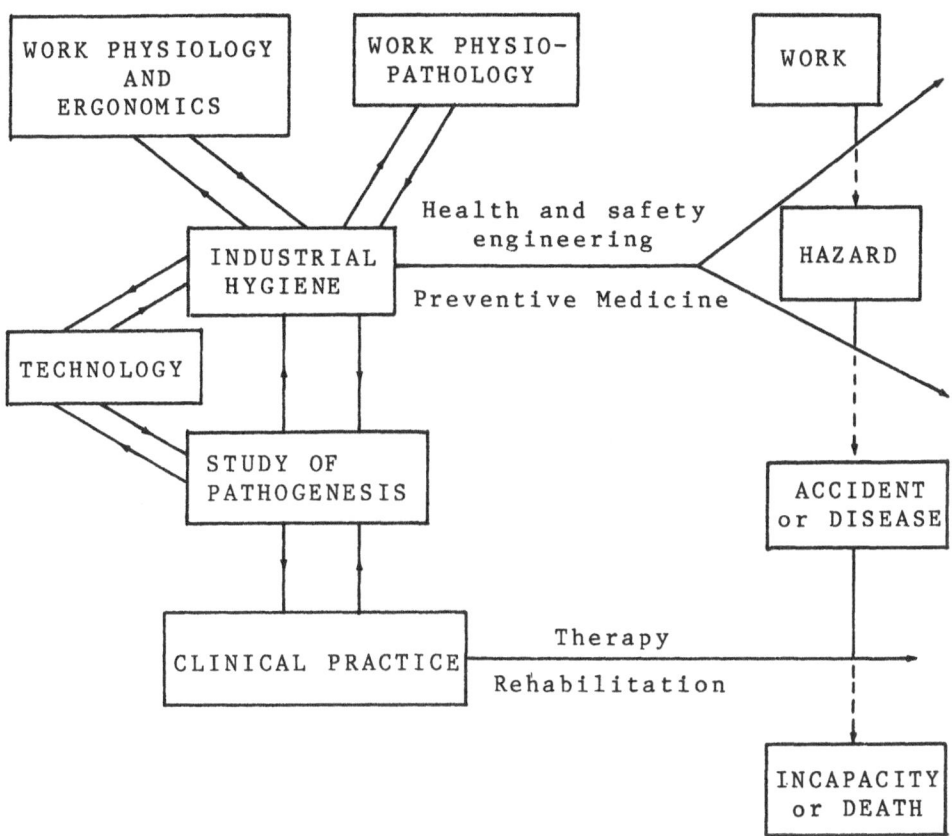

FIGURE 1   INTER-RELATIONSHIPS IN OCCUPATIONAL MEDICINE
(From Candura F. "Remarks on Occupational
Medicine" Pavia University, Publisher, 1961).

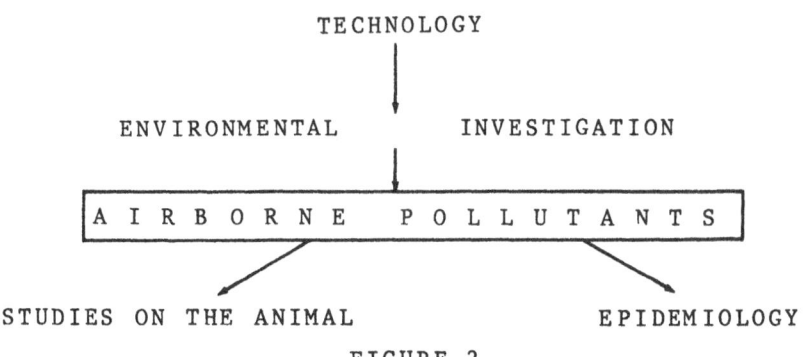

FIGURE 2

TABLE 1

TENTATIVE CLASSIFICATION OF AIRBORNE CONTAMINANTS

1. ASPHYXIANTS
2. IRRITANTS
3. TOXIC COMPOUNDS or POISONS
4. SENSITIZERS

TABLE 2   ASPHYXIANTS

1. DEFINITION. Airborne materials, which, "just" due to their presence cut down the partial pressure of the oxygen in the inhaled air.

2. COMPOSITION.

  2.1 Most common elemental asphyxiants:
      nitrogen-Hydrogen-Helium-Argon

  2.2 Most common compounded asphyxiants:
      carbon dioxide-methane-ethane-propane-ethylane
      acetilene

3. MOST COMMON HAZARDS. Mining; working in shafts, wells, tanks, vats, tuns, silos, holds.

TABLE 3   IRRITANTS

1. DEFINITION. Airborne materials (dusts, fumes, smogs) able to cause inflamation (heat, redness, swelling and pain in reaction to injury) at the exposed surface levels (mucous membranes of the eye, and respiratory tract; skin).

2. COMMONLY ACCEPTED CLASSIFICATION.

  2.1 EYE AND RESIRATORY TRACT IRRITANTS

  2.2 TRACHEO-BRONCHIAL IRRITANTS

  2.3 DEEP AIRWAYS IRRITANTS

relative mechanisms of action. Nevertheless, this appears to be an outline which permits a satisfyingly simple approach to a problem that is both complex and not easily accessible. At any rate, it is essential to state first of all that since the typical means of penetration of the occupational noxious agents is the respiratory tract, a complete listing of all the possible air pollutants of confined environments would require a complete treatment of the whole of occupational pathology, which is impossible, especially in relation to occupational toxic substances. I will then limit myself to a few examples.

The "simple" asphyxiants are aeriforms which function only in so far as they reduce -due to their "simple" presence- the partial pressure of oxygen in the air breathed. It is obvious that for them to function they must be present in high concentration, and it is equally obvious that an environment occupied by "inert" aeriforms and therefore in low or zero concentrations of oxygen, is incompatible with life itself. Despite this fact, countless people have died and continue to die due to the lack of elementary precautions which are explicitly indicated by laws, regulations, orders and specifications (such as the norm which prescribes that anyone who lowers himself into vats, wells and other confined environments should tie himself to a suitable piece of rope).

The irritant air pollutants (Table 3) are, on the other hand, substances (gas, vapors, dust, smoke, smog) capable of injuring the surfaces with which they come in contact (conjunctivas, respiratory mucous, cutis), causing in them an inflammatory process which is more or less extensive and more or less serious.

Commonly, distinctions are made between the irritants of the upper, middle and lower respiratory tracts: Tables 4, 5, and 6 provide some examples of such compounds, and they particularly attempt to identify the major representatives of each category, always with the specification that these schematic indications also suffer from schematicism. For example, acrolein is classified among the oculo-respiratory irritants due only to the fact that its irritant action is so elevated as to often provoke a real escape reaction. It is at any rate certain that in high concentrations, this aldehyde causes pulmonary edema. In this and in other cases, then the increasing gravity of effect is linked to the increase in the dose.

As toxic substances, the occupational doctor legitimately understands a substance (or a mixture of substances) typically devoid of action at the level of the surface of contact and apt to provoke, even in

TABLE 4

EXAMPLES OF THE EYE AND RESPIRATORY TRACT IRRITANTS

1. Ammonia, NH_3 (chemical industry, coal-based chemical industry, petrolchemistry, metallurgical industry, electroplating, freezing industry; production and use of synthetic resins, dyeing industry, duplication procedures and so on).

2. Hydrochloric Acid, HCl (chemical industry, metal working, puckling, polyvinylchloride combustion and so on).

3. Sulphuric Acid, H_2SO_4, miliary stone of the chemical industry (chemical industry, metal working, pickling, battery charge and so on).

4. Fluorine, which in air originates immediately hydrofloric acid, HF which is present in glassworks, works with siliceous products, phosphorus industry, Al, Be, Zn metallurgy, electroplating industry, titanium white production, alkylation of gaseous fractions in oil refining, and in freons production.

5. Phosphoric Acid, H_3PO_4 (in glass industry, supherphosphates industry, petroleum industry, petrolchemistry, pickling of metal surfaces).

6. Nitric Acid, HNO_3 (chemical industry, electroplating, petrolchemistry, industry of artificial and synthetic resins, dyes industry and explosives, photoengraving).

7. Formaldehyde, HCHO (petrolchemical industry, synthetic resins industry, explosives industry; disinfectant; tanning industry) other aldehydes: acetic aldehyde, allylic aldehyde (acrolein: $CH_2=CH-CHO$).

8. Acetic Acid, CH_3COOH (petrolchemical industry, synthetic resins industry, artificial fibres industry, rubber industry, dyes industry, pharmaceutical industry, photographic industry); other organic acids.

9. Kentonic halogen derivatives: chloroacetone, alogeno-aceto-phenons (chemical industry).

TABLE 5

EXAMPLES OF DEEP AIRWAYS IRRITANTS

1. Phosgene, $COCL_2$, (Plastic industry, aromatic nitriles production for resins, elastomers and fibres; polycarbonates production for fibres and various manufactured articles; diisocyanate production for urethane foam; chemical industry as intermediate reagents and unwanted product in chlorinations. Welding, as unwanted product developed by workpieces degreased by trichloethylene.

2. Nitrous vapours, nitric oxide, NO, and dinitrogen trioxide (N_2O_3) originate nitrogen oxide (NO_2) + dinitrogen tetroxide (N_2O_4) (use of NHO_3 in nitrogen industry, in superphosphates industry, in propellers industry, metal working as a reagent and as a pickling agent, welding, chemistry, petrolchemistry, in artificial fibres and resins productions, in elsatomers production, in explosives and dyeing industry, unwanted product from nitrogen and oxygen welding).

3. Ozone, O_3 (arc welding, plama and flash welding due to catalytic activity of ultraviolet rays upon the atmospheric oxygen).

4. Sulphuric anhydride, SO_3 (H_2SO_4 production by catalytic methods, sulphonation for dyes and surface active compounds).

5. Cyanogen chloride (chemical industry).

6. Dimethylsulphate $(CH_3O)_2SO_2$ (chemical industry).

TABLE 6

EXAMPLES OF TRACHEO-BRONCHIAL IRRITANTS

1. Sulphur dioxide, SO_2, airborne pollutant from sulphur containing oils combustion: it may be oxidised to SO_3 which originates H_2SO_4 (metallurgical industry, chemical industry, petroleum industry, freezing industry, textile industry, paper industry, sugar industry, food industry)

2. Chlorine, Cl_2 (brine electrolysis, chemical industry, petrolchemistry, plastics industry, paper industry, textile industry, water conditioning) and other halogens (Br and I: chemical industry, photographic industry, dyes industry)

minimal doses, disturbances of even a mortal nature in a specific target. At times capable of penetrating even the unbroken skin, the occupational toxic substances typically penetrate by means of the respiratory system (and are therefore air pollutants) and, fortunately they are also at times provided with irritant action (Table 7).

As has already been said, a complete treatment of the air pollutants with toxic action would require an analysis of the entire range of occupational pathology, and I will therefore confine myself to a few examples, with particular emphasis on some of the general character-istics peculiar to occupational poisons (Table 8).

Treacherousness is one distinctive characteristic, which has always been recognized by our penal legislators, who - when they outline the aggravating circumstances leading to personal injuries - speak of "poison or other treacherous means", precisely to place poison on the same level as other means capable of injury, lurking in the shadows to catch the victim by surprise. I might mention, among the means capable of injury on the same level as poisons, the explosive packet sent by post and glass intimately mixed into food.

Carbon monoxide, is a typical air pollutant, which owes precisely to the absolute absence of action at the level of the surface exposed, the colourful, journalistic appellative of "invisible assassin", by which it is well-known both within and outside professional circles. At any rate, many poisons are provided with irritant action (Table 7).

The aggressive nature of poisons has always been known, and this is expressed in an injury which is typically of a serious nature, at times being mortal, and which manifests itself at a distance from the exposed surface, in the most diverse organs and systems: the nervous system, liver, kidneys, blood, cardiovascular system, etc.

As to the potency of a poison, the occupational doctor - aligned with the physicist - understands the work carried out by the poison in a unit of time; in practice, in an equal amount of time, the more potent the poison, the less time involved in producing the harmful result.

As regards the degree of danger involved, it has always been known that a poison can kill, but it must be emphasized that the harm is not necessarily acute. In other words: if some poisons kill within a few minutes, others kill after many years, because, for example, they

TABLE 7

PARTICULARY IRRITANT TOXIC COMPOUNDS

1. <u>Sulphydric acid</u> H_2S, (extractive industry, coking industry, hydrogen production; natural gas industry gaseous artificial fuels; petroleum industry, petrolchemistry; elastomers industry; production of sulphur dioxide from H_2S; chemistry, mainly of P and K; metallurgical industry of Fe, Pt, Ti, Ra; pigments industry, mainly of white titanium TiO_2, yellow CdS, chrome greens; gas welding viscose rayon industry; paper industry); irritant of the eye and of the respiratory tract; is an outstanding causative agent of oedema; firstly showing exciting effects on the CNS as well as cytocrome inhibition.

2. <u>Phosphine</u> PH_3 (calcium cyanamide industry; acetalic resins industry; moisture effect on Ca_3P_2 and on Zn_3P_2; gas welding): strongly oedemagenic is also causing important disturbances of the CNS.

3. <u>Epoxides</u>. 3.1 <u>Ethylene oxide</u> $CH_2\overset{O}{-}CH_2$ (propellers industry; petrolchemical industry; ethanolamines production; gycols and surface active agents production): irritant of the skin, eye and of the respiratory tract, is causing pulmonary oedema and neurotoxicity; in addition it may give cutaneous sensitization.

   3.2 <u>Propylene oxyde</u> $CH_3-CH\overset{O}{-}CH_2$ (surface active agents industry); see above, although it is less toxic.

   3.3 <u>Epichlorhydrin</u> $CICH_2-CH\overset{O}{-}CH_2$ (glicerine and exposy-resins production): irritant for the skin, eyes, lungs; it is also toxic for the kidneys.

   3.4 <u>Homologous longer chain epoxides</u>: sensitizers, irritants and neurotoxic.

4. <u>Products from pyrolysis of fluorinated polymers</u>. <u>Polymer fume fever</u> and similar reactions are caused by carbonyl fluoride COF_2 and other related compounds.

TABLE 8

PRIMARY CHARACTERISTICS OF OCCUPATIONAL POISONS

1   Treacherousness

2   Aggressiveness

3   Potency

4   Specificity

TABLE 9

CAUSES OF OCCUPATIONAL TUMORS

PHYSICAL AGENTS

Repeated trauma
Ultra violet irradiation
Ionising radiation

CHEMICAL AGENTS
(a)   Inorganic

Bryllium and its salts
Chromium and its salts
Metallic iron and haematite
Cobalt and its salts
Nickel and its salts
Arsenic and its salts
Codmium and its salts
Radioactive chemicals

(b)   Organic

Alphatic hydrocarbons and their chlorides
Aromatic hydrocarbons
Polycydic hydrocarbons
Aromatic amines
Azo compounds

(c)   Biological

Bihharzia
Herpes virus
DNA virus
Smallpox virus

develop a malignant tumor. Table 9 illustrates some of the most
typical oncotic air pollutants.

As to specificity, it is characteristic that individual poisons strike
individual targets and that workers employed in particular operations
will present particular clinical situations involving certain organs and
systems, just as it is characteristic that individual poisons are not
poisons for the entire living species. As an example, the hemlock
which kills the philosopher spares the parrot. I am in the habit of
warning my students not to turn into parrots to escape the hemlock,
since the parrot has poisons of its own. A pollutant which proves to
be poisonous to experimental animals may prove to be tolerated by man
and vice-versa. The greatest caution is necessary, therefore, in
transferring to man the results of such an experiment, and above all,
as we have said at the beginning, particular attention must be given to
epidemiological studies, especially with regard to poisons and to
oncotic poisons in particular. Tables 10 and 11 classify the major
inorganic air pollutants in which toxic action is recognized. There
are few pure toxic substances; that is, those which show no
manifestations of irritant and/or sensitizing action. As far as the
toxic air pollutants of organic nature are concerned, these are so
numerous that it is possible for me only to refer to them.

Table 12 indicates the occupations most exposed to risk and deals
with cases already well-known. Table 13 illustrates the subject of
bronchial asthma in relation to persons who work with chemicals,
textile workers (Table 14), hair stylists (Table 15) and health
workers (Table 16).

Table 17 deals with agricultural workers and animal experts.
I wish to render homage to my first teacher, my father, a brave
naturalist who was born here in bountiful Sicily and who, since 1932,
has pointed out that some pathogenic actions referring to plants and
vegetal derivatives (hay, wood, etc.) were in reality due to
saprophytes and parasites (acari in particular) as well as to their
relative predators (which are often other acari). In particular he
suggested the possible pathogenicity of excrements and remains of
both types of acari and even the noxiousness of the dust deriving
from such remains some time after death. My late lamented brother
and I, together with our father, took up those distant studies in 1961.
It is well known today that household dust and other types of dust
are sensitizing agents, above all when derived from the remains of
the two well known congeneric species of the genus Dermatophagoides,
the pteronyssinus TRT and the farinae HUGHES.

TABLE 10

ATMOSPHERIC   POLLUTANT (INORGANIC)

Beryllium salts
Inorganic carbon compounds
(eg carbon monoxide, carbon disulphide)
Nitrogen compounds
(eg HF, HCl)
Silicon compounds
(eg asbestos)
Phosphorus compounds
(eg phosphine, parathion, phosphonium compounds)

TABLE 11

ATMOSPHERIC POLLUTANTS (INORGANIC)

Sulphur compounds (e.g. SO_2)
Chromium compounds
Cobalt compounds
Nickel compounds
Zinc compounds
Arsenical compounds (e.g. Leurisite)
Antimony compounds
Mercury compounds
Lead compounds
Bismouth compounds
Radioactive compounds

TABLE 12

WORKERS MOST EXPOSED TO OCCUPATIONAL
AIRBORNE SENSITIZATION DISEASE

Housewives	Grocers
Agricultural workers	Textile industry workers
Millers and bakers	Barbers, hair cutters
Woodworkers and joiners	Peltry industry employees
Pharmaceutical industry employees	Health service employees
	Clerks
Plastics industry and dyes industry employees	Carriers
	Bricklayers

TABLE 13

Principal Agents Causing Occupational Asthma
in the Chemical Industry

Plastics (TDI)	Proteolytic enzymes
Varnishes	Phthalic anyhdride
Colourants	Antibiotes and their intermediates

TABLE 14

Principal Agents Causing Occupational Asthma
in the Textile Industry

Cotton	Silk
Linoleum	Sisal
Wool	Jute

TABLE 15

Principal Agents Causing Occupational Asthma
in Hairdressing

Furfural	Dermatophagoides
Vegetable gums	Vegetable essenses
Hair dyes	

TABLE 16

Principal Agents Causing Occupational Asthma
in Health Care

Antibiotics

Proteolytic enzymes (papain,
trypsa, bronchlin)

Adhesives

Fur in laboratory animals

TABLE 17

Principal Agents Causing Occupational Asthma
in Agriculture and Animal Husbandry

Cereals   Fowl

Insects   Fur

Furfutal

TABLE 18

Principal Agents Causing Occupational Asthma
in Housewives

Dermatophagoides (in house dust)

Proteolytic enzymes (detergents)

Pyrethrum

Table 18 makes another fitting acknowlededement: the homage to
the debt that even in this particular area of occupational pathology
is paid by that incomparable worker: the housewife. In addition to
the ever-present occupational risks - such as the hot-dry microclimate
of the kitchen stove, the most exhaustive imaginable work organization,
with evening and holiday duties, the household dust just-mentioned and
carbon monoxide - the housewife finds herself today having to confront
new risks (the physical risk being those of electricity or micro-wave,
chemical risks being those we have mentioned and still others), as our
houses come to resemble more and more dangerous factories and less
and less the oases of peace of which the poets sang. It seems to me
only fitting, then, to close my talk with due tribute to this unique form
of worker, typically functioning in a confined environment, who,
among other things has the stupefying amiability of not asking for pay.
No wonder that the full-time housewife is a particular form of worker
that is disappearing.

DRUG INDUCED HYPERSENSITIVITY AND PARAQUAT

Paul Smith

Department of Pathology
University of Liverpool
England

To date there are well over 40 drugs or therapeutic agents which are known to produce pathological changes in the lungs of humans (Rosenow, 1972). In addition there are several other agents which will induce changes in the lungs of animals for which a human association has not been demonstrated. In this paper I shall confine myself to five of these, namely busulphan, bleomycin, chlorphentermine, iprindole and paraquat.

BUSULPHAN

Busulphan is an alkylating agent which is used in the treatment of chronic myeloid leukaemia. Pulmonary side effects occur in quite a high proportion of cases, thus in one survey, 6 out of 14 patients receiving busulphan therapy showed histological changes in their lungs (Heard and Cooke, 1968). The most prominent feature of these changes is a fibrinous exudate into the alveoli and respiratory bronchioles. This becomes organised to produce an intra-alveolar fibrosis. Sometimes the organising fibrin becomes incorporated into the alveolar walls to produce interstitial pulmonary fibrosis. In the large majority of cases this pulmonary fibrosis is focal and produces only mild impairment of respiratory function, however, a small minority of patients develop a more extensive and fatal pulmonary fibrosis. The following description concerns such a case (Littler et al., 1969).

The patient was a 61 year old male who had been treated for chronic myeloid leukaemia with busulphan for nearly two years. He developed progressive dyspnoea and died in respiratory failure. On gross examination the lungs were diffusely involved by fibrosis but this was particularly dense

at the apices. Histologically a prominent feature was an eosinophilic debris
within the alveolar spaces. Much of this material was fibrin but it also
contained cellular debris and desquamated alveolar epithelial cells. In some
alveoli this debris had become organised to produce intra-alveolar fibrosis.
In other regions of the lungs, especially at the apices, the fibrosis was
predominantly interstitial in nature, the alveolar walls being greatly thickened by collagen, fibroblasts and chronic inflammatory cells. In these areas
of interstitial fibrosis the alveoli were lined by a continuous, cuboidal
epithelium composed of granular pneumocytes. There were also desquamated
granular pneumocytes within the alveoli, the whole histological picture
conforming to that of fibrosing alveolitis. Hyperplasia of granular pneumocytes is not pathognomonic for busulphan lung but represents a non-specific
reaction of the lung to injury. Moderate numbers of granular pneumocytes
scattered about the alveolar walls are a normal feature of the lung. Their
function is to synthesise and secrete the phospholipid surfactant substance
which is essential to avoid collapse of the alveoli. Under the electron
microscope the cells are seen to contain characteristic lamellated or whorled
structures consisting of this surfactant. These organelles are referred to as
lamellar bodies. When the membranous pneumocytes or type I epithelial
cells are damaged the granular pneumocytes multiply, elongate and recover
the alveolar walls. This regenerative phenomenon occurs in a wide variety
of pulmonary lesions, including busulphan lung.

An unusual feature of the case described above and those described by
Heard and Cooke (1968) was the presence of large bizarre cells lining some
of the alveoli. These had eosinophilic, vacuolated cytoplasm with large,
hyperchromatic nuclei. Electron microscopy has shown them to be greatly
enlarged granular pneumocytes (Littler et al., 1969).

BLEOMYCIN

The pathogenesis of busulphan lung is not known with complete certainty.
However, much more is known about another drug which produces an almost
identical pathology. This drug is bleomycin, a cytotoxic drug isolated from
Streptomyces verticillus. It is used extensively in the chemotherapy of
squamous cell carcinoma and lymphomas. In one study, 12 out of 35 patients
receiving bleomycin showed a pulmonary pathology similar to that of busulphan
lung and in 10 of these patients there had been clinical evidence of respiratory
distress (Luna et al., 1972).

Experiments in which large doses of bleomycin have been given to mice
and dogs have revealed the following pathogenesis of bleomycin lung.
(Fleischman et al., 1971 ; Adamson and Bowden, 1974 ; Aso et al., 1976 ;
Jones and Reeve, 1978). Firstly there is degeneration and swelling of

endothelial cells, both in the alveolar capillaries and muscular pulmonary arteries. This is followed by degeneration, ballooning or even necrosis of the alveolar type I epithelial cells or membranous pneumocytes. As a consequence of this damage a fibrinous exudate accumulates in the alveoli, which, with the passage of time, becomes organised into intra-alveolar and interstitial fibrosis. The damaged alveolar epithelium then regenerates by a hyperplasia of granular pneumocytes, some of which may desquamate into the airways. Some granular pneumocytes undergo dysplasia to produce huge, bizarre cells. This last change appears to be caused by the direct cytotoxic action of bleomycin which is known to interfere with DNA synthesis. It is not a normal regenerative process and appears to be peculiar to bleomycin and busulphan lung.

CHLORPHENTERMINE

During the years 1967 - 1970 considerable interest centred upon the possible pulmonary side-effects of anorexigenic drugs. This is because there was strong epidemiological evidence that one of them, aminorex fumarate (Menocil), was the cause of an epidemic of primary pulmonary hypertension in Austria, Switzerland and Germany. As a result of this apparent association, other anorexigens were suspected of inducing pulmonary hypertension. One of these was chlorphentermine, a drug sold under the name of lucofen but which has now been withdrawn from the market. Experiments with various animal species have failed to demonstrate any evidence of hypertensive pulmonary vascular disease after administration of either aminorex or chlorphentermine. Nevertheless, chlorphentermine is associated with striking changes within the lung parenchyma of rats. At the light microscopic level the most obvious of these changes is the presence of numerous, large mononuclear cells within the alveolar spaces. These cells have a moderately eosinophilic, finely granular or vacuolated cytoplasm and they may be multinucleated (Heath et al., 1973). Under the electron microscope these intra-alveolar cells can be seen to contain numerous lysosomes demonstrating that they are macrophages (Smith et al., 1973). They also contain large numbers of electron-dense lamellar inclusions reminiscent of the lamellar bodies of granular pneumocytes. Some macrophages are so distended with these structures that other cytoplasmic organelles are hard to find. The lamellar inclusions consist of phospholipid, a material that also accumulates in large quantities freely within the air spaces. In this location, however, it is often in a different physical state, adopting a lattice-like array of parallel tubes called tubular myelin (Weibel et al., 1966). This is the phospholipid pulmonary surfactant which is normally responsible for reducing the surface tension of the alveolar lining and hence preventing pulmonary collapse. In normal lungs it is present in small quantities, but following chlorphentermine administration enormous quantities of this surfactant are liberated into the alveoli. Macrophages then migrate into the airways and attempt to phagocytose all this phospholipid. However their cytoplasm becomes distended with this material and they eventually die and disintegrate. Under the light microscope this appears as focal accumulations of faintly eosinophilic granular debris often associated with fragmenting macrophages. Under the electron microscope one sees cellular debris and lamellar inclusions mixed with the tubular myelin surfactant.

It seems reasonable to assume that the origin of all this phospholipid surfactant is the granular pneumocyte since this is the cell in which it is normally synthesised. However, lamellar inclusions are found in a whole variety of cells uninvolved with surfactant metabolism. These include capillary endothelium, membranous pneumocytes, goblet cells, ciliated cells and the Clara cells. This may represent a passive intake of phospholipid from the airways or alternatively it may be that chlorphentermine interferes with the metabolism of these cells. The latter possibility is supported by the fact that chlorphentermine stimulates an excessive production of phospholipid in other organs (Lüllmann et al., 1975). Furthermore when chlorphentermine is added to cultured macrophages in vitro, numerous lamellar inclusions appear in their cytoplasm (Drenckhahn et al., 1976).

There remains the question, what happens to the alveolar debris with the passage of time? Does it simply accumulate within the alveoli, or does it become organised to produce interstitial and intra-alveolar fibrosis like busulphan and bleomycin lung? To answer this question we administered chlorphentermine in the drinking water to rats for periods up to one year (Smith et al., 1974a). In this experiment the alveoli became totally occluded by enlarged macrophages which had a pale vacuolated cytoplasm with a foamy appearance. These cells were so numerous that by 4 months they distended the alveoli and caused consolidation of large areas of the lungs. By 10 months most of the macrophages had disintegrated to produce a granular, faintly eosinophilic debris filling many of the alveoli. The histological picture was identical to that occurring in human lungs in a disease called alveolar proteinosis. There was no organisation of this debris or interstitial fibrosis even after a year. The electron microscope showed that this debris consisted largely of lamellar inclusions of various sizes mixed with small quantities of cellular debris. There was no fibrin (Smith et al., 1974a). This difference in chemical composition of the alveolar debris in chlorphentermine lung from that in busulphan lung may explain the lack of pulmonary fibrosis in the former. Indeed the end-stage of chlorphentermine lung should, perhaps be more aptly described as an "alveolar phospholipidosis".

IPRINDOLE

Similar changes to the above are produced by the drug iprindole. This is chemically and pharmacologically different from chlorphentermine, being a tricyclic anti-depressant sold under the name of Prondol. When this drug is administered to rats it too causes an outpouring of phospholipid-laden macrophages into the alveoli. (Vijeyaratnam and Corrin, 1972). It also causes lamellar inclusions to appear in a variety of other cells (Vijeyaratnam and Corrin, 1974). Also in common with chlorphentermine the end-stage of this process is a histological pattern like that of alveolar proteinosis (Vijeyaratnam and Corrin, 1973). Recent studies indicate that chlorphentermine and iprindole belong to a family of drugs which influence phospholipid metabolism in many types of cells (Drenckhahn et al., 1976). All these drugs share the property of having hydrophobic and hydrophilic groups in close proximity within the molecule; that is they are amphipathic. Collectively these drugs are responsible for producing phospholipidosis in various organs of rats but there is no evidence that either chlorphentermine or iprindole induce similar changes in humans.

PARAQUAT

Unlike the other substances described in this paper, paraquat is not a drug but a herbicide. It is sold in two principal forms. One of these is a concentrated solution called Gramoxone, sold only to farmers and market gardeners. The other is a diluted, granular form called Weedol which is for domestic use. Weedol is comparatively safe but Gramoxone is highly potent when ingested.

People illicitly obtain samples of Gramoxone from horticulturist acquaintances and store it in lemonade, beer or wine bottles which they leave in the garage or garden shed. In this way the weedkiller may be accidentally drunk by children or even unwary adults. One mouthful is often all that is required for a fatal outcome. More recently paraquat in the form of Gramoxone has been used increasingly commonly as a means of committing suicide. Although paraquat produces focal degenerative lesions in the liver and kidney, its main target organ is the lung.

Pathogenesis of Paraquat Poisoning

The pulmonary changes in paraquat poisoning can conveniently be divided into two stages, a destructive phase followed by a proliferative phase. (Smith and Heath, 1976). The sequence of events which occur during these two phases has been studied in detail in experimental rats. In this animal the first change detected by the electron microscope occurs approximately 6 hours after an intraperitoneal injection of paraquat. (Smith and Heath 1974a). This consists of a swelling of the type I alveolar epithelial cells or membranous pneumocytes, associated with a loss of electron density of their cytoplasm. By 18 hours these degenerative changes are pronounced and are associated with balloon-like swellings which project into the alveolar spaces. In these balloons the cytoplasm is very pale with disruption of mitochondria and endoplasmic reticulum (Vijeyaratnam and Corrin, 1971 ; Smith and Heath, 1974a). By 24 hours after paraquat administration, epithelial balloons are numerous and the granular pneumocytes also show early degenerative changes in the form of mitochondrial swelling and loss of the contents of their lamellar bodies. So extensive is the damage to the alveolar epithelium that by 2 days the epithelial balloons have ruptured leaving large gaps in epithelial continuity. Eventually these remaining islands of cytoplasm slough off leaving the capillary endothelial cells completely devoid of an epithelial covering. Surprisingly these denuded capillaries show no evidence of damage (Vijeyaratnam and Corrin, 1971 ; Smith and Heath, 1974a ; Sykes et al., 1977).

, In humans, the destructive phase occurs between 3 and 7 days after ingesting paraquat (Smith and Heath, 1976). Although the early stages have been available for study only on rare occasions, they appear to involve degeneration and disintegration of the alveolar epithelium as in the rat (Toner et al., 1970). In both species this is associated with alveolar capillary congestion, pulmonary oedema, pulmonary collapse and the presence of eosinophilic hyaline membranes within respiratory bronchioles (Smith and Heath, 1976). In man the destructive phase of paraquat poisoning can have three possible sequelae depending upon the dose ingested and the susceptibility of the individual. Firstly it can prove fatal in itself. This is particularly so where large doses have been taken such as in suicides. However, even with relatively small doses, susceptible individuals may die during the first week. Alternatively, the patient may recover with complete resolution of signs and symptoms. Such resolution can occur spontaneously but is much more likely if curative measures are instituted rapidly. All too frequently, however, a short period of apparent well-being deteriorates as dyspnoea of increasing severity culminates in death between 2 and 4 weeks after ingesting the poison. This dyspnoea heralds the onset of the proliferative phase.

The hallmark of the proliferative phase is pulmonary fibrosis. The development of this lesion has been extensively studied in animals and commences with an infiltration into the alveoli of numerous cells with a primitive, non-specialised appearance. Under the electron microscope these cells are seen to be rounded in shape with numerous, long, filamentous pseudopodia at their periphery (Smith et al., 1974b). They have a large nucleus, displaced to one pole, and their cytoplasm contains many small mitochondria and sparsely distributed rough endoplasmic reticulum comprising short, parallel, unbranched membranes with narrow cisternal spaces. At first glance these cells may be mistaken for macrophages, but in fact they contain few, if any, lysosomes. Studies of the behaviour of these cells indicate that they are mesenchymal in nature representing primitive fibroblasts. For this reason the term "profibroblast" has been coined for them (Smith et al., 1974b). Soon after entering the lung the profibroblasts commence differentiation into mature fibroblasts. To achieve this they lose their pseudopodia and their mitochondria enlarge and become more prominent. At the same time the rough endoplasmic reticulum becomes more extensive and branched. These intermediate forms then elongate and the cisternae of their rough endoplasmic reticulum dilates to produce the characteristic features of mature, metabolically-active, fibroblasts. These processes occur exclusively within the alveolar spaces so that the resulting fibrosis is entirely intra-alveolar in nature (Smith, et al., 1974b; Kelly et al., 1978). As the fibrosis progresses the alveolar walls may fragment so that the alveolar capillaries become engulfed by fibroblastic tissue. The resulting histological picture is a dense, cellular, intra-alveolar fibrosis with obliteration of the alveolar architecture (Smith and Heath, 1976).

In humans the pathogenesis of pulmonary fibrosis in paraquat lung is not so clearly established, mainly because we are usually examining the terminal stage of the disease. In a few cases, however, the integrity of the alveolar walls persists thereby acting as markers for the location of the fibrosis. In these cases the fibrosis can be seen to be exclusively intra-alveolar in nature (Copland, Kolin, and Shulman, 1974; Smith and Heath 1974b). In the majority of cases, however, the alveolar walls are so disrupted and the fibroblastic tissue so dense, that the origin of the fibrosis cannot be determined with certainty.

It should also be borne in mind that the above pulmonary pathology may be further complicated by oxygen toxicity. Many patients develop severe dyspnoea and require oxygen therapy to survive. It is now clearly established that prolonged administration of high concentrations of oxygen can induce both interstitial and intra-alveolar fibrosis on its own. This may explain the observation of both these types of pulmonary fibrosis in some cases of paraquat poisoning (Dearden, et al., 1978; Rebello and Mason, 1978).

The pathogenesis of paraquat lung differs from that of busulphan or bleomycin lung in that it does not involve organisation of an alveolar exudate, at least when uncomplicated by oxygen therapy (Smith and Heath 1974b and 1976). Thus the pulmonary fibrosis in paraquat lung is much more cellular than that due to organisation and develops considerably more rapidly.

Secondly, repeated small doses of paraquat to animals result in a negligible alveolar exudate but an extensive fibrosis still ensues. Thirdly inhalation of an aerosol of paraquat causes extensive pulmonary damage and exudate but pulmonary fibrosis never develops after this treatment (Gage, 1968). It may be that a metabolite of paraquat is responsible for stimulating the infiltration of profibroblasts into the lung.

The Biochemistry of Paraquat Toxicity

The biochemical mechanisms underlying the toxic action of paraquat have been extensively studied both in plants and animals. The key to its toxicity lies in its unusual chemical properties.

The paraquat molecule consists of two pyridine rings linked together. It thus belongs to a group of compounds called bipyridyls. The two nitrogen atoms are in the para position and are in a quaternary state of substitution ; hence the common designation of para - quat. In its normal state it is ionised with a positive charge on each of the two nitrogen atoms. This ion is colourless in solution. The paraquat ion will readily accept an electron to produce a free radical which is stable. This reaction is indicated below in which the paraquat molecule is represented by the letters PQ.

$$PQ^{2+} + e^- \rightarrow PQ^+.$$

The paraquat free radical is only permanently stable in the absence of oxygen. In the presence of molecular oxygen it is converted back to the paraquat ion liberating hydrogen peroxide.

$$2PQ^+ + O_2 + 2H^+ \rightarrow H_2O_2 + 2PQ^{2+}.$$

At one time it was thought that the hydrogen peroxide was responsible for the toxicity of paraquat but more recently it has been shown that the oxidation of the free radical has at least two important intermediate stages represented below.

$$PQ^+ + O_2 \rightarrow PQ^{2+} + O_2^- \text{ (superoxide ion).}$$

$$PQ^+ + O_2^- \rightarrow PQ^{2+} + O_2^{2-} \text{ (peroxide ion).}$$

$$O_2^{2-} + 2H^+ \rightarrow H_2O_2 \qquad \text{(hydrogen peroxide).}$$

Both the superoxide and peroxide ions are highly unstable and will react with and polymerise unsaturated fatty acids by a process called lipid peroxidation (Bus et al., 1977). In the cell, lipid peroxidation of cellular

membranes will cause disruption of cytoplasmic organelles and cell death.
There is now good evidence that this process is responsible for the toxic
action of paraquat in both plants and animals. In plants it destroys the
thylakoids of the chloroplast but in animal cells its action seems to be more
generalised, involving a variety of cell membranes.

When this mechanism was first discovered it was not clear why it was
confined almost exclusively to the lungs. It was then discovered that
although the concentration of paraquat in the blood and systemic organs
rapidly declines as it is excreted, its concentration in the lungs steadily
increases (Rose et al., 1974). Furthermore, this process takes place up a
gradient of concentration by what appears to be a mechanism of active
transport (Rose et al., 1974 and 1976). As a result of such experiments it
emerges that paraquat selectively attacks the lungs simply because it is
selectively taken up by the lungs. The mechanism of action of paraquat on
the lung is shown diagramatically in the figure. Once in the alveolar
epithelium the paraquat ion is able to accept electrons from the electron
transport chain and become reduced to its free radical. The alveolar
epithelium is in direct contact with molecular oxygen which reconverts the
paraquat free radical to the paraquat ion. Each time this occurs superoxide
and peroxide ions are formed. These attack the cell membranes by polymerising
and peroxidising their unsaturated fatty acids. As a result the sodium pump is
impaired and the alveolar epithelium swells up with water until it ruptures.

All cells contain an enzyme, superoxide dismutase, which normally
removes the small quantities of superoxide that may be generated during
oxidative metabolism. In the presence of paraquat, however, this enzyme
is saturated permitting the superoxide ion to reach toxic levels of concentration.
Note that in this reaction paraquat is acting merely as a catalyst, being
alternately reduced and oxidised cyclicly. In this way small quantities of
paraquat can generate large amounts of superoxide ion. This explains why
such small doses of paraquat can prove fatal.

An understanding of the biochemical mechanism underlying paraquat
toxicity provides us with ammunition to combat the disease. One such line
of attack would be to increase the level of superoxide dismutase. This has
been done experimentally in rats with a moderate degree of success (Autor,
1974). Unfortunately, to be effective, the enzyme has to be administered
before paraquat which is of no value in human cases of paraquat poisoning.
An alternative would be to reduce the oxygen tension in the lung in an
attempt to supress the oxidation of the paraquat free radical. Once again
this affords some degree of protection in animals (Rhodes et al., 1976) but
appears to be of limited value in humans. It is probable that the partial
pressure of oxygen has to be very low before it limits the rate of the reaction.

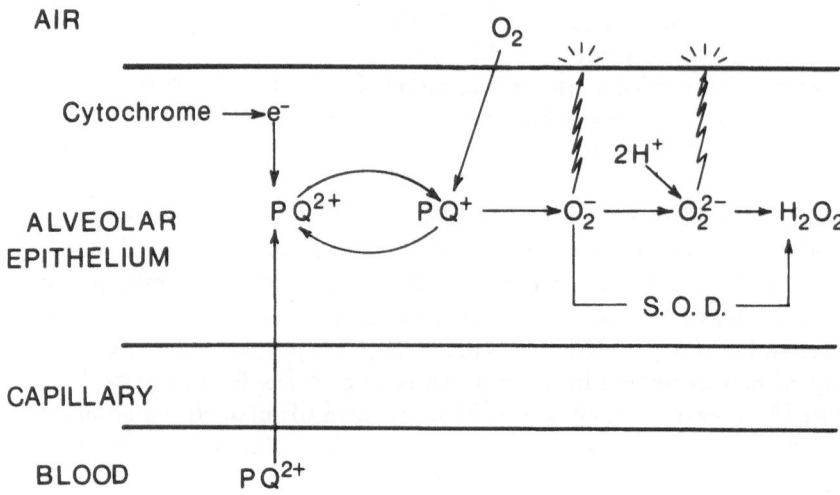

The diagram represents a section through part of an alveolar wall with air on the outside and capillary blood on the inside. Paraquat ions (PQ^{2+}) circulate in the blood stream and are taken up into the alveolar wall by a mechanism of active transport. Once in the alveolar epithelium they are involved in a catalytic oxidation-reduction cycle generating superoxide and peroxide ions as described in the text. The destructive action of these ions on the cell membrane is represented by jagged lines. Superoxide ions are removed by the action of superoxide dismutase (S.O.D.) which converts them to hydrogen peroxide.

A third line of attack would be to slow down the process of lipid peroxidation. Certain endogenous anti-oxidants such as vitamin E, glutathione peroxidase and selenium will do this (Bus et al., 1977) but none of these seems to be of much therapeutic value. The only remaining course of action is to prevent the paraquat being taken up by the lungs in the first instance, and this involves removing it from the body as rapidly as possible, an approach which is now universally adopted. Therapy thus involves a combination of gastric lavage with colloidal suspensions of Fuller's earth, forced diuresis and possibly haemodialysis. Past experience has shown that if this regime is commenced within 24 hours after ingestion a cure may be effected and if this delay is reduced to 10 hours the prognosis is good (Cavalli and Fletcher, 1977).

REFERENCES

Adamson, I.Y.R. and Bowden, D.H., 1974, The pathogenesis of bleomycin-induced pulmonary fibrosis in mice, Am. J. Pathol., 77:185.

Aso, Y., Yoneda, K., and Kikkawa, Y., 1976, Morphologic and biochemical study of pulmonary changes induced by bleomycin in mice, Lab Invest., 35: 558.

Autor, A.P., 1974, Reduction of paraquat toxicity by superoxide dismutase, Life Sci., 14:1309.

Bus, J.S., Aust, S.D., and Gibson, J.E., 1977, Lipid peroxidation as a proposed mechanism for paraquat toxicity, in: "Biochemical Mechanisms of Paraquat Toxicity,"
A.P. Autor, ed., Academic Press, London.

Cavalli, R.D., and Fletcher, K., 1977, An effective treatment for paraquat poisoning, in: "Biochemical Mechanisms of Paraquat Toxicity,"
A.P. Autor, ed, Academic Press, London.

Copland, G.M., Kolin, A., and Shulman, H.S., 1974, Fatal pulmonary intra-alveolar fibrosis after paraquat ingestion, New Eng. J. Med., 291:290.

Dearden, L.C., Fairshter, R.D., McRae, D.M., Smith, W.R., Glauser, F.L., and Wilson, A.F., 1978, Pulmonary ultrastructure of the late aspects of human paraquat poisoning, Amer. J. Path., 93:667.

Drenckhahn, D., Kleine, L., and Lüllmann-Rauch, R., 1976, Lysosomal alterations in cultured macrophages exposed to anorexigenic and psychotropic drugs, Lab. Invest., 35: 116.

Fleischman, R.W., Baker, J.R., Thompson, G.R., Schaeppi, U.M., Illievski, V.R., Cooney, D.A., and Davies, R.D., 1971, Bleomycin induced interstitial pneumonia in dogs, Thorax, 26 : 675.

Gage, J.C., 1968, Toxicity of paraquat and diquat aerosols generated by a size-selective cyclone effect of particle size distribution, Brit. J. Industr. Med., 25: 304.

Heard, B.E., and Cooke, R.A., 1968, Busulphan lung, Thorax, 23:187.

Heath, D., Smith, P., and Hasleton, P.S., 1973, The effects of chlorphentermine on the rat lung. Thorax, 28:551.

Jones, A.W., and Reeve, N.L., 1978, Ultrastructural study of bleomycin-induced pulmonary changes in mice, J. Path., 124: 227.

Kelly, D.F., Morgan, D.G., Darke, P.G.G., Gibbs, C., Pearson, H., and Weaver, B.M.Q., 1978, Pathology of acute respiratory distress in the dog associated with paraquat poisoning, J. Comp. Path., 88:275.

Littler, W.A., Kay, J.M., Hasleton, P.S., and Heath, D., 1969, Busulphan lung, Thorax, 24:639.

Lüllmann, H., Lüllmann-Rauch, R., and Wassermann, O., 1975, Drug-induced phospholipidoses, C.R.C. Crit. Rev. Toxicol., 4: 185.

Luna, M.A., Bedrossian, C.W.M., Lichtiger, B., and Salem, P.A., 1972, Interstitial pneumonitis associated with bleomycin therapy, Am. J. Clin. Path., 58: 501.

Rebello, G., and Mason, J.K., 1978, Pulmonary histological appearances in fatal paraquat poisoning, Histopathology, 2: 53.

Rhodes, M.L., Zavala, D.C., and Brown, D., 1976, Hypoxic protection in paraquat poisoning, Lab. Invest., 35: 496.

Rose, M.S., Lock, E.A., Smith, L.L., and Wyatt, I., 1976, Paraquat accumulation: Tissue and species specificity, Biochem. Pharmacol., 25: 419.

Rose, M.S., Smith, L.L., and Wyatt, I., 1974, Evidence for energy-dependent accumulation of paraquat into rat lung, Nature (Lond.), 252:314.

Rosenow, E.C., 1972, The spectrum of drug induced pulmonary disease. A review, Ann. Intern. Med., 77:977.

Smith, P., and Heath, D., 1974a, The ultrastructure and time sequence of the early stages of paraquat lung in rats, J. Path., 114:117.

Smith, P., and Heath, D., 1974b, Paraquat lung : A reappraisal, Thorax, 29:643.

Smith, P., and Heath, D., 1976, Paraquat, C.R.C. Crit. Rev. Toxicol., 4: 411.

Smith, P., Heath, D., and Hasleton, P.S., 1973, Electron microscopy of chlorphentermine lung, Thorax, 28:559.

Smith, P., Heath, D., and Hasleton, P.S., 1974a, Effects of prolonged administration of chlorphentermine on the rat lung, Path. Europ., 9: 273.

Smith, P., Heath, D., and Kay, J.M., 1974b, The pathogenesis and structure of paraquat-induced pulmonary fibrosis in rats, J. Path., 114:57.

Sykes, B.I., Purchase, I.F.H., and Smith, L.L., 1977, Pulmonary ultrastructure after oral and intravenous dosage of paraquat to rats, J. Path., 121:233.

Toner, P.G., Vetters, J.M., Spilg, W.G.S., and Harland, W.A., 1970, Fine structure of the lung lesion in a case of paraquat poisoning, J. Path., 102:182.

Vijeyaratnam, G.S., and Corrin, B., 1971, Experimental paraquat poisoning: A histological and electron-optical study of the changes in the lung, J. Path., 103:123.

Vijeyaratnam, G.S., and Corrin, B., 1972, Pulmonary histiocytosis simulating desquamative interstitial pneumonia in rats receiving oral iprindole, J. Path., 108:105.

Vijeyaratnam, G.S., and Corrin, B., 1973, Pulmonary alveolar proteinosis developing from desquamative interstitial pneumonia in long term toxicity studies of iprindole in the rat. Virch. Arch. Path. Anat 351:1.

Vijeyaratnam, G.S., and Corrin, B., 1974, Fine structural alterations in
    the lungs of iprindole-treated rats, J. Path., 114:233.
Weibel, E.R., Kistler, G.S., and Töndury, G., 1966, A stereologic electron
    microscopic study of "tubular myelin figures" in alveolar fluids of rat
    lungs, Z. Zellforsch., 69:418.

DISCUSSION

LECTURER: Smith   CHAIRMAN: Cumming

DENISON: The superoxide ion is very destructive and can
                destroy every other molecule that it encounters, and
                I am surprised that there is not a lot of systemic
                damage during the time that the paraquat is getting
                to the lung. Perhaps that implies that the union of
                paraquat and electrons to generate the superoxide ion
                is quite a slow process. Do you know anything about
                that?

SMITH: I do not know the precise time interval that is
                required. There is of course a certain amount of
                systemic damage in paraquat poisoning, particularly
                during the early phase. It is hepatotoxic, not as
                toxic as it is to the lung, but it does produce a
                certain amount of degeneration or even death of cells
                in the liver. As it is being excreted through the
                kidney it does produce damage to the renal tubule,
                and indeed there is recent evidence that it is to
                some extent selectively taken up by the brain, not
                quite as avidly as by the lung, but it does seem to
                have an affinity for paraquat, and this has produced
                cerebral damage as a consequence. It is not entirely
                restricted to the lung, but I assume that is because
                the lung has a high molecular oxygen concentration,
                once paraquat gets into the lung it produces a large
                quantity of super oxide ions, which are, as you say,
                highly destructive. But the precise time interval
                taken in order to generate them I am not familiar
                with.

JEFFERY: I wonder if I might refer back to a point raised by
                Heath on the difficulty of distinguishing between a
                granular pneumonocyte, a type II pneumonocyte, and a
                macrophage. You point out that in the macrophage one
                notes the presence of lysosomes, while in the type II
                cells these are absent. I wonder if I can suggest
                that there is another way of distinguishing these two
                cells, and that is by looking at the mitochondria.
                The mitochondria in the type II pneumonocyte are
                unusual in that they are so large, they are at least
                twice the size of those found in the macrophage, and
                have very well developed crestae, and this I think

might be a rather simpler way of distinguishing between the two cell types. Have you looked at this?

SMITH: I have looked at large numbers of both types of cells, and I would agree with you that, as well as the presence or absence of lamellar inclusions and whether or not there are lysosomes present, there are other differences. There are differences in the mitochondria, there are differences, to some extent, in the Golgi apparatus, and I think that the granular pneumocyte has this sort of microvillous border, whereas the macrophage would have a rather convoluted mass of pseudopodia. These are EM differences, but the problem is with light mocroscopy. You can often be pretty sure what it is that you are looking at, but you can never be absolutely certain, unless you do histochemistry or look at these things with the elctron microscope. I am not saying that pathologists never really know whether a cell is a macrophage or a granular pneumocyte, but to be absolutely categorical about it you really need to look at the thing under the electron microscope, or do enzyme histochemistry.

WIDDICOMBE: A few years ago I wanted to use paraquat in order to produce experimental lung disease in rabbits, and I was told what I should have known and I suspect you know, that if you give it to rabbits they do not get lung disease because they die of kidney failure long before the lung is sufficiently affected. This leads to a double question: firstly, do any of the other drugs that you have described this morning also have specific effects in the sense of acting differently on different species, as paraquat seems to do. We know that in some species paraquat acts more on the kidney as in the case of rabbits, on the liver, as in the case of other species and on the lung in the case of rats and men, does this give you any indication about how the drug might be acting? Because there is not a great difference in oxygen tension between these organs, so presumably there must be some difference in some kind of membrane pumping system which you described.

SMITH: The rabbit is, as you say, unusual in this respect, in that paraquat seems to affect the kidney much more severely and earlier than other species. I do not know, to be quite honest with you, why that is. It is perhaps just an intrinsic property of the rabbit kidney that it is not easily damaged by certain toxic substances, I am not really sure. Certainly it is almost impossible to produce fibrosis in the lungs of rabbits. As far as the reason for this idiosyncratic difference is concerned, I really do not know.

WIDDICOMBE: Could I just make a very quick statement. It is possible to produce fibrosis in the lung of a rabbit with, for example fatty acid emboli, but you do not seem to be able to do it with paraquat.

SMITH: I think all attempts failed miserably. In fact, if you give a reasonable dose calculated to produce fibrosis they die suddenly and in some cases rather inexplicably, with rather little changes in the lung, and one assumes that this is due to damage elsewhere in the systemic circulation, perhaps cerebral damage.

SANT'AMBROGIO: All the substance that you have considered were either injected or ingested, and you also considered inhaled paraquat All these substances eventually reach the lungs from the blood. The lung, which means intrapulmonary airways, alveoli and all the corresponding vascular structure, is very well perfused. Do you have any data on the extrapulmonary airways, in which there is oxygen on the luminal side but not much perfusion, at least not as much as in the lungs themselves? Is there a specific pathology for the extrapulmonary airways?

SMITH: As far as I know, paraquat, unless inhaled, does not produce damage to the extrapulmonary airways. If it is inhaled then it does, but by the normal route, which is the oral route or occasionally by injection, it is not the extrapulmonary airways which show damage.

COHEN: A few comments and then questions. Firstly, with regard to paraquat poisoning, I believe the kidney toxicity to be particularly in man, because if there is kidney toxicity then the paraquat cannot be cleared and then there is more uptake into the lung thus exaggerating the pulmonary toxicity. The kidney toxicity is therefore an important facet of the overall situation. In relation to the mechanism proposed, if it is possible to put on the last slide some work from ICI, as well as other evidence in the literature, suggests that perhaps superoxide is not the mechanis of paraquat toxicity and I do not think that we should necessarily have a closed mind on that subject. What they have shown is that there is an accumulation of paraquat into the lung and probably into the type II cell, and that is an extremely important facet in the toxicity. When paraquat is taken into the lung there is an immediate increase in pentose phosphate shunt activity which will lead to an increase in NADPH production, which will in turn stimulate the reduction of paraquat and production of superoxide ions. What they have suggested, and have some data to support, is that it is the ratio of NADPH to NADP level, which might be important in this toxicity. Two of the points which argue against the mechanism proposed on that slide is that there is degeneration of the superoxide anion, O_2^- which then causes lipid peroxidation. The two pieces of data against this being the mechanism of toxicity are that in vitamin E deficient animals there is no exaggeration of pulmonary toxicity when given paraquat, or a very very small increase. The major mechanism of dealing with peroxides is by glutathione peroxidase, and in animals with altered glutathione peroxidase there is no increase in paraquat toxicity. Another aspect which has to be considered is that there are differences between oxygen toxicity and paraquat toxicity. If we assume that oxygen toxicity is mediated by superoxide anions, one might expect more similarity between the toxicities of oxygen and paraquat.

SMITH: I was interested to hear you talking about vitamin E, incidentally. This has been tried experimentally in animals in an attempt to abolish the lipid peroxidation mechanism, and it has appeared to produce some ameliorating effect. It certainly has not completely blocked this reaction at all, it has only suppressed it to some extent. Not being a biochemist, I do not know how effective vitamin E ought in theory to be and whether one would anticipate that it would completely prevent that reaction or whether a slight reduction is all that one could expect. As far as oxygen toxicity is concerned, there are some slight differences between the two. Paraquat, as we have seen, affects the alveolar epithelium, the capillary showing no evidence of damage at all, and this may well be why we see a rather watery, edematous exudate as water is drawn out of the capillary by osmotic forces. In oxygen poisoning one gets damage both to the alveolar epithelium and the capillary; one can see both being damaged at the ultrastructural level. Consequently there is much more fibrin liberated, but both cells are involved. The reason why this occurs in paraquat poisoning, just in this cell, might simply be because that is where it is being localised, whereas in oxygen toxicity both epithelium and capillary being bathed in a high oxygen tension give rise to damage in both types of cell. Certainly there does seem to be this difference in distribution in the cellular damage between paraquat poisoning, affecting only alveolar cells and oxygen toxicity which tends to effect both.

CUMMING: Turning to the chlorphentermine story, I was very interested in looking at what you call microtubular bodies and I would like to suggest an alternative hypothesis which may result in this. It is well known to colloid chemists that if you take a test tube and fill it with a gel containing sodium chromate and you put onto the top of this solid gel silver nitrate solution, then after a short interval, there is a precipitation of brown silver chromate. Then it stops precipitating and there is a blank, and then a second layer, and so on, the phenomenon of periodic precipitation, which has the characteristic that the intervals slowly increase. This is a characteristic of proteinaceous material through which there is diffusing another substance and I suggest that this may be a possible explanation for what you called microtubules.

SMITH: I think that is very interesting. At the moment there is only one paper, so far as I know, which actually describes or studies these things quantitatively to find out whether they are a lattice or tubes, and this was a paper by Weibel. It was he who coined the term tubular myelin. So this alternative hypothesis should perhaps be more widely promulgated.

CUMMING: As the silver ions diffuse down and the chromate ions diffuse up they form a precipitate, and that makes a depletion layer of chromate molecules within the diffusion pathway of the chromate, and when the silver ions get down there are no chromate ions to react with it. When the depletion layer has been traversed to this point the same thing happens again. Because of depletion the time interval is gradually increased. The characteristics of the diffusion front is what determines the increased time interval. It is a simple expression of physical diffusion laws in a hydrophylic sol.

CORRIN: Might I also comment on that lamellar structure,
               which is a feature of any complex lipid, because of
               hydrophobic and hydrophilic groupings taking up
               alternate positions. I wanted to stick with the
               chlorphentermine material. There is no chemical
               analysis of this, your assumption is that it
               represents surfactant. It might be conceivable that
               type I endothelial cell in the alveolus and the
               ciliated goblet and Clara cells in the airways are
               secreting surfactant. It does not seem conceivable
               that the endothelial cells are secreting pulmonary
               surfactant or the many cells in other organs of the
               body. I think we should look at an alternative
               explanation. One could be that there is widespread
               focal cytoplasmic degredation with autophagocytosis
               and then the lamellar bodies represent residual
               bodies, the end result of autophagocytosis. I would
               like to comment on your criticisms of my hypothesis
               in paraquat that the fibrosis is the result of the
               organisation of exudates. I accept your criticisms
               as very valid arguments, but I will leave with the
               comment that you have not substituted an alternative
               hypothesis to explain the fibrosis.

SMITH: First, with regard to the chlorphentermine, I did not
               think that I suggested that all these cells have
               started actively secreting surfactant in the same way
               as the granular pneumocyte does. Indeed, as you say,
               there is good evidence nowadays that chlorphentermine
               affects lysosomes, in particular in a whole variety
               of cells, not necessarily in the lung, and that the
               phospholipid which is normally present in lysosomes
               is produced at an excessive rate, mainly because it
               is not capable of being broken down as it is
               produced. It accumulates because it cannot be
               catabolised and these residual structures occur as a
               consquence of this. This takes place not only in the
               lung, but in the kidney, the adrenals and in any
               other organ you would care to name, by a list of
               about twenty odd chemical compounds that have since
               been discovered. As for the mechanism of fibrosis of
               paraquat poisoning, I agree that if one suggests that
               it is not a process or organisation then it is not
               quite so straightforward to suggest what does occur.
               Certainly the appearances are not those of an
               exudate, and since the generation of fibrosis does
               not seem to require the presence of an exudate at
               all, or if the exudate is due to inhalation of the
               poison there is no fibrosis. This does imply that

the fibrogenic effect of paraquat may be exerted
somewhere else, that it is not just a purely
destructive action on the lung epithelium, this being
the first phase. What follows it may not be a simple
direct consequence of it such as we have in
Busulphan, where there is an exudate organising to
fibrosis; one follows as the inevitable consequence
of the other. In paraquat I do not think that this
is the case. One gets alveolar damage, oedema, but
what follows is not inevitably a consequence of this.
It may be that paraquat has an influence on the
connective tissue surrounding the bronchi or
something like this, stimulating fibroblasts to
proliferate and to emigrate. I agree that having
dispenses with the organistion theory it is not as
easy to find one to put in its place. I still
believe that it is not a process of organisation.

THE METABOLISM OF CHEMICALS BY THE LUNG

Gerald M. Cohen

Department of Biochemisty, University of Surrey

Guildford, GU2 5XH, Surrey, England

SUMMARY

The lung, due to its position between 'le milieu interieur' and the outside environment, is exposed both to a wide variety of environmental chemicals including atmospheric pollutants and cigarette smoke and also to chemicals in the systemic circulation. Many of these chemicals may be metabolised in the lung possibly modifying their pharmacological, therapeutic or toxicological actions. Many xenobiotics (foreign chemicals) are metabolised by pulmonary microsomal mixed function oxidases, with very broad substrate specificities. The oxidation products of these pulmonary mixed function oxidase may be altered by a wide variety of environmental factors, in particular diet and cigarette smoke. Such variations in enzyme activities could lead to alterations of the pharmacology, toxicology or carcinogenicity of the xenobiotic. the localisation of high concentrations of cytochrome P-450 in the Clara cell may be the molecular basis for the organ specificity of several pulmonary toxins.

INTRODUCTION

The lung is the major portal of entry into the body of all inhaled compounds, such as those contained in polluted city air, tobacco smoke and medicinal and household aerosols. The very large surface area of the alveoli, (about 70 square meters) besides being ideally suited for the function of gas exchange also facilitates the penetration into the body of these potentially toxic agents. As the entire cardiac output passed through the lungs, all chemicals present in the circulation must also traverse the lungs. Not surprisingly, many of the foreign chemicals and

429

biological agents to which the airways are exposed have been
implicated in a wide variety of diseases of the respiratory tract,
for example, lung cancer (tobacco smoke, benzo(a)pyrene,
nitrosamines, polonium-210, chromates, nickel and arsenic),
chronic bronchitis and emphysema (cigarette smoking and air
pollution), diffuse pulmonary fibrosis (busulphan, bleomycin,
nitrofurantoin, methysergide, paraquat, methylphenylethylhydantoin
and diphenylhydantoin) and phospholipidoses (chlorphenteramine).

The lung is exposed to a wide variety of exogenous chemicals,
which may exert profound pharmacological, therapeutic or
toxicological actions. These chemicals may be metabolised by the
lung, possibly increasing or decreasing their toxicity or
pharmacological action (1-4). The enzymes responsible for the
pulmonary metabolism of these foreign chemicals (xenobiotics)
include both mixed function oxidases and enzymes catalysing
conjugation reactions which are discussed in the text. A
remarkable property of these mixed function oxidases is their very
wide substrate specificity. In spite of their potential
significance, there are relatively few detailed studies of the
pulmonary metabolism of xenobiotics. With the use of appropriate
examples, I shall try to illustrate both the versatility and
importance of these pulmonary metabolising enzymes in the
metabolism and biological activity of some chemicals.

However, before giving a more detailed description of the
wide variety of metabolic pathways of xenobiotics present in
respiratory tissue, I should like to emphasise some of the
problems associated with assessing the contribution of pulmonary
metabolism to the overall metabolism of a chemical in vivo.
Pulmonary metabolism studies have been carried out with many
different preparations including isolated perfused lungs, short-
term organ cultures and different subcellular fractions. The
possible difficulties of the extrapolation of in vitro data
obtained from some of these studies in particular those using
various subcellular fractions is well illustrated in several
studies using both endogenous and exogenous chemicals (Table 1).
Thus, histamine is not metabolised by the pulmonary circulation of
dogs and rats in vivo or by the isolated dog lung. However, lung
slices or homogenates from many species including man and rat
rapidly metabolise histamine (5). Similarly angiotensin II is
metabolised by lung homogenates but not by isolated perfused
lungs (5).

These examples illustrate that the dangers of extrapolation of
in vitro metabolic data to in vivo applies to both endogenous and
exogenous compounds as well as to polar and non-polar compounds.
thus a greater reliance may be placed on metabolic studies which
have maintained either or both cellular integrity and cellular

Table 1.
Pulmonary Metabolism of these Compounds in vivo or in
Perfusion is not in Agreement with theri Metabolism by
Subcellular Fractions

Histamine
Angiotensin II
Imipramine
Chlorcyclizine

architecture. Some major advantages of such systems are that they
will (i) generate their own cofactors at levels similar to those
found in vivo, (ii) be subjected to all the control mechanisms of
the cell and the organ, (iii) simulate in vivo substrate
availability to the metabolising enzymes, (iv) maintain the
in vivo relationship of both phase I and phase II metabolising
enzymes such that the final products observed resemble more
closely the in vivo situation. The advantages and uses of
isolated cell suspensions and cell cultures in drug metabolism
studies has been reviewed (6). Thus the advantages of use of
isolated perfused lung, cell suspensions and cultures as well as
short-term organ cultures are apparent. However, in spite of the
above limitations valuable information can also be obtained from
studies employing subcellular fractions. These experiments are
generally simpler to perform because (i) the technical expertise
is less demanding (ii) the results are less complicated by the
absence of active conjugating enzymes (in general these require
addition of appropriate cofactors in order to exhibit enzyme
activity, for example uridine diphosphoglucuronic acid is required
for UDP-glucuronosyltransferase activity) and (iii) better control
of the xenobiotic concentration and its accessibility to the drug
metabolising enzymes are obtained.

Pulmonary Drug Metabolising Enzymes

Whilst the liver has been recognised as the major site of
drug or xenobiotic metabolism, recently much attention has also
been paid to the metabolism of these xenobiotics by extrahepatic
tissues in particular at portals of entry into the body, i.e. the
lung, intestine and skin. the metabolism of these xenobiotics at
their portals of entry is markedly influenced by the environment,
dietary status and prior exposure to other chemicals (7). Owing
to a deficiency of good data, it is difficult as stated earlier to
assess the contribution of pulmonary metabolism to the overall

metabolism of a compound. In most cases, in vitro enzymic
activities of liver are significantly higher than those of lung
but these differences may be offset in vivo by other factors such
as blood flow and distribution. Thus with compounds such as
amines which are concentrated in lung tissue (8), the contribution
of the lung to the overall metabolism of the compound may be far
greater than can be determined from in vitro enzymic activities.

The metabolic transformations, which most xenobiotics undergo
in the body, are numerous and diverse but they may be conveniently
classified into two main phases (9). In Phase I, compounds may
undergo either oxidation, reduction or hydrolysis. In general,
Phase I reactions result in the addition or exposure of functional
groups, such as -OH or COOH, which may then undergo Phase II or
synthetic (conjugation) reactions. The net result of such Phase I
and II metabolism is often the production of more polar, less
lipid-soluble, more readily excretable and less biologically
active compounds. This overall scheme is illustrated below:-

```
                                OH
        Phase I                 Phase II
DRUG                METABOLITE                    CONJUGATE
        oxidation,          glucuonidation  Increased Polarity
        reduction              sulphation      More Readily
                                                Excreted
```

Many chemicals which already possess functional groups may
undergo Phase II reactions directly, for example, the conjugation
of isoprenaline to 3-0-methylisoprenaline. The enzyme, catechol
o-methyl transferase, catalysing this reaction is present for the
metabolism of structurally related endogenous catecholamines but
will also o-methylate exogenous catechols such as isoprenaline.

Phase I Reactions

Many of the Phase I reactions, in particular the oxidations,
are carried out in the endoplasmic reticulum (microsomes) of the
cell. These reactions are catalysed by a microsomal mixed-
function oxidase system, which requires NADPH and molecular
oxygen. The terminal oxidase of this system is cytochrome P-450
(so called because of the light absorption peak at 450 nm when
reduced microsomes are bubbled with carbon monoxide). This enzyme
system appears to be unique in having such a remarkably broad
substrate specificity. The activity of this enzyme system may
also be markedly enhanced by prior administration of other
xenobiotics. Whilst much of what is known about the xenobiotic
metabolising enzyme systems comes from studies on the liver,
comparable work on the lung emphasises the similarities rather
than the diversities of this enzyme system in the different

tissues (10). However, with some compounds, such as (9)-tetrahydrocannabinol important qualitative and quantitative differences, in the metabolites formed and the inducibility of these enzyme systems, are found (1,2). Various reductases and esterases have also been found in different subcellular fractions of lung often in the cytosol.

Phase II Reactions

Phase II synthetic or conjugation reactions generally involve the addition of small endogenous compounds to the xenobiotic or its metabolite(s). Such endogenous compounds include uridine diphosphate glucuronic acid (UDPGA), 3'-phosphoadenosine 5'-phosphosulphate (PAPS), S-adenosylmethionine, acetyl-CoA and glutathione in glucuronidation, sulphation, methylation, acetylation and glutathione conjugation reactions respectively (9).

All the above types of Phase I and Phase II conjugation reactions have been demonstrated in pulmonary tissues (1) and the interested reader is referred to recent reviews for further details (1-4).

Factors affecting pulmonary metabolism

Whilst many factors such as age, sex and diet are known to effect both Phase I and Phase II reactions in the liver, little is known about the effect of such variables on pulmonary metabolism. Two points seem particularly important to stree:- a) effects on the pulmonary metabolism of one substrate do not necessarily mean similar effects on other substrates, b) many of these effects have only been demonstrated as yet on in vitro preparations and for reasons described earlier in the article require verification by other methods such as perfusion.

Many of these studies on the variables affecting pulmonary metabolism have been carried out using the environmental carcinogen benzo(a)pyrene. This is due to the possible relationship of benzo(a)pyrene, a constituent of cigarette smoke, to human lung cancer and also to the availability and relative ease of a highly sensitive fluorometric assay for the measurement of benzo(a)pyrene 3-monooxygenase (E.C.1.14.14.2 - also called benzo(a)pyrene hydroxylase or aryl hydrocarbon hydroxylase, AHH). The wide use of this assay in very many studies necessitates a few words about the assay and its limitations. The assay measures the production of monohydroxybenzo(a)pyrenes predominantly 3- and 9-hydroxybenzo(a)pyrene. Benzo(a)pyrene is also metabolised to other metabolites, including 4,5-dihydro-4,5-dihydroxybenzo(a)pyrene, 7,8-dihydro-7,8-dihydroxybenzo(a)pyrene and 9,10-dihydro-9,10-dihydroxybenzo(a)pyrene, which are not measured by the fluorimetric assay. The failure of the assay to measure these metabolites may be particularly important as it has been suggested that such dihydrodiol metabolites might be the precursors of the active diol-epoxide carcinogenic metabolite(s) of benzo(a)pyrene, for example, the conversion of 7,8-dihydro-7,8-dihydroxybenzo(a)pyrene to its diol-epoxide 7,8-dihydro-7,8-dihydroxybenzo(a)pyrene 9,10-oxide (11). However, even with these limitations the high sensitivity of the fluorimetric assay has enabled much valuable information to be acquired.

The results of several of these studies are summarised in Table 2 together with a comparison of the effect of similar experimental protocols on hepatic benzo(a)pyrene 3-monooxygenase activity. Several differences, in particular with regards, inducibility and inhibition, between the properties of the pulmonary and hepatic enzymes are apparent (Table 2). Particularly important would appear to be the observations of Wattenberg on the effects of diet on pulmonary and other extra-hepatic metabolism. Replacement of standard laboratory diet with a control purified diet results in a dramatic decrease in basal levels of benzo(a)pyrene 3-monooxygenase activity to almost zero (12). Wattenberg has also found potent inducers of pulmonary benzo(a)pyrene 3-monooxygenase activity in various members of the Brassicaceae family (e.g. brussel sprouts, cauliflower, cabbages) as well as in other naturally occurring compounds such as tangeritin and flavone (12). These marked effects of diet on extra-hepatic metabolism may have important pharmacological and toxicological implications as well as explaining both some of the very wide variation in basal levels of pulmonary benzo(a)pyrene 3-monooxygenase activity observed in the literature and also some of the reported discrepancies of the effects of exogenous agents on this activity. The different responses of the lung compared to the liver to many of the factors studied suggest that xenobiotic metabolism in the different organs may be under individual control and not interrelated. In support of this are the differences observed in the development of lung and liver microsomal enzymes during the post-natal period (13) (Table 2).

The differential effects of inducers and inhibitors on the activity of pulmonary microsomes to metabolise various substrates suggested the presence in both hamster and rat of at least two different monooxygenases present in lung resembling the several forms of cytochrome P-450 found in liver (Table 2). Further support for at least two forms of rabbit pulmonary cytochrome P-450 have also come from more recent studies using partially purified cytochrome P-450 in a reconstituted monooxygenase system. The pulmonary monooxygenase system has been resolved into three components, NADPH-cytochrome P-450 reductase, phospholipid and cytochrome P-450. At least two forms of rabbit pulmonary

Table 2.

Comparitive Effects of Various Agents on Benzo(a)pyrene 3-Monooxygenase Activity in Lung and Liver

Property	Inducing Agent	Lung	Liver	Species
	3-Methylcholanthrene	+	+	Rat
	3-Methylcholanthrene	-	-	Rabbit
Inducibility	3-Methylcholanthrene	+	-	Hamster
	Phenobarbitone	-	+	Rat
	Phenothiazine	+	+	Rat
	β-Napthoflavone	+	+	Rat
	Cigarette Smoking	+++	+	Rat
Sex		♂ = ♀	♂ > ♀	Rat
Ozone		Inhibition	No change	Hamster
Purified Diet		Marked Inhibition	No effect	Rat
In vitro Addition of 7,8-Benzoflavone		Inhibition	Enhancement	Control rats
Development		Gradual increase to adult levels	Sharp increase between 14-30 days	Rabbit

cytochrome P-450, i.e. $P-450_I$ and $P-450_{II}$ have been isolated, purified and differentiated by their monomeric moleculur weights, substrate specificities and immunochemical properties (14). Of interest was the observation that only one of these purified pulmonary cytochrome P-450's, but not the other, was capable of metabolising the pulmonary carcinogen benzo(a)pyrene to reactive metabolites which covalently bound to exogenous DNA (14).

Possible Role of Pulmonary Metabolism in Mediating Pulmonary Toxicity

It is well known primarily due to the work of the Millers, that many chemical carcinogens, and other toxic chemicals, have to be metabolically activated before exerting their carcinogenic or toxic actions (15) (Miller and Miller, 1966; Miller, 1970). Although it is possible that active metabolites may be formed in the liver and then transported to the lung, it is more likely that such reactive metabolites are formed in susceptible tissues. Thus recently in attempts to explain organ-specific toxicity much attention has focussed both on the extrahepatic metabolism of chemicals and environmental factors, which by affecting this metabolism, may ultimately alter the carcinogenic or toxic response of the organism.

In Table 3, I have compiled a list of chemicals whose pulmonary toxicity may be mediated by reactive metabolites generated in situ rather than in the liver.

Table 3
Pulmonary Toxicity of these Compounds may be Mediated by Reactive Metabolites Generated in situ

Compound	Source	Toxicity
Benzo(a)pyrene	Cigarette Smoke	Lung Cancer
4-Ipomeanol	Moldy Sweet Potatoes	Lung Oedema, Clara Cell Necrosis
3-Methylfuran	Environmental Pollutant	Bronchiolar Necrosis
Parathion	Insecticide	Respiratory Failure
-Naphthylthiourea	Pulmonary Toxin	Pulmonary Oedema
Carbon Tetrachloride	Industrial Solvent	Clara Cell Necrosis
Nitrofurantoin	Antibacterial Drug	Pulmonary Oedema
Napthalene	Moth Balls	Clara Cell Necrosis

It is very interesting to note the high percentage of these pulmonary toxins which cause bronchiolar or Clara cell necrosis. Boyd has presented extensive evidence that the Clara cell is a major site of cytochrome P-450 dependent enzymes for the metabolic activation of 4-ipomeanol (16). Autoradiographic studies of animal lungs dosed in vivo with radiolabelled 4-ipomeanol showed specific localization of covalently bound radioactivity in the Clara cells (16). This was followed by Clara cell necrosis and then pulmonary oedema. A very good correlation has been observed between the amount of pulmonary alkylation and the lung toxicity of 4-ipomeanol (17). A very recent study, using immunofluorescence to localise the cytochrome P-450, has shown that at least one form of the cytochrome P-450's (i.e. $P-450_1$) found in rabbit lung was actually concentrated in the Clara cell (18). Although, the endogenous role of this P-450 is at present unclear, although it may well be involved in lipid metabolism, the concentration of a particular type of cytochrome P-450 in a specific cell type in the lung may well provide an explanation for the pulmonary toxicity of several structurally diverse chemicals.

Other Cell Types Responsible for Xenobiotic Metabolism

In addition to the above evidence implicating the Clara cell as one cell type in the lung important in the metabolism of xenobiotics the role of other cell types is not very clear. This is in part due to the complexity of the lung which is composed of at least forty cell types. Using a histochemical method Watternberg and Leong (7) demonstrated benzo(a)pyrene 3-monooxygenase activity in the alveolar walls but not in the tracheal or bronchial mucosa. However, activity was detected in the tracheal mucosa, using a more sensitive quantitative assay. Using a different histochemical technique aniline 4-hydroxylase activity was found in both alveolar tissue and bronchial epithelium and in this case the enzyme activity appeared greater in the bronchial epithelium (19).

Whilst in general it is difficult to separate the various cell types in lung and thus discern which particular type or types are important for metabolism of xenobiotics, one particular group of cells, i.e. alveolar macrophages, are fairly readily available. The mixed function oxidase activity (as measured by biphenyl 4-hydroxylase, benzo(a)pyrene 3-monooxygenase, benzphetamine N-demethylase and metabolism of bromobenzene) of alveolar macrophages of rats and rabbits was very low and did not contribute significantly to overall metabolism in the lung (20,21). Glutathione S-aryl transferase and UDP-glucuronosyl transferase were absent from rabbit alveolar macrophage soluble and microsomal fractions respectively but these macrophages readily acetylated p-aminobenzoic acid (20). However, in marked contrast to the above results, human pulmonary macrophages have been shown to metabolically activate benzo(a)pyrene to mutagenic metabolites (22).

Other cells or cell types which appear to be capable of metabolism of xenobiotics are those of the upper respiratory tract, i.e. the trachea and the bronchus. The majority of studies on lung have concentrated primarily on alveolar tissue, whilst relatively little work has been done on the trachea and bronchus. Metabolic activation of chemicals by the human bronchus and rodent trachea may be of great importance as the majority of lung cancers in man rise from a squamous metaplastic differentiation of the bronchial epithelium and in the animal model for bronchogenic carcinoma, tumours histologically similar to those found in man arise in the trachea. Thus several studies have been carried out on the metabolism and binding of benzo(a)pyrene by these tissues. Benzo(a)pyrene is metabolised by short-term organ culture of rodent trachea and human bronchus and by stripped human broncheal epithelium to 4,5- 7,8- and 9,10-dihydrodiols and 3-hydroxybenzo(a)pyrene (23,24). Human bronchial mucosa was able to activate, benzo(a)pyrene, 7,12-dimethylbenz(a)anthracene, 3-methylcholanthrene and dibenz(a,h)anthracene, into metabolites which covalently bound to DNA. In a study of 37 patients, a 75-fold interindividual variation was found in the binding of benzo(a)pyrene to cultured human bronchus (25). Following culture of human bronchus with benzo(a)pyrene the product bound to DNA is indistinguishable from that formed when the diol-epoxide 7,8-dihydro-7,8-dihydroxybenzo(a)pyrene 9,10-oxide was reacted with DNA (24,26). From these studies it is clear that the area of the respiratory tract most susceptible to develop 'lung' cancer, i.e. bronchial epithelium is capable of covalently binding to DNA. The rates of formation and removal of reactive metabolites in these cells may well be critical factors in determining individual susceptibilities to lung cancer.

Thus it is apparent from the above that there are different cells in the respiratory tract capable of metabolically activating foreign chemicals to highly reactive toxic and carcinogenic metabolites. The significance of these findings to respiratory disease in man awaits much further investigation.

ACKNOWLEDGEMENT

Part of this work was carried out by grants from the Medical Research Council and Chemical Research Centre of Great Britain.

REFERENCES

1. G.M. Cohen, Pulmonary metabolism of inhaled chemicals and irritants, in Scientific Foundations of Respiratory Medicine, eds. J.G. Scadding and G. Cumming. W. Heinemann Ltd. In Press.

2. G.M. Cohen, Pulmonary metabolism of inhaled sustances and possible relationship with carcinogenicity and toxicity, in Lung Metabolism, eds. A.F. Junod and R. de Haller, Academic Press, New York, 1975.

3. G.E.R. Hook and J.R. Bend, Pulmonary metabolism of xenobiotics, Life Sci. 18:279 (1976).

4. R.M. Philpot, M.W. Anderson and T.E. Eling, Uptake, accumulation and metabolism of chemicals by the lung, in Metabolic Functions of the Lung, eds. Y.S. Bakhle and J.R. Vane Marcel Dekker Inc. New York, 1977.

5. Y.S. Bakhle and J.R. Vane, Pharmocokinetic function of the pulmonary circulation, Physiol. Rev. 54:1007 (1974).

6. J.R. Fry and J.W. Bridges, in Progress in Drug Metabolism, Vol.2. eds. J.W. Bridges and L.F. Chasseaud, Wiley and Sons, London, 1977.

7. L.W. Wattenberg and J.L. Leong, Tissue distribution studies of polycyclic hydrocarbon hydroxylase activity, in Handbook of Experimental Pharmacology, Vol.28/2, eds. B.B. Brodie and J. Gillette, Springer-Verlag, Berlin, 1971.

8. E.A.B. Brown, The localisation, metabolism and effects of drugs and toxicants in lung, Drug Metab. Rev., 3:33 (1974).

9. R.T. Williams, The metabolism and detoxication of drugs, toxic substances and other organic compounds, Chapman and Hall Ltd., London, 1959.

10. T.E. Gram, Comparitive aspects of mixed function oxidation by lung and liver of rabbits, Drug Metab. Rev., 2:1 (1973).

11. P. Simms, P.L. Grover, A. Swaisland, K. Pal and A. Hewer, Metabolic activation of benzo(a)pyrene proceeds by a diol-epoxide, Nature, 252:326 (1974).

12. L.W. Wattenberg, Dietry modifications of intestinal and pulmonary aryl hydrocarbon hydroxylase activity, Tox. Appl. Pharmacol., 23:741 (1972).

13. J.R. Fouts and T.R. Devereux, Developmental aspects of hepatic and extrahepatic drug-metabolizing enzyme systems: Microsomal enzymes and components in rabbit liver and lung during the first month of life, J. Pharmacol. Exp. Ther., 183:458 (1972).

14. C.R. Wolf, B.R. Smith, L.M. Ball, C. Serabjit-Singh, J.R. Bend and R.M. Philpot, The rabbit pulmonary monooxygenase system: catalytic differences between two purified forms of cytochrome P-450 in the metabolism of benzo(a)pyrene, J. biol. Chem., 254:3658 (1979).

15. E.C. Miller, Some current perspectives on chemical
     carcinogenesis in humans and experimental animals:
     Presidential address. Cancer Res., 38:1479 (1978).
16. M.R. Boyd, Evidence for the Clara cell as a site of
     cytochrome P-450-dependent mixed-function oxidase activity in
     lung, Nature, 269:713 (1977).
17. M.R. Boyd and L.T. Burka, In Vivo studies on the relationship
     between target organ alkylation and the pulmonary toxicity of
     a chemically reactive metabolite of 4-ipomeanol, J.
     Pharmacol. Exp. Ther., 207:687 (1978).
18. C.J. Serabjit-Singh, C.R. Wolf, R.M. Philpot and
     C.G. Plopper, Cytochrome P-450: Localization in rabbit lung,
     Science, 207:1469 (1980).
19. P.Grasso, M. Williams, R. Hodgson, M.G. Wright and S.D.
     Gangolli, The histochemical distribution of aniline
     hydroxylase in rat tissues, Histochem. J., 3:117 (1971).
20. W.D. Reid, J.M. Glick and G. Krishna, Metabolism of foreign
     compounds by alveolar macrophages of rabbits, Biochem.
     Biophys. Res. Commun., 49:626 (1972).
21. G.E.R. Hook, J.R. Bend and J.R. Fouts, Mixed function
     oxidases of the alveolar macrophage, Biochem. Pharmacol.,
     21:3267 (1972).
22. C.C. Harris, I.C. Hsu, G.D. Stoner, B.F. Trump and J.K.
     Selkirk Human pulmonary alveolar macrophages metabolise
     benzo(a)pyrene to proximate and ultimate mutagens, Nature,
     272:633 (1978).
23. G.M. Cohen, S.M. Haws, B.P. Moore and J.W. Bridges,
     Benzo(a)pyren-3-yl hydrogen sulphate, a major ethyl acetate-
     extractable metabolite of benzo(a)pyrene in human, hamster
     and rat lung cultures, Biochem. Pharmacol., 25:2561 (1976).
24. H. Autrup, F.C. Wefald, A.M. Jeffrey, H. Tate, R.D. Schwarz,
     B.F. Trump and C.C. Harris, Metabolism of benzo(a)pyrene by
     cultured tracheobronchial tissues from mice, rats, hamsters,
     bovines and humans, Int. J. Cancer, 25:293 (1980).
25. C.C. Harris, H. Autrup, R. Connor, L.A. Barrett,
     E.M. McDowell and B.F. Trump, Interindividual variation in
     binding of benzo(a)pyrene to DNA in cultured human bronchi,
     Science, 194:1067 (1976).
26. P.L. Grover, K. Pal, A. Hewer and P. Sims, The involvement of
     a diol-epoxide in the metabolic activation of benzo(a)pyrene
     in human bronchial mucosa and mouse skin. Int. J. Cancer,
     18:1 (1976).

DISCUSSION

LECTURER: Cohen   CHAIRMAN: Bonsignore

SPINA: Many years ago people spoke of a carcinogenic action
                of isoniazide. I would like to know what is your
                opinion on this.

COHEN: Isoniazide, which is an antituberculous agent, is a
                very potent carcinogen in a particular animal model
                of carcinogenesis, which utilises a strain of mice
                very susceptible to pulmonary adenoma formation.
                They exhibit spontaneous adenoma formation and when
                you give chemicals, such as the hydrazines,
                isoniazide being a member of that class, you get a
                higher incidence of pulmonary adenoma. I do not
                think we can say because of that, that isoniazide is
                a carcinogen in man. I believe there are no data, as
                far as I am aware, that isoniazide is carcinogenic in
                man. It will cause hepatic damage or hepatic
                necrosis, in some patients, but there is no evidence
                that it is a carcinogen in man. It is a potent
                carcinogen in the strain A mice.

JEFFERY: Two questions, one short one, one a little bit
                longer. Did I understand it correctly that P 450 was
                selectively localised to Clara cells or is it also
                found in some amounts in the other cell types?

COHEN: It is a complicated question. Because there are
                multiple cytochrome P 450's present in the lung. One
                form of cytochrome P 450 is localised in the Clara
                cell.

JEFFERY: Is that the form that would be significant in terms
                of carcinogenesis?

COHEN: It is perhaps important in activating some
                carcinogen, but it is the form which activates
                chemicals such as the 4 epoxides. The whole thing
                has been taken a step further, in that someone is
                looking to see how many human tumors are derived from
                the Clara cell. Unfortunately, from what Corrin has
                said to me that is going to be a very small number,
                but if you had a tumor which was derived from this
                Clara cell you could direct a cancer chemotherapeutic
                agent to that tumor with selective toxicity, because

it would activate the chemical to a reactive alkylating agent, which would then selectively destroy the tumor because it would be only the tumor cells which would activate that chemical in such a specific manner.

JEFFERY: The second question is that there are some interesting reports in the literature that people with low circulating levels of vitamin A might be more susceptible to bronchial carcinoma, and then, conversely, there have been claims of success of treatment with high levels of vitamin A in therapy for bronchial carcinoma. Is there any link between vitamin A and the systems that you described?

COHEN: There is a definite relationship between vitamin A and the systems I have described, a lot of work has gone on in that particular area. I think most of it is erroneous. Though vitamin A can modify the system, its effect is in my opinion at the second stage of carcinogenesis. Vitamin A affects the differentiation of cells, it almost certainly affects the development of the tumour rather than acting at what I would call the initiating step. So I think that vitamin A comes much later in the story, and its effects are very complex. I think in most cases they inhibit tumour formation, there are cases where they enhance tumour formation, probably by causing lysosomal instability, and then that would perhaps lead to the release of DNAse which ultimately could enhance tumour formation.

CORRIN: What I said with regard to Clara cell neoplasms is that very few neoplasms consist of Clara cells, but in many adenocarcinomas, large cell undifferentiated carcinomas, even the common squamous carcinomas, we do not know from which particular cell type they are derived originally. They could all conceivably be derived from Clara cells. I would like to ask a question about these various groups. How far are we from identifying, in our smokers, which of them are at risk those who are particularly susceptible?

COHEN: It is a very difficult question. A few years ago, in 1973, Kellerman produced a so-called risk test, based on lymphocyte hydrocarbon hydroxylase, which is the metabolism of benzopyrene to 3-hydroxybenzopyrene. He suggested that people with a high induceability in their lymphotytes for coverting benzopyrene to 3-hydroxybenzopyrene were 36 fold more at risk than people with a low induceability. It seemed that perhaps there was beginning to be some hope in that particular area relating to identifying subpopulations. The National Cancer Institute put 6 million dollars into following that up, and this has led absolutely nowhere. I wanted to develop the point even further. My personal conviction is that by describing the metabolic profiles, as we have done, we in fact might still not recognise those people who are most at risk. Obviously, if someone forms masses of reactive metabolites and that all binds to DNA, perhaps they would be much more at risk. I think that very nice experiments are going to come out from Chester Beatty in London, where they have used mice of different susceptibilities to skin carcinogenesis. Not only have they shown that the metabolism of the carcinogen in the skin of those mice is the same, but so is the actual product bound to the DNA in mice of very different susceptibilities to skin carcinogenesis. If the same applies to the human situation we could waste a lot of time looking at metabolism, perhaps we should concentrate more on the binding site and even more on the steps after the binding. It is perhaps only in those steps, in terms of tumor development and tumor promotion, which are critical in determining why one individual is susceptible and why another is not.

WIDDICOMBE: How do you think that Clara cell necrosis leads to pulmonary oedema?

COHEN: I think perhaps our pathologists could help me.

WIDDICOMBE: It is interesting, because on Peter's evidence you can punch lots of holes in the airway epithelium and not get oedema, as far as I know. Can somebody else answer?

HEATH: I think it is hard, I do not see how Clara cells lead to pulmonary oedema, but it is hard to ask me to explain somebody else's hypothesis, when I do not see a connection. One thing I do know, this has been discussed. I just wonder what were the criteria for diagnosing necrosis of Clara cells? Because the normal physiological activity of Clara cells looks extremely like necrosis to the uninitiated and I am just wondering if all this Clara cell necrosis which has appeared in your tables is genuine. I think the distrinction between the physiological appearances needs to be worked out very carefully.

COHEN: I have actually brought papers with diagrams,
                    basically I think they considered as evidence the
                    ballooning of the Clara cell, often with dilatation,
                    and then actually necrosis. But I know that may be
                    normal function.

HEATH: It is interesting, the two histological features
                    which you have just described are both physiological
                    features of Clara cell activity.

SANT'AMBROGIO: Do you think that the lung is better at deactivating
                    foreign chemicals when these chemicals are given from
                    the air side or when they are given into the systemic
                    circulation? Is it made to fight against air-borne
                    pollutants or is it better prepared to fight
                    pollutants carried by the blood? Another way to look
                    at this question is to, compare lesions which are
                    caused by the same substance when they are given into
                    the air or into the blood.

COHEN: This has in fact been done by Boyd in the United
                    States with carbon tetrachloride. The carbon
                    tetrachloride, in addition to causing some damage in
                    the peripheral lung, will also cause Clara cell
                    necrosis, and it is more active at causing Clara cell
                    necrosis following inhalation than following
                    injection.

SANT'AMBROGIO: That means that it is better prepared to fight
                    air-borne pollutants?

COHEN: No, I think what it may mean in many cases is a pure
                    pharmacokinetic phenomenon, in that the lung, if you
                    like, gets first shot at the chemical, as opposed to
                    having a large fraction of a chemical which if it is
                    given intraperitoneally or orally will be metabolised
                    in the liver and deactivated; so that the effective
                    dose in the respiratory tract will be much higher. A
                    second point would be if the chemical, on its first
                    passage through the lung was taken up into the lung,
                    where there would be a concentration of the chemical
                    which would then act as a reservoir for reactive
                    metabolites. Some work of the group at the
                    Hammersmith Hospital has shown different metabolic
                    pathways of some bronchodilators when given
                    intrabronchially rather than given in the systemic
                    circulation. Which suggests that some metabolic
                    pathways are only active when the chemical is passing
                    into the airways via the airways and then through the

cells, rather than from the systemic circulation into the respiratory tract.

CANDURA: I wanted to make a short comment, and then two questions, if I may. The comment is that this morning we tried to say that, based on the experience of the occupational physician, poisons are not only organ specific, that is they have peculiar clinical patterns in workers exposed to specific poisons, they are also species specific and strain specific. In other words, they are active specifically on certain species, on certain strains. So we can explain that a particular strain of rats develops a malignant tumor when it is given a drug which is beneficial to man, such as the hydrazide of isonicotinic acid. There are many other examples, for instance the dog is not sensitive to a powerful carcinogen for man, that is beta naphthylamine. The question I would like to ask is: don't you think that in between phenol, that is hydroxybenzene and benzene, there is an intermediate compound, an epoxide one, which may be the one which is active. I see that you have already answered because you are writing the formula of benzene epoxide. The last question then: among the various factors which have an influence, age, environment, disease etc., don't you think we should also mention drugs? Because we occupational doctors spend a lot of our time in trying to convince practicing physicians to be very careful when they prescribe drugs to people exposed to industrial poisons. We have great perplexities about the possibility that some very powerful enzyme inducers, such as barbiturates and so on, can promote certain changes, but in a negative direction. In other words, it is possible that enzyme induction favours the metabolism of toxic substances, but it is also possible that they help the metabolism of carcinogens.

COHEN: The first question, yes there is good evidence for an epoxide intermediate in the metabolism of benzene to a phenol. Part of that evidence is that there is excreted in the urine a glutathione conjugate, which means that there must be an epoxide intermediate. The finding of a glutathione conjugate in the urine is good evidence for a reactive metabolite, and that could be the reason for the toxicity mediated by benzene. The second question. In terms of drugs affecting metabolism toxicity,

of course the answer is yes. I just felt there was a limit to how
big and how much writing I shall put on the slide. So I put some of
the factors. On a later slide in fact, where I have factors
affecting pulmonary metabolism, there were several drugs listed.

THE LUNG AT HIGH ALTITUDE

Donald Heath

Department of Pathology
University of Liverpool
England

When man ascends to high altitude or lives permanently at great heights, such as the High Andes of Peru and Bolivia, he becomes acclimatized to the conditions characteristic of the mountain environment. Virtually every system of the body is involved in this process. Since the major adverse environmental factor inherent in life at high altitude is hypoxia due to the diminished barometric pressure, one would expect that modifications in pulmonary structure and function would be prominent in acclimatization. Such is the case but we are still ignorant about many aspects of the condition of the lung at altitude that might be thought to be common knowledge.

Chest size

The shape and size of the chest of the native highlander have attracted much attention in the past and this is not surprising since it is in this area that one might anticipate finding the greatest structural adaptation to provide an increased ventilatory reserve compatible with a healthy and active life at high altitude. The classic field work of Hurtado in the Andes in 1932 established the view that the chest is big in highlanders. His studies were made on Quechua and Aymara Indians and he thought that the adults, in particular "gave the impression that practically all the body mass lies on the chest". It seems very likely that this overstates the case and takes no account of the considerable variation in the size and shape of the chest that is to be found in native highlanders. On a visit to La Paz (3800 m) in 1979 we came across a young Indian native of the city whose massive chest appeared to confirm the classical view put forward by Hurtado. Unfortunately close questioning revealed that his spare-time was devoted to weight-lifting! In contrast to

this other natives of the Bolivian capital have a flat chest. The early views
of Hurtado were based on physical anthropological indices such as the so-
called 'Chest Index' defined as $\frac{depth}{width} \times 100$, but these physical measurements
were somewhat crude. Thus 'width' was defined as the distance between the
midaxillary lines at the level of the sixth ribs. 'Depth' was the distance from
the middle of the sternum to the vertebral column at the level at which width
was determined. Hurtado found that in the child at sea level this index falls
so that the chest changes from being rounded to being ellipsoidal. At high
altitude the chest in childhood maintains a more rounded shape. There is a
widening of the costal angle.

Careful comparisons of child-growth surveys in Peru and mountain
communities in other countries make it clear that patterns of growth are
population-specific and one cannot extrapolate the effects of environmental
stress, such as the hypoxia of high altitude, from one population to another.
Thus in Ethiopia children native to the mountains grow and mature faster than
their counterparts from the lowlands (Frisancho, 1978). Studies in the
Himalayas and Tien Shan mountains also demonstrate that the enlargement of
the thorax is not a general characteristic of all highland populations. Hence
it would appear that the factors associated with an increase in chest size in
the Andean region are not the same as those in Asiatic populations (Frisancho,
1978). All we can say is that in general highlanders tend to have big chests.

The nature of the enlargement of the lung at high altitude

Early studies by Hurtado (1932) suggested that the big chests of highlanders are associated with increased vital capacity compared to lowlanders.
Studies carried out many years later by Frisancho (1975) showed that the
physiological characteristics of the lung at high altitude are acquired during
childhood. Thus the forced vital capacity was the same in lowlanders who
acclimatized to high altitude during growth, as in native highlanders. It was,
however, appreciably smaller in lowland subjects who acclimatized to the
hypoxic environment in adult life. Experimental evidence from Burri and
Weibel (1971) confirms that young rats exposed to chronic hypoxia show an
increase in the number of alveolar spaces, in alveolar surface area and in
lung volume. Hyperoxia has the reverse effect on the developing lung.

The nature of the enlargement of the lung at high altitude is suggested
by the work of Brody and his colleagues (1977). Their studies reveal that,
while whole lung volume in Peruvian highlanders between 17 and 30 years of
age is 30 to 35 per cent greater than that of lowlanders of similar race, age,
and body size, bronchial flow rates expressed as a fraction of lung volume
are less in highlanders than in lowlanders. Such results suggest that the high-

altitude lung represents a combination of a bronchial tree comparable to that of a lowlander with no increase in cross sectional area or volume of the airways, with the increased alveolar numbers and hence increased lung volume of a highlander.

The internal surface area of the lung

The implication of the early anthropological observations of Hurtado is that the lungs as well as the chest of the highlander are also unduly large and thereby also their breathing surface, the internal surface area of the lung. However, proof of this has still not been obtained. There are now available to histopathologists several valid methods of tissue morphometry which can determine accurately the internal surface area of the lung, the number of alveolar spaces, and the numbers and cross-sectional area of bronchial airways. Unfortunately, these precise morphometric studies have not been applied to the lungs of highlanders. This investigation is long overdue and will determine in terms of square metres precisely how the internal surface area of the lung, the breathing area, is altered in native highlanders with 'big chests'.

Acclimatization and adaptation

Acclimatization is a reversible, non-inheritable change in the anatomy or physiology of an organism which enables it to survive in an alien environment. Its essence is the maintenance of high partial pressures of oxygen from alveolar spaces to the vicinity of the mitochondria and this is achieved by a variety of mechanisms including a shift to the right of the oxygen-haemoglobin association curve. Adaptation is the development of biochemical, physiological and anatomical features that are heritable and of genetic basis, enabling the species to explore the environment of high altitude to its best advantage. It is characterized by increased oxygen-carrying capacity of the blood by a leftward shift of the oxygen-haemoglobin dissociation curve and mechanisms for increased oxygen utilization in the tissues. As we have already seen an increased internal surface area of the lung seems to be one feature of acclimatization. Another appears to be hyperventilation so that deeper breathing elevates partial pressure of oxygen in the alveolar spaces.

Diffusion across the alveolar-capillary membrane

In some way the alveolar-capillary membrane appears to be modified to facilitate diffusion across it of oxygen. In highlanders the A - a difference falls to about 2 mm Hg (Hurtado, 1964). This fall in A - a difference is related to the increased internal surface area of the lung and to the pulmonary hypertension inherent in life at high altitude. Recent studies by Pearson and Pearson (1976 and 1979) have demonstrated ultrastructural changes in the alveolar-capillary wall in mice (Phyllotis darwini) in the Peruvian Andes. The thickness of the air-blood barrier was found to be the same in high and low altitude. However, there was a significantly greater volume of tissue components. There was an increase in size of membranous and granular pneumocytes and pulmonary endothelial cells. In particular the granular pneumocytes were found to be larger at high altitude with larger nuclei, a greater volume of mitochondria, and more and larger lamellar bodies resulting in increase in surfactant secretion. We shall return to this point again.

The pulmonary vasculature

In spite of the modifications in the lung associated with acclimatization there remains persistent chronic alveolar hypoxia at high altitude. This exerts a pronounced effect on the terminal portion of the pulmonary arterial tree causing it to constrict. Chronic hypoxia exerts opposite effects on the pulmonary and systemic circulations inducing constriction in the former and dilatation in the latter. Thus the Quechua Indians of the Andes and mining engineers of Caucasian extraction living for a long time at high altitude show an appreciable fall in the level of systemic blood pressure (Marticorena et al. 1969) but mild to moderate pulmonary hypertension (Peñaloza et al. 1962). At first the smooth muscle cells of the pulmonary arterioles constrict forming muscular evaginations which appear as clear cyst like extrusions of the myocytes (Smith et al. 1978). Subsequently this initial vasoconstrictive effect is followed by hyperplasia and hypertrophy of smooth muscle. The normal pulmonary arteriole has a wall consisting of a single elastic lamina. In contrast in native highlanders, in those acclimatizing to high altitude and in rats subjected to simulated high altitude in a decompression chamber, circularly-orientated smooth muscle develops internal to the original elastic lamina. Internal to the newly-formed muscle a second much thinner elastic lamina forms. Such muscularized pulmonary arterioles are very characteristic of Quechua Indians and sufferers from Monge's disease. Ultrastructural studies confirm the presence of many new smooth muscle cells between two elastic laminae. In effect thin-walled pulmonary arterioles are transformed into something resembling a systemic arteriole capable of elevating pulmonary vascular resistance.

The pulmonary hypertension of high altitude never becomes associated with the coarse intimal fibroelastosis so characteristic of the raised pulmonary arterial pressure of congenital cardiac shunts. Hence the pulmonary hypertension of the high-altitude lung is largely reversible. This is not to say, however, that such reversibility is immediate. The administration of oxygen to highlanders will reduce their pulmonary hypertension by about a fifth but total relief of the pulmonary hypertension takes a full two years when the

Quechuas descend to the coast (Peñaloza et al. 1962). No doubt this is
associated with the time needed for regression of muscle in the terminal
portions of the pulmonary arterial tree.

The mechanism by which chronic alveolar hypoxia induces constriction
of the terminal portion of the pulmonary arterial tree is controversial and there
is no time to dwell on this important but perplexing problem. Suffice it here
to say that the view of Haas and Bergofsky (1972) that the hypoxia stimulates
perivascular mast cells to release histamine and related substances with a
constrictive effect on the smooth muscle of the arteriolar wall no longer enjoys
the vogue it had recently (Williams et al. 1977).

The pulmonary trunk

The changes which take place in the small pulmonary blood vessels in
those living at high altitude increase resistance to the flow of blood through
the lungs and hence lead to pulmonary hypertension. The elevation of pulmonary
arterial pressure influences the structure of the pulmonary trunk and the large
conducting elastic pulmonary arteries. The medial coat of these large arteries
does not show the same transition of elastic tissue patterns so characteristic of
sea-level subjects. In lowlanders during fetal life the elastic tissue of the
media of the pulmonary trunk has an aortic pattern associated with physiological
pulmonary hypertension. In infancy there is stick-like fragmentation of elastic
fibrils to produce a 'transitional pattern'. Finally, after the age of two years,
an 'adult pulmonary pattern' is established in which there is an open network
of branched, irregularly shaped elastic fibrils (Heath et al. 1959). These
changes are associated with, and produced by, the precipitous fall in pulmonary arterial pressure which occurs on birth at sea level. At high altitude
this fall in pressure does not occur in infancy and childhood and indeed there
is moderate pulmonary hypertension. Hence the usual atrophic changes in
the elastic tissue of the media do not take place and instead a 'persistent'
type of elastic tissue pattern occurs in which many long thick fibres are found
in association with fewer fragmented ones. This pattern, somewhat reminiscent of the aorta, occurs in association with the pulmonary hypertension of
high altitude. At elevations between 4 040 and 4 540 m the aortic type of
elastic tissue pattern exists up to the age of 9 years and is then followed by
the 'persistent configuration' for the rest of adult life (Saldaña and Arias-
Stella, 1963). At somewhat lower altitude the aortic pattern persists for a
shorter period but it is again followed by the persistent configuration. It is
thus a salutary fact that the histological structure of all classes of pulmonary
blood vessel depends on the partial pressure of oxygen of the air we breathe.

Chemoreceptor tissue in the lung

It is now well established that the carotid bodies undergo characteristic changes at high altitude. Thus they are enlarged at high altitude in man (Arias-Stella and Valcarcel, 1973) and animals (Edwards et al. 1971). The enlargement is due to hyperplasia of the light type of chief cell showing vacuolation (Heath et al. 1970) while at ultrastructural level there is micro-vacuolation around neurosecretory vesicles (Edwards et al. 1972). The functional significance of these changes is as yet not understood. The electron microscopic changes in the neurosecretory vesicles may be concerned with chemoreception or even with the secretion of a hormone.

It is not so widely recognized that similar ultrastructural changes occur in the Feyrter cells of the lung at simulated high altitude (Moosavi et al. 1973). Feyrter cells are argyrophilic cells which are distributed throughout the bronchioles and to a less extent in the alveolar walls. Their function is unknown. Nevertheless, they are 'A.P.U.D. cells' and closely resemble the carotid body in ultrastructure. It is conceivable that they also comprise chemoreceptor tissue in the bronchial tree. Glomic tissue is also to be found around the pulmonary veins (Edwards and Heath, 1972) but the response of this tissue to the chronic hypoxia of high altitude has not been studied.

Alveolar surface tension

I have already referred to studies which suggest that at high altitude the granular pneumocytes become larger and produce more surfactant. There has been one report that the acute exposure of sea level-mice to an altitude of 4270 m induces increased alveolar surface tension (Castillo and Johnson, 1969). We have found that in llamas (Lama glama) adapted to high altitude the bronchiolar Clara cells are hyperactive and extrude the apical portions of the cell into the bronchial tree, perhaps carrying a secondary source of pulmonary surfactant overcoming the tendency of the environment to increase surface tension in the lung. In llamas born and living at sea level in Chester Zoo in England on the other hand the bronchiolar Clara cells are inactive and passive (Heath et al. 1979).

High altitude pulmonary oedema

Serious or fatal oedema of the lung may occur in those ascending to high altitude who indulge too early in vigorous exercise such as rock climbing or skiing. It also occurs in those who live and work permanently at high altitude, especially if they are faced with repeated re-entry into the hypoxic environment after visits to lower altitude. It is a disease of considerable practical importance to those living in mountainous areas and is likely to develop in healthy active

young men or children. If it is not treated promptly by treatment with oxygen
or descent to lower altitude, it may prove fatal. We have considered the
nature of this condition at length elsewhere (Heath and Williams, 1977).

We have studied the ultrastructural changes in the lungs of rats exposed
for twelve hours in a hypobaric chamber to a subatmospheric pressure of 265
mm Hg which simulates an elevation roughly corresponding to the summit of
Mount Everest (Heath et al. 1973). Under these conditions there is a formation
of multiple endothelial vesicles which protrude into the pulmonary capillaries.
They arise by pedicles from localized widened areas of the fused basement
membrane of the alveolar wall where there seems to be an accumulation of
fluid. In longitudinal section the vesicles have an elongated shape which
accommodates itself to the lumens of the pulmonary capillaries into which
they project. They appear large enough to occlude these capillaries.

Obviously one must be cautious before ascribing functional significance
to structural change in the absence of physiological data. However, these
vesicles are composed largely of oedema fluid and could form rapidly and
thus account for the very sudden onset of the syndrome of high altitude
pulmonary oedema. Likewise they could shrink equally rapidly on return to
low altitude or on the administration of oxygen. These oedema vesicles
could produce a haemodynamic effect by protruding into venous capillaries.
This would meet the requirements laid down by Fred et al. (1962) to explain
the coexistence of acute pulmonary arterial hypertension and acute pulmonary
oedema in the presence of normal left atrial and pulmonary venous pressure.

The lung in acute decompression

When mammalian lungs are suddenly subjected to the greatly diminished
atmospheric pressure of extreme altitudes, there ensues a very different set of
ultrastructural changes from those seen in simulated high altitude pulmonary
oedema. We have studied these events in the lungs of rats exposed suddenly
to a barometric pressure equivalent to that of the summit of Mount Everest
(Mooi et al. 1978). Under these circumstances there is swelling and destruction of the granular and membranous pneumocytes with exposure of the denuded
fused basement membrane of the alveolar-capillary wall. At the same time
there is what amounts to a suction effect on the pulmonary capillaries so that
erythrocytes are sucked from them into the interstitial tissues of the lung. It
is of interest that the administration of heavy doses of frusemide to such rats
prevents these deleterious effects on the pneumocytes but induces a cystic
change in the endothelial cells of the pulmonary capillaries. Pilots of aircraft occasionally have to eject themselves into the rarefied air of extreme
altitude and expose themselves to the danger of sustaining lung damage of
this type.

REFERENCES

Arias-Stella, J., and Valcarcel, J. (1973)
The human carotid body at high altitudes.
Pathologia et Microbiologia, 39, 292 - 297.

Broday, J.S., Lahiri, S., Simpser, M., Motoyama, E.K., and
Velasquez, T. (1977)
Lung elasticity and airway dynamics in Peruvians native to high altitude.
American Journal of Physiology:
Respiratory, Environmental and Exercise Physiology 42, 245 - 251.

Burri, P.H., and Weibel, E.R. (1971)
Environmental oxygen tension and lung growth.
Respiratory Physiology, 11, 247 - 264.

Castillo, Y., and Johnson, F.B. (1969)
Pulmonary surfactant in acutely hypoxic mice.
Laboratory investigation, 21, 61 - 64.

Edwards, C., and Heath, D. (1972)
Pulmonary venous chemoreceptor tissue.
British Journal of Diseases of the Chest 66, 96 - 100.

Edwards, C., Heath, D., and Harris, P. (1972)
Ultrastructure of the carotid body in high-altitude guinea-pigs.
Journal of Pathology, 107, 131-136.

Edwards, C., Heath, D., Harris, P., Castillo, Y., Krüger, H., and
Arias-Stella, J. (1971)
The carotid body in animals at high altitude.
Journal of Pathology, 104, 231 - 238.

Fred, H.L., Schmidt, A.M., Bates, T., and Hecht, H.H. (1962)
Acute pulmonary edema of altitude.
Clinical and physiologic observations.
Circulation, 25, 929-937.

Frisancho, A.R. (1975)
Functional adaptation to high altitude hypoxia.
Science, 187, 313-319.

Frisancho, A.R. (1978)
Human growth and development among high-altitude populations.
In: Baker, P.T. (Editor)
The Biology of High-Altitude Peoples.
Cambridge, Cambridge University Press, p. 117.

Haas, F., and Bergofsky, E.H. (1972)
Role of the mast cell in the pulmonary response to hypoxia.
Journal of Clinical Investigation, 51, 3154 - 3162.

Heath, D., Edwards, C., and Harris, P. (1970)
Postmortem size and structure of the human carotid body.
Thorax, 25, 129-140.

Heath, D., Moosavi, H., and Smith, P. (1973)
Ultrastructure of high altitude pulmonary oedema.
Thorax, 28, 694-700.

Heath, D., Smith, P., and Biggar, R. (1980)
Clara cells in llamas born and living at high and low altitudes.
British Journal of Diseases of the Chest, 74, 75-80.

Heath, D. and Williams, D.R. (1977)
Man at High Altitude.
Edinburgh, Churchill Livingstone.

Heath, D., Wood, E.H., DuShane, J.W., and Edwards, J.E. (1959)
The structure of the pulmonary trunk at different ages and in cases of
pulmonary stenosis.
Journal of Pathology and Bacteriology, 77, 443-456.

Hurtado, A. (1932)
Respiratory adaptations in the Indian natives of the Peruvian Andes.
American Journal of Physical Anthropology, 17, 137-165.

Hurtado, A. (1964)
Some physiologic and clinical aspects of life at high altitudes.
In: Cander, L. and Moyer, J.H. (Editors)
Aging of the Lung.
New York, Grune and Stratton p. 257.

Marticorena, E., Ruiz, L., Severino, J., Galvez, J., and Peñaloza, D.,
                                                                    1969,
Systemic blood pressure in white men born at sea level ; changes after long
residence at high altitudes.
American Journal of Cardiology, 23, 364 - 368.

Mooi, W., Smith, P., and Heath, D. (1978)
The ultrastructural effects of acute decompression of the lung of rats ; the
influence of frusemide.
Journal of Pathology 126, 189-196.

Moosavi, H., Smith, P., and Heath, D. (1973)
The Feyrter cell in hypoxia.
Thorax, 28, 729-741.

Pearson, A.K., and Pearson, O.P. (1979)
Granular pneumocytes and altitude:
a stereological evaluation.
Cell and Tissue Research, 201, 137-144.

Pearson, O.P. and Pearson, A.K. (1976)
A stereological analysis of the ultrastructure of the lungs of wild mice living
at low and high altitude.
Journal of Morphology, 150, 359-368.

Peñaloza, D., Sime, F., Banchero, N., Gamboa, R. (1962)
Pulmonary hypertension in healthy man born and living at high altitudes.
Medicina Thoracalis, 19, 449-460.

Saldaña, M., and Arias-Stella, J. (1963)
Studies on the structure of the pulmonary trunk.
II. The evolution of the elastic configuration of the pulmonary trunk in
people native to high altitudes.
Circulation, 27, 1094 - 1100.

Smith, P., Heath, D., and Padula, F. (1978)
Evagination of smooth muscle cells in the hypoxic pulmonary trunk.
Thorax, 33, 31-42.

Williams, A., Heath, D., Kay, J.M., and Smith, P. (1977)
Lung mast cells in rats exposed to acute hypoxia, and chronic hypoxia with
recovery.
Thorax, 32, 287-295.

D I S C U S S I O N

LECTURER: Heath   CHAIRMAN: Bonsignore

DENISON: Donald Heath was bold enough to venture into
              physiology, and there was one point, and only one
              point, with which I disagree with him. It is not an
              important one and it concerns the interpretation of
              shifts in the haemoglobin dissociation curve. I am
              first going to draw the normal haemoglobin
              dissociation curve. We are born with a curve that
              lies over here. In the first few weeks of life it
              moves like that. We have this type of dissociation
              curve to cope with the hypoxia of intrauterine life.
              When man is taken to high altitude his curve moves
              further to the right. This has an advantage in the
              tissues but a disadvantage in the lung. Llamas have
              a curve over in this direction. These changes were
              discovered independently and teleologists saw the
              hand of God in both changes as being wise adaptations
              to high altitude.

HEATH: Could I just interrupt you and say, it is interesting
              and perhaps you could say it, that the Americans saw
              the advantage with the move to the right and Barcroft
              saw the hand of God in it moving to the left.

DENISON: Now, I agree entirely with that. There is a very
              interesting paper by Hebble in the Journal of Applied
              Physiology in late 1978 which points out that this is
              a very unwise adaptation to altitude. It is the
              correct move to make if you have got arterial hypoxia
              or ischaemia, and it is the proper adaptation to
              circulatory insufficiency. He had a very nice
              demonstration of this, because he found twins with a
              very rare haemoglobin pathology, haemoglobinopathy, a
              variant called Minneapolis, which moves in that
              direction, and these twins were able to survive very
              well at high altitude. I think the physiologist
              would say: this is truly an adaptation to hypoxia,
              and this one is a false one. It is because man does
              not normally go to high altitude, but he very
              commonly becomes ischaemic.

HEATH: What I would say is this; I would still like to use
              these words, if I may. The hand of God moving it to
              the left is the sign of adaptation, and the one

coming over here acclimatisation. I agree that
acclimatisation is an imperfect thing compared to
adaptation. But I would just like to make a little
point I referred to in the paper. Some work which I
think came from Italy, from the University of Rome,
in which three years ago some workers showed that the
Sherpas of the Himalayas have their haemoglobin
dissociation curve moved over to the left, and so it
would appear that the Sherpas are much more like
adapted animals, like the llama, than acclimatised.
Again this is a thing that morbid anatomists very
frequently take into account and that is the factor
of time, because whereas the Quechuas have been there
for about 35,000 years, the Sherpas have been there
about half a million years. That is the difference,
I think.

BONSIGNORE: How much does the polycythaemia intervene in this
displacement to the left? There are two ways of
adaptation. This adaptation is because of alkalosis
following on hyperventilation. This is a natural
displacement to the left because of polycythaemia.

HEATH: Can I say that animals which show adaptation do not
show increased levels of haemoglobin. If you have a
look at the llama and the alpaca, the thing that
characterises all of them is their very low levels of
haematocrit. This is again another one of the
sharply contrasting phenomena to acclimitisation,
which of course is associated with gross
polycythaemia. So the feature of these high altitude
camilids is that they have low haematocrits. So they
are adapted animals, not acclimatised.

BONSIGNORE: Yes. Human haemoglobin never changes. We have a
haemoglobin F in Indians, but one does see these F
normal haemoglobins in llamas and alpacas.
Incidentally at this point, when I heard Cumming this
morning introducing his talk and talking about the
long nose of the camel and saying that this was an
adaptation to heat, of course he did not mention that
there are about five or six species of the camel
family which live in freezing conditions on the top
of mountains. I forgot to ask him that.

FAGAN: It is not really a question, more a sort of comment.
It was very interesting to hear Donald with his
stimulating paper. I am sorry he got the idea that I
criticised Weibel because I do not. The length of
time that it has taken to catch up with what Weibel

pointed out and to discover the flaws in just blindly applying his method is an indication of the superiority of his work above the comprehension of the rest of us. What is happening now is that we are catching up with the special situations in which Weibel's methods cannot be followed blindly. The same holds true for stereology in special situations like the high altitude lung, that if you blindly follow these methods you will get exactly the answer that you expect to get due to the use of certain constants. What we have to do is go back to the beginning and redefine the constants in each of these situations. For myself I should at least point out that I have tried stereology. I did not see any stereological results in Heath's paper, and another thing I would like to say is that I have got a sabbatical coming up in 1981, so if he is going, I will come too.

BONSIGNORE: Since discussion has been so stimulating I have not terminted it, but we are now in the position that I must ask you to relinquish the break and proceed directly to our next paper.

PATHOGENESIS OF PULMONARY FIBROSIS DUE TO

ASBESTOS AND SILICA

R.J. Emerson*, E. Bateman and P. Cole
Host Defense Unit, Dept. of Medicine
Cardiothoracic Institute, Brompton Hospital
Fulham Road, London SW3 6HP
*Funded by the Chest, Heart and Stroke Assoc.

INTRODUCTION

Inhalation of silica or asbestos for a sufficient time and at a sufficient dose can lead to irreversible damage of the blood/gas barrier and pleura. As well as these particles being fibrogenic, asbestos has been shown to be associated with increase in bronchial carcinomas and mesotheliomas (Becklake, 1976). The effects of these particles on lung is dependent on a number of variable factors - puriety, dose, physical and chemical composition and host factors which alter normal pulmonary defense mechanism: for example, existing acute or chronic inflammation, cigarette smoke, immunological status and lung anatomical variations which may cause alterations in the site of particle retention between individuals.

Inhaled particles of appropriate size which are not eliminated by the mucociliary apparatus and other methods as discussed by Corrin in this volume can reach the alveoli where they are phagocytosed by macrophages. Those particles escaping phagocytosis are able to interact with Type I and Type II epithelial cells and fibroblasts. Consequently the effects of these particles on macrophages and other resident cells of the alveolar region are relevant to the pathogenesis of mineral dust induced pulmonary fibrosis.

In vitro models of fibrogenesis are used to study the interactions of particles with specific cell types (e.g. asbestos-macrophage, asbestos-fibroblast interactions). These models are also used to study cell -cell interactions and to analyse the effects of soluble cell products on other cells; for example, soluble macrophage products on fibroblasts.

461

Ideally, mechanisms observed by in vitro experimentation
should be evaluated by parallel in vivo experiments. However, these
models are more difficult to interpret because of the complexity of
in vivo systems. In vivo models also allow study of various host
factors that may be partly responsible for fibrosis, other than the
fibrogenic agent itself.

PHYSICOCHEMICAL PROPERTIES RELATED TO FIBROGENECITY

Silica

     The three main crystalline forms of pure mineral silica (silicon
dioxide) which are cytotoxic to macrophages in culture and fibrogenic
in mammals are quartz, cristobalite and tridymite. A rarer form,
coesite also has cytotoxic and fibrogenic activity but not to the
extent of the former three varieties. A fifth type, stishovite
is neither cytotoxic nor fibrogenic (Allison et al, 1968).
Stishovite differs from the other crystalline forms of silica by
virtue of its octahedral configuration, whereas the other types
have a tetrahedral configuration (Chao et al, 1962). Therefore, it
appears that the tetrahedral configuration is a requirement for the
fibrogenic activity of silica.

     Since silica particles are nearly symmetrical their diameter is
important when determining which particles are able to reach the
terminal bronchioles and alveoli. Generally, particles less than
5μm in diameter can penetrate these terminal regions of the lung,
and it is here that the silica-macrophage interactions take place
which can eventually lead to fibrosis.

Asbestos

     Asbestos (fibrous silicates) is divided into two classes
according to its chemical configuration. Serpentine asbestos is a
sheet silicate whereas amphibole asbestos is a double chain
tetrahedral structure. Chrysotile is the major serpentine form
while all other asbestos minerals belong to the amphibole class
(crocidolite, amosite and anthophyllite). The physical
characteristics of the asbestos fibres, rather than their chemical
composition appear to be the important factor in their ability to
induce fibrosis. All asbestos minerals are potentially fibrogenic.

     Since fibres are phagocytosed by macrophages, which have finite
diameters, fibre length is an important factor in their ability to
induce fibrosis. Allison (1973) has shown that short fibres
(< 5 μm) are completely phagocytosed whereas long fibres (> 30 μm)
were never completely ingested. Fibres of intermediate size
(5 - 20 μm) were sometimes completely phagocytosed but not at other
times. It is widely believed by most investigators that short fibres

(<5 μm) are not fibrogenic, possibly because they are completely
phagocytosed and rapidly eliminated. However, interstitial
pulmonary fibrosis in rats has been produced following the
inhalation of asbestos in which 84% of the fibres were 5 μm or
less in length (Holt et al, 1964).

SILICA-INDUCED FIBROGENESIS

It is generally accepted that silica-induced fibrogenesis
takes place via a two-stage mechanism (Allison et al, 1968). Stage
I involves the interaction of particles with macrophages in such a
way that a factor or factors are released which stimulate collagen
biosynthesis (Stage 2) by fibroblasts.

Although evidence has been put forward to support a two-stage
mechanism of fibrogenesis, it has proved difficult to isolate and
characterise the macrophage factor (s) which stimulate fibroblasts.
Allison et al (1977) provided support for a two-stage mechanism
in experiments using diffusion chambers limited by Millipore filters
(pore size 0.8μm) implanted in mouse peritoneum. They observed the
following: unstimulated mouse peritoneal cells alone, or silica
alone, within the chamber, produced only a slight reaction which
was no different from that elicited by chambers containing saline
solution. By contrast, the combination of small amounts of silica
and peritoneal cells produced fibrosis of parietal and visceral
peritoneum. Larger amounts (>50μg) of silica which resulted in
cell death of most of the macrophages in the chamber, produced a
lesser reaction.

In similar experiments Bateman et al, 1979, used diffusion
chambers limited by Nuclepore membranes of pore diameter 0.05 μm.
Although these experiments strongly support a two-stage mechanism of
fibrogenesis, their results were significantly different from
Allison's. The smaller pore size was used to prevent the escape
from the chamber of the particulate Dorentrup No.12 quartz used
in these experiments, and to prevent cell contact by cytoplasmic
processes through the membrane filter pores. Electron microscopy
revealed that more than 75% of the silica particles were less than
0.8 μm in diameter and only 10.7% were less than 0.2 μm. The pore
size used in Allison's experiments was 0.8 μm. Direct contact
between cells on the outside and inside of the chamber was shown
to be absent by electron microscopy and this confirmed the findings
of others (Wartiovaara et al, 1972). Millipore nitrocellulose
filters, because of their spongy three-dimensional structure, allow
cytoplasmic ingrowths and direct contact through the membrane
(Lehtonen et al, 1973). The exclusion of cell contact is important
when confirmation of diffusible macrophage products or factors is
sought.

Using Nuclepore filters the following was observed:

1.   Silica alone in the chamber produced no significant fibrosis,
2.   Unstimulated mouse peritoneal macrophages alone caused a slight
     but significant fibrotic response on the outer surface of the
     diffusion chamber filter; and
3.   The combination of silica and peritoneal macrophages failed
     to produce fibrosis over a range of silica from 20 $\mu g/10^6$
     macrophages to as little as 5 $\mu g/10^6$ cells.

The presence of lymphocytes did not appear to influence these
results. The reason for this result is the rapid cytotoxic effect
of silica even at low concentrations, such that insufficient numbers
of macrophages survive for long enough to produce fibrogenic factor
(s) in sufficient quantities, and for adequate duration, to produce
fibrosis. This was confirmed by very low viable cell counts at two
weeks compared with identical experiments performed with asbestos,
where fibrosis was produced and viable cell counts of up to 25% were
observed at this time.

These results highlight the difference between silica and
asbestos, the former being more cytotoxic than the latter. Since
asbestos is less cytotoxic for macrophages and its presence can
be tolerated for longer, it allows a more prolonged stimulation of
the macrophage, which in turn can produce and release fibrogenic
factor (s) for a longer period. However, unlike the in vivo
situation, the diffusion chamber has a non-replaceable pool of
macrophages. Consequently any fibrosis observed is the result of
fibre interaction with only one generation of macrophages. Silica,
therefore, appears to require a continuous recruitment of
macrophages for its effect. Also, the asbestos induced fibrosis in
this model was maximal at two to four weeks and then resolved
completely over the next two months. This suggests that a sustained
production of fibrogenic factor(s) is required for progressive
fibrosis. This is achieved in vivo as silica and asbestos are able
to react with successive generations of macrophages over many months
or years

ASBESTOS-INDUCED FIBROGENESIS

Experiments using asbestos in diffusion chambers as discussed
above established that macrophages are required for fibrosis and
supports the theory that asbestos-induced fibrogenesis also operates
via a two-stage mechanism. As with silica, stage I involves fibre-
macrophage interactions and stage 2 is a macrophage-fibroblast
interaction. It is thought that during stage I stimulation of
macrophages by asbestos fibres results in chronic lysosomal enzyme
release. Stage 2 (macrophage-fibroblast interaction) may occur via
at least three pathways, two of which involve lysosomal enzymes.
The enzymes might have a direct effect on fibroblast or fibroblast

stimulation might result from products of tissue destruction or mediators released from the inflammatory response.

The enzymes could have an indirect effect on fibroblast by degrading connective tissue components, cell membranes and proteins of the complement system. In this way other systems become involved which can influence fibroblast activity. It is also possible that other macrophage factor (s) as yet undescribed may be produced following their interaction with asbestos.

HOST FACTORS IN FIBROGENESIS

Despite their complexity, in vivo models of fibrogenesis are most appropriate for assessment of host factors that may be partly responsible for fibrotic lesions. Although several in vivo models of fibrogenesis exist, it emerges that:

1. The majority of in vivo models have assessed the fibrogenicity of agents or reproduced particular disease entities, concentrating on histopathological changes rather than mechanisms of fibrogenesis.
2. Where mechanisms have been examined, insufficient parameters have been used to enable a complete understanding of the events that occur in complicated in vivo models.
3. Many studies have not examined events occurring in vivo shortly after insult before obvious pathology occurs; generally, also, too few time points have been included in many studies.
4. Many of the insulting agents have been administered in high concentrations ("unphysiological dose") preventing the assessment of host factors upon which fibrogenesis may partly depend.

An animal model which would overcome some of these criticisms has recently been developed (Emerson, 1980).

Assessment of host factors was achieved by exposing the lungs of SPF guinea-pigs (whose systemic mononuclear cells had been sensitized and whose mononuclear phagocytes had been activated), to asbestos, a known fibrogenic agent which appears to require interaction with cells in order to produce fibrosis, at a dose that by itself did not cause fibrosis.

Asbestos fibres and asbestos body formation were prominent throughout the entire study (1 year) even though the animals were only exposed for two eight hour periods. Histologically the reaction to asbestos began as a minor peribronchiole mononuclear cell inflammatory reaction at a few scattered individual respiratory bronchioles, but gradually more and more bronchioles became affected with the reaction spreading to the surrounding

alveolar septa. The minor increase in reticulin fibres never
developed into collagen fibres and the reaction was short-lived with
complete resolution occurring even though asbestos bodies were
continually present.

Sensitization of lymphocytes with Freund's complete adjuvant
and/or M.tuberculosis (H37Ra) was achieved as demonstrated by
lymphocyte blast transformation to PPD. The pulmonary reaction to
these agents was a diffuse mononuclear cell infiltration of the
interstitium and air space. The severity of the reactions to the
various insults was dependent on the treatment given. Although there
was no histological evidence of increased collagen, there were
diffuse increases in reticulin, particularly in animals with
sensitized mononuclear cells which were exposed to asbestos.
However, even in the case where severe pulmonary inflammatory
changes occurred, complete resolution of the response took place
within one year.

Immunological activation of the mononuclear phagocytes was
confirmed by criteria based on electron microscopy observation of
the alveolar lumen cells, morphology of glass adherent mononuclear
cells and measurement of enzyme activity of pulmonary mononuclear
phagocytes. It was concluded from these criteria that :-
1. there was a heterogeneous population of mononuclear phagocytes
   in the alveolar lumens, a majority of which were mature and
   appeared activated, however monocytes and immature mononuclear
   phagocytes were also present in much smaller numbers;
2. the morphological studies of guinea-pig pulmonary glass adherent
   mononuclear cells gave inconclusive results as to the state of
   activation possibly due to the inability of the cells to spread
   on a glass substrate compared to activated peritoneal
   macrophages;
3. enzyme activity of the pulmonary mononuclear phagocyte provided
   the most sensitive method for demonstrating activation of the
   cells.

The results show that systemic mononuclear cell activation does
not enhance establishment of fibrosis in guinea pigs that have
inhaled asbestos for a short time. Although in this model the
results indicate that mononuclear cell activation may not be an
important host factor in fibrogenesis, one must consider that the
amount of asbestos capable of interacting with activated cells may
be too small to cause fibrosis no matter how much or how the cells
were activated. Dose/response studies in a similar model would be
needed to answer this question.

Based on observations of human pulmonary fibrosis, Carrington
(1968) proposed a set of anatomic criteria that should be met by
the ideal experimental model. These included a mixed cellular
exudate in the interstitium, protein exudate in air spaces with or

without leukocytes, proliferation of epithelial cells, gradual progression to fibrosis and honeycombing, continuing cellular activity even when partly fibrotic, and widespread focal lesions. Among the criteria that are considered favourable but not essential are the presence of hyaline membranes, smooth muscle cell proliferation, multiple species for the same agent and multiple agents for the same species. Features that the ideal model must not show are abundant polymorphonuclear leukocytes, at least after the first few days, prolonged state of extensive oedema, early respiratory failure and contamination by ordinary bacterial disease.

Although these criteria may meet the ideal animal model, it is not critical that the model be an exact replica of human pulmonary fibrosis. Rather the model should be used as a system to enable investigators to distinguish abnormal from the normal. In this manner a model is helpful in understanding the inability of the human lung to return the fibrotic to normal. Also, animal models have the advantage over human studies of :-
1. pairing control and fibrotic animals;
2. enabling dose response studies of the insulting agent;
3. studying drug therapy; and
4. obtaining the entire lung for comparative morphological, physiological and biochemical studies.

Perhaps it is not surprising that the present model did not meet all the requirements discussed by Carrington.

In man and experimental animals, the progression of interstitial fibrosis from oedema, mononuclear cell infiltration through organization to honeycombing is not always seen after diffuse alveolar damage. Thus it is possible that pulmonary injuries sometimes organize and sometimes resolve. The question of what tips the balance in favour of organization or resolution is yet unanswered.

This study was partially based on the premise that an accumulation of mononuclear cells in the lung is an essential prelude to pulmonary fibrosis. Therefore, keeping the asbestos exposed lung packed with stimulated mononuclear cells, particularly mononuclear phagocytes, should eventually lead to fibrosis. Richerson et al (1978) have taken a similar approach by immunizing rabbits systemically with Freund's complete adjuvant via the foot-pads and followed this with repeated intravenous injections of killed BCG. Their study as with these experiments produced large accumulations of interstitial and intra-alveolar mononuclear phagocytes. However, resolution of lesions occurred unless BCG was given continuously and in either case, fibrosis was scanty. Butler (1975) also thought that the macrophage-laden lung would produce a more intense fibrosis. Following systemic sensitization of mononuclear cells, hamsters were given paraquat intra-tracheally.

Results indicated that the macrophage-laden lungs were protected against paraquat injury, since the degree of fibrosis was less in animals given BCG plus paraquat compared to paraquat treatment alone.

It appears that unless there is a continuous administration of irritant to the lung resolution of the lesions without fibrosis occurs. It should also be noted that although mononuclear phagocytes may appear to be a prelude to fibrosis their presence alone, even in activated states is insufficient as a stimulus for progressive fibrogenesis. Indeed, the influx of large numbers of mononuclear phagocytes to the lung following insult is usually considered to be a protective mechanism and the presence of concurrent or subsequent fibrosis may be either due to a functional defect of the cell or continued persistence of irritant at concentrations so large that the mononuclear phagocyte is overwhelmed and is no longer able effectively to carry out its function of phagocytosis.

The common problem of complete resolution of the inflammatory and fibrotic response occurring in animal models appears to be overcome only by continual administration of insulting agents throughout a study. This is in direct opposition to what occurs in human s, particularly following exposure to fibrogenic inorganic dusts since once fibrosis has occurred, progression will continue after removal from exposure. It is not known whether this progression is due to continual activity either by the persisting fibrogenic agent or to some unknown self-perpetuating host reaction (i.e. virus). In the case of inorganic dust, persistence of cytotoxic dust in the lung may in itself be enough to cause continuing fibrosis but in the case of soluble agents such as avian protein, the persistence of antigen is more difficult to accept.

In all cases where exposure to fibrogenic agents occurs (human and animal) there is the question of individual host susceptibility. It has been shown that even among heavily exposed asbestos workers only a proportion developed clinical asbestosis and of those affected the extent of lung involvement varied greatly (Selikoff and Lee, 1978).

Although the results of this model are far from satisfactory in answering the questions as to how far fibrosis is occurring as the result of persisting activity of initiating agent (due either to persistent exposure or to persisting activity of retained particles in the lung) and as to how far the perpetuating fibrosis depends on host factors and no longer on the initiating agent, they show enough promise to justify further investigations.

The pulmonary response to insult, preceeding fibrosis is similar to that seen in man and other experimental animal models of

fibrogenesis. However, to obtain full benefit from this model dose/
response relationships and other methods of macrophage,activation
should be studied.

References

Allison, A.C., Experimental methods - cell and tissue culture:
    effects of asbestos particles on macrophages, mesothelial cells
    and fibroblasts, in:"Biological Effects of Asbestos,
    Proceedings Working Conference International Agency for
    Research on Cancer", Lyon, France, October, 1972, P.Bogovski,
    J.C.Gilson, V. Timbrell and J.C.Wagner, eds., I.A.R.C.
    Scientific Publication, No.8, Lyon, 1973.
Allison, A.C., Clark, I.A., and Davies, P., 1977, Cellular
    interactions in fibrogenesis, Ann.Rheum.Dis., 36:Suppl., 8-13.
Allison, A.C., Harington, J.S., and Birbeck, M., 1968, An
    examination of the cytotoxic effects of silica on macrophages,
    J.Exp. Med. 124:141.
Bateman, E.D., Emerson, R.J., and Cole, P., The use of diffusion
    chambers to examine the biological effects of mineral dusts,
    in: "Proceedings of the International Workshop on the in vitro
    Effects of Mineral Dusts", Cardiff, Wales, September 4-7, 1979,
    in press.
Becklake, M.R., 1976, Asbestos related diseases of the lung and
    other organs: their epidemiology and implications for clinical
    practice, Am. Rev. Respir. Dis., 114:187.
Butler, C., 1975, Pulmonary interstitial fibrosis from paraquat in
    the hamster, Arch. Pathol., 99:563.
Carrington, C., 1968, Organizing interstitial penumonia, definition
    of the lesion and attempts to devise an experimental model,
    Yale J.Biol.Med., 40:352.
Chao, E.C.T., Fahey, J.J., and Littler, J., 1962, Stishovite, SiO_2,
    a very high pressure new mineral from Meteor Crater, Arizona,
    J.Geophys. Res., 67:419.
Emerson, R.J., 1980, Cellular Interactions in Pulmonary Fibrogenesis,
    Ph.D. Thesis, University of London, London, England.
Holt, P.F., Mills, J., and Young, D.K., 1964, Experimental asbestosis
    in the guinea-pig, J.Pathol.Bact., 92: 185.
Lehtonen, E., Nordling, S., and Wartiovaara, J., 1973, Permeability
    and structure of ethanol sterilized nitrocellulose (Millipore)
    filters, Expl.Cell.Res., 81: 169.
Richerson, H.B., Seidenfield, Jj., Ratajeyak, H.V., and Edwards, D.W.,
    1978, Chronic experimental interstitial pneumonitis in the
    rabbit, Am. Rev. Respir.Dis., 117: 5.
Selikoff, I.J., and Lee, D.H., Asbestos and Disease, Academic Press,
    London, 1978.
Wartiovaara, J., Lehtonen, E., Nordling, S., and Saxen, L., 1972,
    Do membrane filters prevent cell contacts? Nature 238: 407.

D I S C U S S I O N

LECTURER: Emerson   CHAIRMAN: Bonsignore

BONSIGNORE: We have time for one or two questions.

PRODI: Do you have any data on the biological activity as a
                 function of fibre length for asbestos fibres, because
                 I think your technique could be very suited to that
                 study and, as I understand it, it would be a very
                 important piece of information.

EMERSON: You are thinking of the interaction of asbestos with
                 the cell itself, what does the asbestos do to the
                 cell membrane to cause a release of factors?   I
                 myself have not looked at that, but it has been shown
                 by Davis that macrophages in vitro would phagocytose
                 asbestos particles and would selectively release
                 glycolytic enzymes, glucosaminidase, galactosidase
                 and a number of others, without any release of LDH,
                 the cells remaining viable.   What I think is
                 happening is that it is perpetuating an inflammatory
                 response that may lead to a fibroblastic response.
                 As far as the chemistry of asbestos interacting with
                 the cell membrane, I myself do not have any
                 information.

NEWMAN TAYLOR: Can I ask you two quick questions. Firstly, you have
                 shown us the predominantly mononuclear cell
                 inflammatory reaction to short periods of asbestos
                 inhalation, on your final slide you make the point
                 that one may get a resolution with elimination of the
                 irritant causing the reaction.   But yet I believe I
                 am correct in saying that you demonstrated that the
                 asbestos fibres were still to be found in the lung at
                 the end of the experiment, when in fact there had
                 been resolution, and I wonder whether you have any
                 explanation to offer for this.

EMERSON: I was very surprised seeing asbestos sitting in the
                 lung. It may be individual host susceptibility to
                 the agent. Clinically, I think one could think about
                 patients who are exposed to asbestos that do not
                 develop fibrosis, while others do. This is the big
                 question that we are trying to answer; why do some
                 people get pulmonary fibrosis following asbestos
                 exposure and others do not. Yet one can see the

asbestos sitting in these lungs. Whether there is latent virus or these cells have to be triggered in some way to go back and react with the asbestos particles, whether they are coated some way or protected.

NEWMAN TAYLOR: The second point was that there is some evidence from epidemiological studies that asbestosis may be a rapidly progressive fibrosis which is self limiting, and I wondered if you have some evidence from experimental animals which would either support or refute that hypothesis.

EMERSON: Yes, there is experimental evidence to support that. I have always thought that asbestos exposure, once the fibrotic lesion develops and fibrosis begins, removal of an individual from exposure does not prevent progressive fibrosis. Animal studies have shown that the development of fibrosis depends on the species. There can be progressive fibrosis in one species and complete resolution in another. It is most likely due to host variation, host susceptibility.

CARINI: Have you tried to culture the macrophage in vitro and then look for a soluble factor.

EMERSON: I have not. I wanted to in the beginning and then tried but I was doing this work with only one technician. I would like to go back and repeat it. When I think of setting up experiments like that I would like to have parallel in vivo and in vitro experiments, but you need a few hands to perform this type of experiment. It is an excellent point and I have considered it.

RESPIRATORY DISEASES IN FARMERS

Massimo Crepet, Giuseppe Mastrangelo, Bruno Saia,[+] and
Paolo Paruzzolo[*]

+Istituto di Medicina del Lavoro
*Istituto di Calcolo Scientifico
Università di Padova, Italy

INTRODUCTION

Few epidemiological data have been reported regarding the
health status of rural workers in relation to environmental or
occupational hazards. This study deals with respiratory diseases
in farmers. A cross-sectional epidemiological survey was carried
out to ascertain the prevalence of chronic respiratory diseases
and the various associated factors.

POPULATION AND METHODS

The area surveyed is situated in the Pianura Padana, near
Cremona, Italy. The total population (about 35,000 inhabitants) is
scattered in 26 towns. As there are only 20 to 22 small factories
in the 50,000 hectares covered, atmospheric pollution is not
significant.

The population chosen consisted entirely of agricultural lab-
orers working and living in the area. Their names and addresses
were taken from the Local Authority Registers (Ufficio di Collo-
camento e Cassa Mutua Coltivatori Diretti). Each subject received
a written request to attend an interview and undergo examination.
In cases of noncompliance, a second letter was sent. Table 1 shows
how effective this second attempt was: 3225 subjects were invited
and 2932 agreed to be examined, a response rate of 90%.

The questionnaire used was suggested by the "Società Italiana
di Medicina del Lavoro e d'Igiene Industriale."[1] Questions on respiratory symptoms were taken from the E.C.C.S. questionnaire.[2]

Table 1. Population surveyed

	Number	%
Subjects invited	3225	100.0
Subjects examined		
after 1st attempt	2663	82.6
after 2nd attempt	2932	90.9

Modifications were made in order to establish whether the subject had attacks of breathlessness associated with fever or shivering. Questions were added so that information could be obtained concerning family history of rhinitis, chronic bronchitis, and bronchial asthma, environmental factors, and living conditions. A comprehensive occupational history was also included. There were questions about the length of work in agriculture, size of farms, and type of duty or duties undertaken by the subjects. Specific questions were asked to assess possible contact with animals and plants or dust containing fungal spores or allergy-producing substances. The questions were asked by six trained interviewers.

A chest x-ray was taken and seen by a physician to establish whether the subject had pulmonary tuberculosis or lung cancer. Subjects with lung cancer, active pulmonary tuberculosis, or diffuse scarring of the lung were excluded from the subsequent statistical analysis.

Simple spirometry was performed by trained technicians, using a VICA-TEST spirometer (Hellige). The one-second forced expiratory volume (FEV_1) and the forced vital capacity (FVC) were measured. An electronic device provided the maximum mid-expiratory flow (MMF). Each man made three exhalations; the maximum reading was used; the values were corrected to BTPS.

Diagnostic Criteria

Chronic bronchitis (CB): cough with phlegm almost daily for a period of three months in each year for at least two consecutive years.

Bronchial asthma (BA): 'Yes' answers to the following questions:
1. When resting, have you ever had attacks of breathlessness?
2. Are/were these attacks accompanied by any wheezing or whistling in your chest?

Farmer's lung (FL): 'Yes' answer to the question: Have you ever had breathlessness accompanied by cough and a temperature several hours after handling moldy hay?

Statistical Analysis

A multiple regression analysis was used in order to discover which of a set of variables x_1, x_2, x_3,... exerted the greatest influence on the dependent variable y. For example, by relating FEV_1 simultaneously to a large number of variables describing the subject's age, height, smoking habits, living conditions, occupational history, etc., it is possible to find which factors exert a particular influence on FEV_1. The F test of significance for a regression coefficient verifies whether the corresponding predictor variable can be excluded from the equation of multiple regression without any significant effect on the variation of y.[3]

Apart from FEV_1, FVC, MMF, age, and height, all other variables were 0 or 1 quantities, 1 for subjects who reported having had a characteristic and 0 otherwise. Eighty-one predictor variables were forced in each multiple regression equation.

The statistical handling of the data was done in the Computer Center of Padua University by means of SPSS computer programs.[4]

Our results were obtained with male subjects. At the beginning of the study period the number was 2626, but subjects with pulmonary tuberculosis or lung cancer were excluded, and the results obtained refer to the remaining 2579 males with a mean age of 47.24 $\pm$ 12.54; 57.9% of the subjects were smokers.

RESULTS

Few of the 81 predictor variables showed a significant influence; we only considered the variables influencing an increase of prevalence of the respiratory syndromes or a reduction of mean level of respiratory function. Such predictor variables are shown in the rows of Tables 2 and 3. The intersections of rows and columns give the level of significance of the F test described earlier.

The dependent variables of three multiple regression equations are reported in the columns of Table 2. We can see that FEV_1 depends on age, smoking habits, and family history of chronic bronchitis or bronchial asthma. A significant association exists between the FEV_1 and several aspects of agricultural work: milking and fodder cultivation or its storage in silos. Apart from these variables, which effect a reduction of the mean FEV_1 level, other variables, not shown in the table, are significantly associated with an FEV_1 increase. These are height, use of protective

Table 2. Multiple regression analysis: significance
of the test for deletion of variables
(Dependent variables: FEV_1, MMF, FVC)

	FEV_1	MMF	FVC
Smoking	.000	.000	--
Age	.000	.000	.000
Family history of CB	.014	.056	.034
Family history of BA	.001	.000	--
Silo on the farm	.047	.011	--
Fodder cultivation	.031	.011	--
Milking	.021	.009	--

CB = Chronic bronchitis; BA = Bronchial asthma.

Table 3. Multiple regression analysis: significance
of the test for deletion of variables
(Dependent variables: BC, BA, FL)

	CB	BA	FL
Smoking	.013	--	--
Age	.031	.004	--
Family history of CB	--	.014	--
Family history of BA	--	.001	.009
Moldy hay handling	--	--	.012
Farm less than 5 hectares	--	--	.003
Hayloft on the farm	--	--	.014
Rye cultivation	--	--	.016

CB = Chronic bronchitis; BA = Bronchial asthma;
FL = Farmer's lung.

clothing during pesticide spraying, and use of fertilizers and agricultural machinery.

MMF exhibits the same epidemiological behavior as FEV_1, while
CVF appears to depend only on age and family history of chronic
bronchitis.

In the whole group of 2579 male farm workers, the prevalence
of chronic bronchitis, bronchial asthma, and farmer's lung was
11.85, 3.96, and 2.71% respectively. Table 3 shows the factors that
significantly increase the prevalence: age and smoking habits for
chronic bronchitis; age and family history of respiratory disease
for bronchial asthma; family history of bronchial asthma and handling of moldy hay for farmer's lung. Farmer's lung was common in
farms with a hayloft, rye cultivation, and a surface of less than
5 hectares.

DISCUSSION AND CONCLUSION

In planning the survey, factors that could affect the response
rate were carefully considered. The subjects were informed of the
impending study through assemblies held at work, and were paid for
time lost. Political authorities, unions, and employers all cooper-
ated and the response rates was high (90%).

One aspect of agricultural production is that the type of work
done by a person frequently changes, especially on small farms.
This precludes a strict distinction between duties and renders analysis of the health problems involved in each job difficult. We
must also consider the confounding effects of nonoccupational risk
factors (i.e., factors already known or strongly suspected to be
related to the disease), if unbiased statistical tests of significance are desired. This problem is solved when we have an inter-
pretive tool to evaluate the relative importance of risk factors.
With this in mind we used the multiple regression analysis. In the
multiple regression equation the partial regression coefficient
stands for the expected changes in y with a change of one unit in
x(j) when other x(i)'s are held constant. An equally important and
obvious interpretation is that the combined effects are additive.

A control group of persons living in the same area but not
working in agriculture is being examined; unfortunately these data
are not yet available.

Few epidemiological studies on respiratory symptoms in agricultural communities have been published, and comparison between the
data reported in the literature is difficult because of differences
in age, smoking habits, or definition of the particular disease.
For example, in the present study chronic bronchitis is defined as

"persistent cough and phlegm for at least two consecutive years,"
whereas Higgins,[5] Higgins and Cochrane,[6] Morgan,[7] and Huthi[8] have
defined chronic bronchitis as "persistent phlegm production." This
may explain the lower prevalence of the disease in our population
in comparison to that found in agricultural communities by these
authors.

For bronchial asthma, our estimate of prevalence (3.96%) agrees
with the 3.7% prevalence found in a sample of British farmers,[7] with
the 3.4% found in a random sample of a town population in South
Wales,[9] and with the 3.0% prevalence reported by the U.S. National
Center for Health Statistics in a U.S. national sample of 116,000
people.[10] However, a lower prevalence (1.4%) of bronchial asthma
was found in a rural community in Finland.[8]

In our population we found a significant association between
the disease and a family history which was positive for bronchial
asthma and chronic bronchitis. This finding confirms other data reported which suggest that certain asthmas depend on genetic factors.[11]
However, there is increasing evidence confirming that asthma is a
multifactorial disease. A larger variety and increasing number of
occupational exposures have been associated with either the onset
or the aggravation of the disease. It must be kept in mind that
often the most sensitive individuals are not in the work force being
studied, since they do not remain in a job which causes severe symptoms. Therefore the lack of association between bronchial asthma
and occupational risk factors in our population may be a bias due
to "internal migration."

The diagnosis of acute farmer's lung usually requires a history of exposure to moldy hay or grain, typical clinical symptoms,
abnormal chest x-ray, and serological evidence of precipitins to
M. faeni or vulgaris. However, for epidemiological purposes, these
criteria may be too restrictive. It is well known that abnormalities observed on chest x-rays may clear completely after the acute
attack, and serological studies may become negative during a period
of two to three years following the initial illness.[12] Thus, in
retrospectively evaluating the occurrence of farmer's lung one must
rely on clinical symptoms as the diagnostic criteria. In the farm-
ing population covered by this study, the prevalence of the farmer's
lung syndrome (2.7%) is compatible with the 2.3% found in a rural
population of Somerset,[13] but it is 2- to 3-fold lower than that
reported elsewhere. In 93 Devon farmers, Morgan[7] found that 9.6%
of subjects had a typical history of farmer's lung. A similar prevalence has been reported recently in a western Wyoming population.[12]
In a study carried out on 655 farm workers in Scotland, Grant[14]
found that the prevalence of subjects who had experienced farmer's
lung ranged from 2.3 to 8.6%. Both meteorological and economic
factors may account for these differences. In a retrospective
study of 82 cases of farmer's lung[15] a clustering of cases was

observed in cattle breeding areas, which are regions of extensive hay cultivation and high humidity. The annual distribution of cases was strongly linked to the rainfall during previous hay-making. The working methods were also important factors: prevalence was higher when the work was done manually rather than mechanically. Our data confirm these observations. They show an increase in the prevalence of farmer's lung in small farms and in cases of traditional work organization. As farmer's lung occurs in the minority of exposed workers, some authors have suggested the influence of genetic factors.[16] Our results show that the prevalence of the disease, which has an immunologic pathogenesis, was higher in subjects with a family history of bronchial asthma.

As a cross-sectional study only examines individuals at one point in time, it is extremely difficult to assess the etiological role of factors which exert an influence over a considerable period of time. In general, cross-sectional studies serve as a quick exploration prior to conducting a retrospective or prospective study.

SUMMARY

A population of 2932 farmers (90.9% response rate) was studied in order to ascertain the prevalence of respiratory diseases and the various associated factors.

A questionnaire was used to obtain informations about respiratory symptoms and various risk factors (family history of respiratory disease, environmental factors, living conditions, and occupational history). A chest x-ray was taken and simple spirometry performed.

The results refer to 2579 male subjects. It is shown that chronic bronchitis depends upon age and smoking, bronchial asthma upon age and a family history of respiratory diseases, farmer's lung upon a family history of bronchial asthma and moldy hay handling. This syndrome was found to be more common in subjects working on farms with hayloft, rye cultivation, or an area of less than five hectares.

The mean levels of FEV_1 and MMF were lower in subjects with a family history of respiratory disease, in milkers, and in subjects working on farms with fodder cultivation or its storage in silos.

BIBLIOGRAPHY

1. D. Casula, Questionario per le broncopneumopatie in agricoltura, in: "La salute in agricoltura," Ed. C.S.Z., Cremona (1979)

2. Commission of the European Communities, "Notes on the 1967 E. C.C.S. questionnaire for studying chronic bronchitis and pulmonary emphysema," A. Minette, D. Brille, R. Van Der Lende, F. Sanna-Randaccio, and U. Smidt (Eds.) Industrial Hygiene and Medicine Texts No. 14, Office for Official Publications of the European Communities, Luxembourg (1973)

3. P. Armitage, "Statistical methods in medical research," Blackwell Scientific Publications, Oxford and Edinburgh (1971)

4. N. Nie, D.H. Bent, and C.H. Hull, "Statistical package for the social sciences," McGraw-Hill, New York (1970)

5. I.T.T. Higgins, Respiratory symptoms, bronchitis, and ventilatory capacity in random sample of an agricultural population, Brit. Med. J. 2: 1198 (1957)

6. I.T.T. Higgins and J.B. Cochran, Respiratory symptoms, bronchitis and disability in a random sample of an agricultural community in Dumfriesshire, Tuberche 39: 296 (1958)

7. D.C. Morgan, J.T. Smyth, R.W. Lister, and R.J. Pethybridge, Chest symptoms and farmer's lung: a community survey, Brit. J. Ind. Med. 30: 256 (1973)

8. E. Huthi, J. Takala, J. Nuutinen, and A. Poukkula, Chronic respiratory disease in rural men, Ann. Clin. Res. 10: 87 (1978)

9. M.L. Burr, A.S. St. Leger, C. Bevan, and T.G. Merritt, A community survey of asthmatic characteristics, Thorax 30: 663 (1975)

10. U.S. Dept. Health, Education and Welfare, PHS, Health Resources Administration, "Prevalence of selected chronic respiratory conditions, United States, 1970," Vital and Health Statistics Series 10: No. 84 (1973)

11. F.E. Speizer, Epidemiological aspects of asthma, Triangle 17: 117 (1978)

12. D. Madsen, L.E. Klock, F.J. Wenzel, J. La Mar Robbins, and C. Du Wayne Schmidt, The prevalence of farmer's lung in an agricultural population, Am. Rev. Resp. Dis. 113: 171 (1976)

13. J.V.S. Pether and F.B. Greatorex, Farmer's lung in Somerset, Brit. J. Ind. Med. 33: 265 (1976)

14. I.W.B. Grant, W. Blyth, W.E. Wardrop, R.M. Gordon, J.C.G. Pearson, and A. Mair, Prevalence of farmer's lung in Scotland: a pilot survey, Brit. Med. J. 1: 530 (1972)

15. P. Leophonte, A. Rongieres, M. Durand, and M. Mirman, La maladie du poumon du fermier dans le départment de l'Aveyron, Poum. et Coeur 34: 219 (1978)

16. C. Giuliano, Alveoliti allergiche estrinseche o pneumopatie da ipersensibilità, in: "Immunologia clinica ed allergologica," U. Serafini, Ed., Uses Edizioni Scientifiche, Florence (1978)

INJURY BY ISOCYANATES

Emilio Sartorelli

Institute of Occupational Medicine
University of Siena
Siena (Italy)

The incidence of occupational asthma is continuously increasing in many working populations: in workers engaged in polyurethane-foam production and in other plastics industries, in the wood and furniture industries, in drug production and other chemical plants, in agriculture and the food industries (for instance, among bakers), and so on.

A well defined risk is related to the manufacture and use of polyurethane compounds obtained by reacting organic isocyanates with various polyfunctional alcohols (polyols). Among the isocyanates we find first of all toluene diisocyanate (TDI) and methylene bis-phenyl diisocyanate (MDI), and also hexamethylene 1,6-diisocyanate, naphthylene 1,5-diisocyanate, and others. Very important products are flexible and rigid polyurethane foams, polyurethane paints and varnishes, and polyurethane insulating products (for instance, thermal insulation for refrigerators).

Isocyanate vapors and aerosols exert a primary irritant effect on respiratory airways even at low atmospheric oncnentrations. The threshold limit value (TLV) of TDI, proposed by ACGIH of USA, is only 0.04 mg/m^3 or 0.005 ppm, which is a surprisingly low permissible atmospheric concentration in work places. Furthermore, in some workers exposure to isocyanates induces a "sensitization" at the respiratory level, with onset of a typical asthma-like syndrome. In these subjects even extremely low atmospheric concentrations of isocyanates can induce severe attacks of asthma, and therefore this condition constitutes and important disability.

Another type of reaction to isocyanates in some chronically exposed working populations is an impairment of ventilatory function, elicited by longitudinal studies

The pathogenesis of isocyanate sensitivity is not yet completely explained. The possibility of immunologic mechanisms has been
pointed out on the ground of different immunological approaches
(see Nava et al., 1975). Recently Karol et al. (1978, 1979) reported
the results of investigations carried out by the radioallergosorbent
test (RAST) using the antigen TDI-HSA (TDI conjugated with human
serum albumin). In sera of some TDI-sensitive workers these authors
found specific IgE antibodies. Butcher et al. (1979), however, believe that a direct pharmacological mechanism may induce TDI-asthma;
in fact, TDI seems to inhibit the stimulatory action of isoproterenol
and prostaglandin E_1 on lymphocyte AMPc in vitro. Many other
authors, including myself, found no correlation between the TDI-
induced asthma and atopic condition.

The diagnosis of asthma induced by "sensitization" to isocyanates can be obtained mainly by bronchoprovocation challenge tests
(Pepys et al., 1972; Sartorelli et al., 1976; Zedda et al., 1976;
Butcher, 1979), but a careful evaluation of clinical pattern is
obviously very important.

According to our experience, among the respiratory function
tests the FEV_1 is very suitable, but we may also calculate systematically the indices derived from flow/volume curves.

Different types of reaction (early, dual, and late) can be
obtained: therefore it is necessary to follow the subject for 24
hours in order to detect the pattern of airway obstruction. In some
cases the asthmatic reactions induced by TDI challenge test can be
inhibited by prior administration of disodium cromoglycate.

The data collected by the Institute of Occupational Medicine
of Siena in the last five years concern a lot of workers engaged in
a refrigerator manufacturing plant (polyurethane foam used as
thermal insulation) and in several small furniture factories (using
polyurethane varnishes).

Among 147 subjects exposed to TDI vapors for 1 to 10 years in
the refrigerator plant (where the atmospheric concentrations of TDI
were very often up to the TLV of 0.04 mg/m^3), 14 quit the job
because of onset of asthma syndrome, which was evidently related to
TDI sensitization (bronchoprovocation test positive). Indeed,
clinical and functional data demonstrated that 30 other workers
developed chronic bronchitis. Irritative symptoms of the eyes and
upper airways were common. A longitudinal study carried out on 40
workers at the same plant over a period of 3 years showed on the
average an annual decrement of the FEV_1 of 125 ml, while in control
populations the annual decrement of the FEV_1 was only 21-33 ml
(Franzinelli et al., 1978).

We have also observed 61 cases of TDI-induced asthma arising
from occupational exposure in the furniture factories. In some
plants we studied the workers on the job and were able to detect
the onset of bronchial obstruction during the day (one-day effect).
Figure 1 demonstrates the fall of the FEV_1 during the afternoon in
two females workers.

The clinical course of isocyanate asthma is very different
depending on the exposure that follows the onset of the respiratory
syndrome. In all reactive subjects who continue to work with even
minimal exposure to isocyanates, the asthma attacks are repeated
unremittingly. In no case did we observe the appearance of a
specific tolerance. On the other hand, in most subjects who left
the dangerous job we observed the disappearance of asthma attacks:
in fact, follow-up of 37 cases for 1-5 years after the end of
exposure showed that only 5 subjects had persistent asthma. However,
control functional tests performed on TDI-reactive subjects who were
no longer exposed nevertheless showed in most cases a persistent
impairment of ventilatory function.

Figures 2, 3, 4, and 5 show typical bronchoprovocation
tests with early, dual, or late positive reactions to TDI exposure.
In all cases the control curve was obtained by exposure to the other
components of the varnish (no TDI).

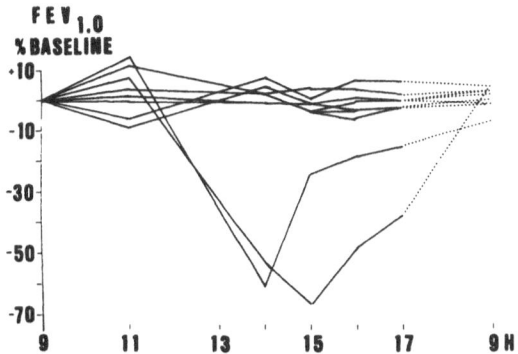

Fig. 1.   Behavior of the FEV_1 during work in 9 subjects (5 males
          and 4 females) exposed to TDI vapors in a furniture
          factory. Two females present a late bronchial obstruc-
          tion. (On another day the prior administration of 40 mg
          of disodium cromoglycate inhibited this late reaction to
          TDI in both subjects completely.)

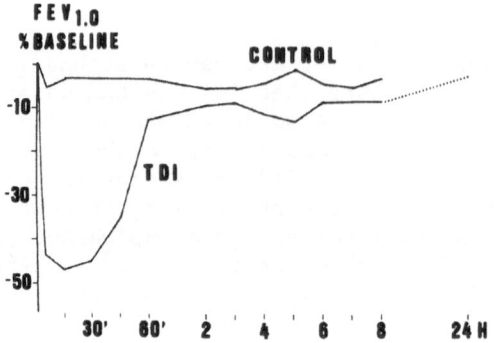

Fig. 2.   Early reaction to TDI challenge test. Asthmatic 52-year-
old male, joiner, not atopic. Serum total IgE 92 U/ml.
Baseline FEV$_1$ -5% predicted.

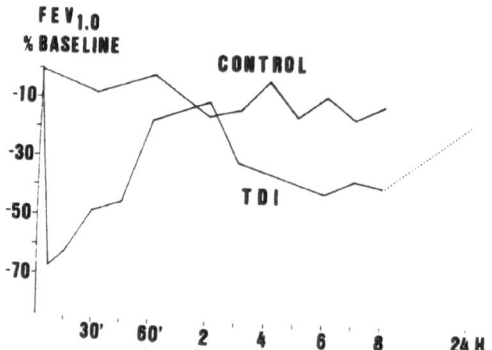

Fig. 3.   Dual reaction to TDI challenge test. Asthmatic 30-year-
old female, furniture maker, atopic. Serum total IgE
80 U/ml. Baseline FEV$_1$ +10% predicted.

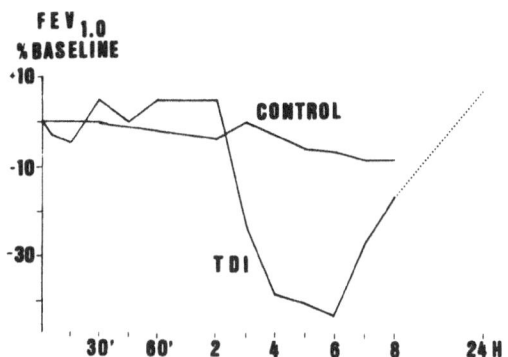

Fig. 4. Late reaction to TDI challenge test. Asthmatic 45-year-
old female, furniture maker, not atopic. Serum total
IgE 35 U/ml. Baseline FEV$_1$ -24% predicted.

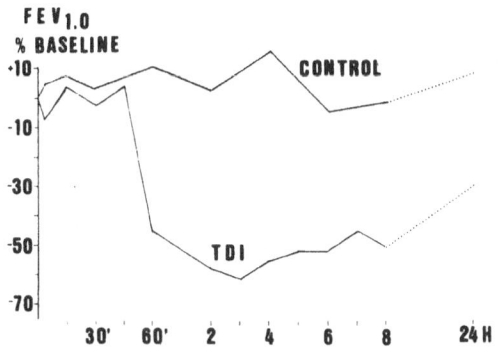

Fig. 5. Late reaction to TDI challenge test. Asthmatic 41-year-
old male, furniture maker, not atopic. Serum total
IgE 5 U/ml. Baseline FEV$_1$ +9% predicted.

In summary, the asthma-like syndrome induced by "sensitization" to isocyanates, mainly of occupational origin, is the most important feature of respiratory injury by isocyanates. A chronic impairment of ventilatory function is another consequence of continuous exposure to low atmospheric concentrations of these highly reactive compounds. Sensitive workers must be removed from the dangerous job, and in any case it is imperative that the atmospheric concentration of TDI, MDI, or other isocyanates be kept below TLV.

REFERENCES

Butcher, B.T., 1979, Inhalation challenge testing with toluene
    diisocyanate, J. Allergy Clin. Immunol., 64, 655.

Butcher, B.T., Karr, R.M., O'Neil, C.E., Wilson, M.R., Dharmarajan,
    V., Salvaggio, J.E., and Weill, H., 1979, Inhalation challenge
    and pharmacologic studies of toluene diisocyanate (TDI)-
    sensitive workers, J. Allergy Clin. Immunol., 64, 146.

Franzinelli, A., Mariotti, F., and Innocenti, A., 1978, Patologia
    respiratoria da isocianati in un'industria di frigoriferi,
    Med. Lavoro, 69, 163.

Karol, M.H., and Alarie, Y.C., 1978, Serologic test for toluene
    diisocyanate (TDI) antibodies, J. Occup. Med, 20, 383.

Karol, M.H., Sandberg, T., Riley, E.J., and Alarie, Y.C., 1979,
    Longitudinal study of tolyl-reactive IgE antibodies in
    workers hypersensitive to TDI, J. Occup. Med., 21, 354.

Nava, C., Arbosti, G., Briatico-Vangosa, G., Cirla, A., Marchisio,
    M., and Zedda, S., 1975, Pathology produced by isocyanates:
    methods of immunological investigation, La Ricerca, 5, 135.

Pepys, J., Pickering, C.A.C., Breslin, A.B.X., and Terry, D.J.,
    1972, Asthma due to inhaled chemical agents--toluene
    diisocyanate, Clin. Allergy, 2, 225.

Sartorelli, E., Franzinelli, A., Catalano, P., Innocenti, A., and
    Severi, A., 1976, Patologia respiratoria da vernici poliure-
    taniche, Atti 39° Congresso Società Italiana Medicina del
    Lavoro, Fiuggi.

Zedda, S., Cirla, A., Aresini, G., and Sala, C., 1976, Occupational
    type test for the etiological diagnosis of asthma due to
    toluene diisocyanate, Respiration, 33, 14.

D I S C U S S I O N

LECTURER: Sartorelli   CHAIRMAN: Cumming

CANDURA: First of all I would like to congratulate you on your beautiful presentation, but I also have an objection. Whilst speaking of furniture factories in Tuscany you neglected our furniture factories in Lombardy. Apart from this, I wanted to ask a question. Several years ago I advanced a hypothesis on the possible action of isocyanate. Taking Methyl Phenyl Isocyanate as an example, this substance hydrolyses very readily, and the reason why foams form is that carbon dioxide bubbles produce many holes, because two molecules of CO_2 are released, with the consumption of two molecules of H_2O and a diamine is formed and these are strongly sensitising substances, in the skin and possibly also in the bronchi. My question to Sartorelli is whether he thinks that this hypothesis can be applicable to some of these compounds in respect of asthma.

SARTORELLI: It is a very interesting question. The problem of the demonstration of the immunological mechanism with chemicals presupposes the presence of an antigen, so the problem would be that of making the antigen with the amines. Although the immunological capacity of amines is well-known, no one appears to have studied this and immunologists tend to stay away from these problems, which are of extreme difficulty. The problem of the 'construction' of antigens with chemical substances, which have to be conjugated with proteins in order to become complete antigens and to be stable, homogeneous and active, is certainly a difficult problem. To give an answer we should require an antigen of this type. One could use provocation tests, but nobody has done them so far. It would be an intelligent thing to do.

CUMMING: It would be perfectly easy to test Candura's hypothesis by giving a provocation test with a diamine directly. Perhaps someone could consider doing that.

SANT'AMBROGIO: Maybe I already have the answer, but I am not sure. Polyurethanes in the form of foam or solid, are they dangerous? Also in the solid form, in the case you describe of the refrigerator manufacturing plant where you found a number of asthmatics or subjects with altered VMS, was the contact with foam or other solid polyurethanes?

SARTORELLI: No the danger is at the moment when the polyurethane foam is generated, that is from the contact of isocyanate with the polyalcohol, when isocyanate vapours are released into the air. When the polyurethane resin is complete there is no danger from it at ambient temperature. In some cases, if the resin is heated, then it can release a certain amount of isocyanic monomer, which is present inside the polyurethane foam. Under these conditions there can be reactions, but it is much more difficult. A fire would produce these conditions.

SANT'AMBROGIO: Another question. When you showed those flow volume curves, in the subject after six hours of exposure to isocyanate, that was a paint, a liquid, wasn't it? Besides the obstruction, was there also an increase in end expiratory volume in that patient?

SARTORELLI: That was not measured.

PAOLETTI: When you spoke of FEV_1, you said you used this test to discriminate between before and after the provocation. Did you choose this test because you did not find significant reductions in flow at 50 and 75%? Because in the slide you showed there seemed to be a remarkable reduction in flow at 50 and 75%. The second question is whether, after exposure, you found subjects where FEV_1 was unaltered and flow t 50 and 75% was reduced, because in our experience, for instance in exercise asthma, this is what we find. Finally, third question, with regard to the reduction in diffusion capacity, don't you think that this is due more to a decrease in ventilation than to a direct impairment of diffusion?

SARTORELLI: I did not say that the indexes from flow volume curves cannot be significant in a high proportion of cases. I said that bearing in mind the conventional significance of index variation, that is VEMS more than 15% reduction and at least 30% reduction for flow-volume curve indexes, because we know the variability of these tests during the circadian rhythm, then taking the average from a large number of subjects we found that VEMS was a more satisfactory index and more discriminating. In individual cases you can have stronger responses with the flow-volume curve. In this particular series all cases, by definition, had altered VEMS. This was also to have diagnostic uniformity. As to CO transfer, with persistent bronchoconstriction it is very likely that the change in CO transfer is due to distribution of ventilation. The radiographs, after 24 hours, were all normal. So I think that alveolar problems are still to be demonstrated. Also from the clinical point of view there is an asthmatic aspect and not an alveolar one.

CUMMING: Thank you Sartorelli. Other questions?

CANDURA: This is not a question, it is a comment. I would like to say that it is very important to bear in mind the differences between the danger involved in finished compounds and the danger involved in compounds which are modified due to various influences. We occupational doctors know very well that many substances, for example when they are burned, can become very dangerous. If, for example, a polyvinyl resin is burned, phosgene can be released, which is a very powerful irritant for the deep lung. If resins like teflon are burned, that is a polyfluoroethylene, they release substances like perfluorobutylene, which is twenty times more powerful than phosgene. I just wanted to stress that we must be very careful when we say that a certain product is 'harmless' because under particular conditions, emergency conditions such as a fire, it may become extremely dangerous.

SARTORELLI: Candura is perfectly right. Another case is that of polyethylene, normally totally harmless, but if it is heated to 200 or 300°C it may release hydrochloric acid vapours.

DETECTION OF CIRCULATING IMMUNE COMPLEXES IN ATOPIC PATIENTS

Claudio Carini and Jonathan Brostoff

Department of Immunology
Middlesex Hospital Medical School
London, England

INTRODUCTION

Circulating immune complexes play a pathogenic role in several diseases by virtue of their immunological and biological properties. On the one hand, they promote an inflammatory reaction by complement activation (1), an action which largely explains their capacity to damage various tissues such as renal glomeruli, blood vessels and joints.

On the other hand, they possess tolerogenic properties and can, to some extent, decrease the immune response by interferring with cellular immunity. They act on T and B lymphocytes and on the collaboration between these cells and macrophages thus hindering the synthesis of specific immunoglobulins. Circulating immune complexes may also play an important role in food allergy as shown by Brostoff et al.(2) in patients with asthma, atopic eczema and hay fever.

In an attempt to elucidate the nature of these immune complexes, we performed oral challenges with a specific allergen in patients with asthma and eczema, to investigate the true course of the entry of antigen, the production of immune complexes containing IgG as well as antigen specific IgE and have related the immunological findings to the clinical symptoms. We then looked for a beneficial effect on symptoms and formation of immune complexes by orally given sodium cromoglycate (3), to see if such treatment blocked the entry of antigen and subsequent formation of immune complexes.

Patients

We selected a number of patients with either having asthma, atopic eczema, hay fever and also allergic to egg. After withdrawal of egg from the diet the symptoms on a following challenge with egg were studied. At the same time, renal peak expiratory flow and blood samples were obtained before and after challenge. On a further occasion, oral sodium cromoglycate was given to the patients at a dose of 500mg 2 hrs before and 500mg a quarter of an hour before the challenge with egg, and similar tests performed. The interval between the two challenges was at least 2 months.

Normal Subjects

Healthy non-atopic adults were used as controls.

METHODS

A refined ultracentrifugation technique was used as previously described (4). Serum samples (0.4ml) were centrifuged on an iso-kinetic sucrose density gradient with a top sucrose concentration of 5% (w/v). The gradient was centrifuged for 16 hrs at 40,000 cpm at 4°C on a Beckman SW41 Ti rotor in a Beckman L2-65 preparative ultracentrifuge. The eluate was separated in 5 drop fractions with an LKB Ultrarac. Each fraction was assayed for total and specific IgE and other classes of immunoglobulins. We used an empirical relation between sedimentation coefficient and molecular weight to estimate the molecular weight of IgE containing components. A similar ultracentrifugation profile was applied to the measurement of ^{125}I labelled IgE myeloma in which the fractions were counted directly in a gamma counter.

Radioallergosorbent Test (RAST)

This method was performed according to Wide et al.(5).

Assay of Haemolitic Complement

Standard amount of monoclonal serum was incubated with ultra-centrifuge gradient fractions in volumes of 20ul at 37°C for half an hour in buffer containing either 5mmol/l calcium and 5mmol/l mag-nesium or 5mmol/l magnesium EGTA. A titration of total haemolytic complement was performed on 20ul aliquots of the incubation mixture using sensitized rabbit red cells in buffer containing magnesium and calcium ions. Consumption of complement was then calculated.

Conglutinin Column Assay

On the basis of experimental work described by other authors
(6) conglutinin columns were used to isolate immune complexes from
serum. Serum samples were freed of lipids by treatment with liqui-
fied trichlorofluoroethane (Frigen 113 TR-T). Cleared serum was then
mixed with an equal volume of 5% polyethylene glycol (PEG) in veronal
buffered saline (VBS). After 1 hr in ice the mixture was centrifuged
at 1000G. The precipitate was gently washed with cold 2.5% PEG in
VBS and was then solubilized in complement fixation diluent (CFD/
Ca^{++}) for further assay on a conglutinin column. The samples were
first activated by incubation with an equal volume of fresh monoclonal human serum diluted 1:2 in phosphate buffered saline (PBS)
for a period of 7 mins at $37^{o}C$ and the activated samples were then
applied to the PMMA-K (Polymethymethacrylate-conglutinin column).
The column was then washed with three column volumes of CFD/Ca^{++}
and complexes were eluted with 1/10 column volume fractions of CFD/
EDTA. Fractions were tested for the presence of specific IgE by
RAST.

Rheumatoid Factor Assay

Discs of Wothman 54 were activated using cyanogen bromide (8).
The discs were then sensitized with an established concentration of
a column purified human IgE myeloma in 0.1M sodium bicarbonate and
incubated for 12 hrs at $4^{o}C$ on a rocking platform. The discs were
then washed many times and blocked with 50mM of ethanolamine and
washed successively with 0.1M sodium bicarbonate and phosphate
buffered saline (PBS).   Each disc was placed in the well of a micro-
titre tray and 50ul of sample added and then incubated for 14 hrs at
$4^{o}C$ on a rocking platform. After incubation the discs were washed
4 times with 1% Tween in PBS and then incubated for 24 hrs at $4^{o}C$
on a rocking platform with 100ul ^{125}I rabbit antihuman IgG (9) at a
dilution to give 100,000 cpm. The discs were then washed 6 times
with 1% Tween in PBS, dried at $37^{o}C$ and counted in a gamma counter.

RESULTS

Fig. 1 shows that immune complexes can be produced as a result
of allergen challenge in the allergic subject. Such complexes can
be isolated as soon as half an hour after challenge, and a peak of
ovalbumin specific IgE complexes reached a maximum over the next
3 hrs. The approximate use of the complexes is 15S (10).

Pre-treatment with oral sodium cromoglycate prevented the
appearance of both the symptoms and complexed IgE. In an attempt to
better understand the nature of the complex, the complexed IgE peak
was analysed for further components

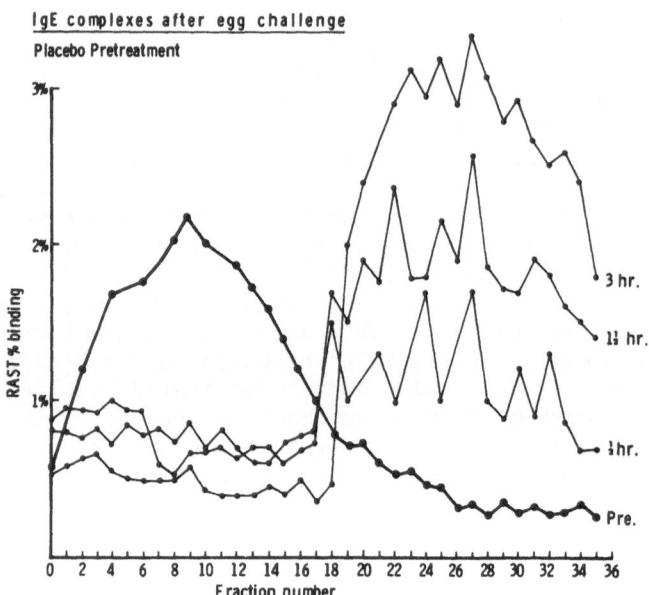

Fig. 1.   After challenge, IgE complexes were produced in the circulation after half an hour (No SCG) and a further rising was
seen at 1½ and 3 hrs.

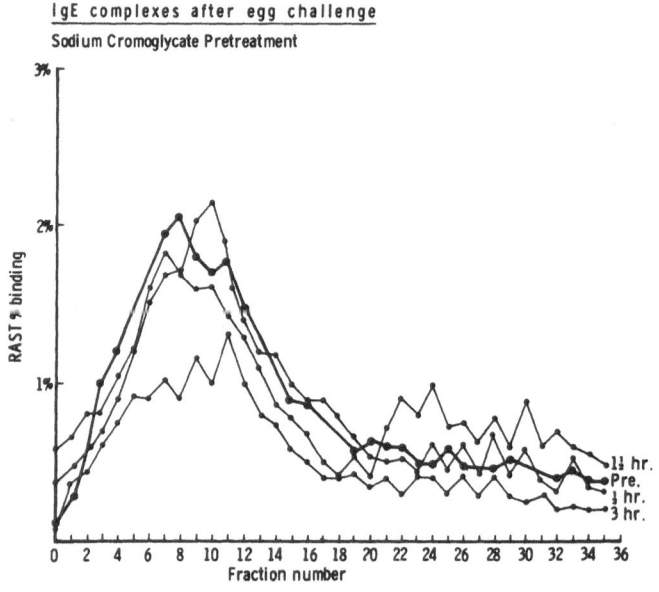

Fig. 2.   Pre-challenge sample with SCG showed a monomeric peak and
a small peak of heavy IgE was seen at 1½ hrs but not later.

Complement Activation by the Complexed IgE Peak

The peak of complexed IgE was examined for its ability to fix complement. Classical pathway activation was seen in the polymeric peak, but not in the area where monomeric IgE are found.

Conglutinin Column

Complement reacting immune complexes bind to conglutinin through fixed C3bi, which has reacted with conglutinogen activating factor (KAF) (11). We were able to show that specific IgE was eluted from the column and assayed by RAST showed an increase of activity in those fractions obtained after challenge with specific allergen. Normal serum and an IgE myeloma were used as controls for non-specific binding. Thus the presence of IgG (as well as specific IgE) in the complexes was confirmed by complement fixation and the conglutinin column.

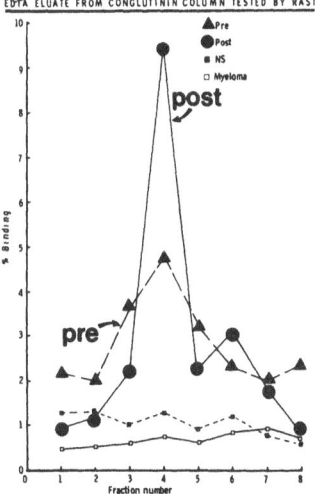

Fig. 3. IgE eluted from the column and assayed by RAST showed an increase of activity in those fractions obtained after challenge with specific allergen.

Rheumatoid Factor: IgG anti-IgE

Further evidence for the presence of IgG in the complex was obtained by the finding of rheumatoid factor (IgG anti-IgE) in the

complexes. Preliminary results indicate that sera from patients with
atopic eczema, asthma and hay fever show high binding in the assay
for rheumatoid factor in comparison to the control sera. Interestingly, patients with systemic lupus erythematosus and rheumatoid
arthritis show some binding in this assay. The specificity of the
atopic rheumatoid factor (IgG anti-IgE) was shown by the blocking
binding by an IgE myeloma but not by an IgG myeloma.

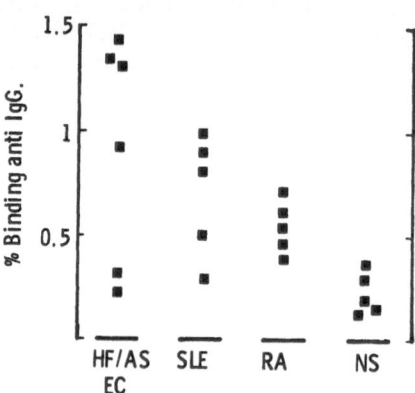

Fig. 4.   Patients with atopic eczema, asthma and hay fever show high
          binding in this assay for R.F. in comparison to the control
          sera.

DISCUSSION

     It is obvious that the intestinal mucosa is not a complete
barrier to the passage of food with the circulation (12).
Sufficient amounts of macromolecular food antigens do cross the
mucosa to stimulate the immune system with the appearance of specific antibodies in the intestinal secretions and serum (13-14).
Although immune complexes containing various immunoglobulin classes
were demonstrated in the non-atopic subjects, they were rapidly
cleared from the circulation and were apparently not damaging.
Formation of immune complexes in the healthy subject is presumably
a normal physiological method of eliminating antigen. In the
subjects sensitive to egg, not only were IgG, IgA and C1q binding

complexes detected after eating egg, but also ovalbumin specific IgE complexes were found. These latter complexes containing IgE were not found in the normal subjects. In food allergy it is envisaged that local production of specific IgE by the intestinal mucosa plays a central role in the causation of food allergy. Even low levels of sensitization can cause the mast cells to degranulate upon challenge with allergen; the increased vascular permeability allows further entry of antigen with subsequent formation of immune complexes. Any class of immunoglobulin may be involved in the formation of these immune complexes. Moreover, in the serum of these patients, IgE complexes which activated complement through the classical pathway have been demonstrated. Although aggregated IgE may possibly activate the alternative pathway, monomeric IgE does not activate complement (15). This evidence supports the idea that the IgE complex is likely to be a mixed complex containing IgG (IgG-allergen-IgE). Support for this comes from conglutinin binding and by the presence of rheumatoid factor. However, we have no direct evidence that any of these immune complexes actually cause the symptoms of broncospasm and itching in these patients. Preliminary experiments with passive cutaneous anaphylaxis suggest that the complexes have strong mast cell degranulating properties. Moreover, pre-treatment with oral sodium cromoglycate prevents the symptoms in atopic subjects following egg challenge, reduces antigen entry and the formation of immune complexes (16). This supports the idea of the "gatekeeper" role of the sensitized mucosa. This study showed a beneficial effect on symptoms by pre-treatment with sodium cromoglycate.

Sodium cromoglycate is poorly absorbed and its effect would seem to be by local inhibition of degranulation of mast cells in the gut, thereby reducing vascular permeability and entry of the antigen. Thus sodium cromoglycate may have a place in the treatment of food allergy in patients, especially who have multiple allergies and whose symptoms continue despite elimination diets.

REFERENCES

(1) P.L. Masson, Ric. Clin. Lab. 6:69 (1976).
(2) J. Brostoff, P. Johns and D.R. Stanworth, The Lancet,ii:741 (1977).
(3) A. Dannaeus, T. Foucard and S.G.O. Johansson, Clin. Allergy 7:109 (1977).
(4) P. Johns and D.R. Stanworth, J. Immunol. Methods 10:231 (1976).
(5) L. Wide, H. Bennich and S.G.O. Johansson, The Lancet ii/1105 (1967).
(6) P. Casali, G. Brighouse and P.H. Lambert, Prot.Biol.Fluids(1978).
(7) P. Casali and P.H. Lambert, Clin. exp. Immunol. 37:295 (1979).
(8) L. Wide, Acta Endocrinol.(KBH), 63 suppl.142:207 (1969).
(9) M. Ceska and V. Lundkvist, Immunochem. 9:1021 (1972).

(10) J. Brostoff, C. Carini, D.G. Wraith and P. Johns, The Lancet
      ii:1268 (1979).
(11) P.J. Lackmann, D.E. Elias and A. Moffett, "Biological Activi-
      ties of Complement", S. Karger, Basel, p.102 (1973).
(12) G. Volkheimer, H. John and F.H. Scholtz, Munch med. Woch.
      107:2293 (1965).
(13) R.M. Rothbery, J. Pediat. 75:391 (1969).
(14) W.A. Walker, K.J. Issel Bacher and K.J. Bloch, Science 177:608
      (1972).
(15) G.O. Solley, G.I. Gleich, R.E. Jordan and A.L. Schroeter,
      J. Clin. Invest. 58:408 (1976).
(16) J. Brostoff, C. Carini, D.G. Wraith, R. Paganelli and
      R.J. Levinsky, "The Mast Cells", J. Pepys, p.380 (1979).

INDEX

499